AI and ML Techniques in IoT-based Communication

IEEE Press
445 Hoes Lane
Piscataway, NJ 08854

AI and ML Techniques in IoT-based Communication

A Path to Sustainable Development Goals

Edited by

Sumita Mishra
Amity University Uttar Pradesh
India

Nishu Gupta
VTT Technical Research Centre of Finland, Oulu
Finland

Polat Goktas
Sabancı University
Türkiye
University College Dublin
Ireland

IEEE Press
WILEY

Published by John Wiley & Sons, Inc., Hoboken, New Jersey.

Library of Congress Cataloging-in-Publication Data has been applied for:

Hardback ISBN: 9781394337231
ePDF ISBN: 9781394337255
epub ISBN: 9781394337248
oBook ISBN: 9781394337262

Cover Design: Wiley
Cover Image: © Evannovostro/Shutterstock

Sumita Mishra dedicates this book to
Her Husband Gaurav Mishra

Nishu Gupta dedicates this book to
His Father Professor K. M. Gupta and Mother Smt. Rita Rani Gupta

Polat Goktas dedicates this book to
His Wife Aycan,
for her steadfast support and inspiration,
and to all who pursue knowledge to serve humanity.

Contents

Preface

The combination of artificial intelligence (AI), machine learning (ML), and the Internet of Things (IoT) is a game changer in a society that is becoming more and more interconnected and technologically advanced. *AI and ML Techniques in IoT-based Communication: A Path to Sustainable Development Goals* is a book written to examine this synergy and its enormous potential to advance global sustainability.

This book is based on the belief that intelligent algorithms can enable linked gadgets to provide creative answers to some of the most important problems in our society today. IoT integration with AI and ML is not only a technical development but also a strategic requirement for creating societies that are smarter and more resilient. This convergence makes data-driven decision-making, real-time responsiveness, and sustainable practices possible in a variety of fields, including healthcare, agriculture, energy systems, and smart cities.

The United Nations' Sustainable Development Goals (SDGs) offer a thorough road map for a brighter future. This book demonstrates how IoT systems improved by AI and ML can significantly aid in accomplishing these objectives. It functions as a guide and a call to action for academia, researchers, and students by presenting real-world examples and providing doable tactics in a variety of fields, such as urban development, environmental monitoring, education, and industrial communication. Every chapter explores a crucial aspect of the subject, ranging from fundamental ideas to moral dilemmas and upcoming developments. The future landscape of intelligent, sustainable communication systems will be envisioned by readers, who will also get insights into emerging trends and wrestle with the issues of ethical and secure deployment.

We hope that this book will inspire as much as it informs. We want our readers to actively participate in this vital and dynamic sector by shedding light on the revolutionary possibilities of AI, ML, and IoT in sustainable development. We think that technology can be a real ally in creating a more sustainable and egalitarian world if people work together and innovate with knowledge.

The book is arranged in a reader-friendly manner, with different parts, each having several chapters. Professionals such as university professors, graduate and doctoral students, industry professionals, and researchers – particularly those working in fields like ML, sensor networks, smart healthcare, intelligent transportation, and sustainable environments – are the target audience for this book.

As a textbook and reference, the book is appropriate for technical education and research at all levels as well as for novices looking for a general overview of the future applications of AI, ML, and IoT technologies, which have the potential to make many things smarter, more intelligent, and self-adaptive. Contributions to the book came from all around the world, and we are grateful to each and every author for sharing their work. We are also grateful to the reviewers for their timely and insightful comments. We express our gratitude to the publishers and authors of the books mentioned in the references.

Dr. Sumita Mishra
Dr. Nishu Gupta
Dr. Polat Goktas

About the Editors

Dr. Sumita Mishra is a distinguished academic and researcher with over 20 years of experience. She currently serves as an Assistant Professor at Amity University, Uttar Pradesh, Lucknow Campus. She has authored more than 70 research papers in reputed journals and international conferences. She holds several patents and copyrights and is actively engaged in interdisciplinary research initiatives. A Fellow of the Institution of Electronics and Telecommunication Engineers (IETE) and a former Senior Member of IEEE, she plays an active role in global research collaborations and technical program committees. Her current research interests include machine learning, optical and wireless communication, and IoT. She has successfully supervised several PhD scholars and continues to contribute significantly to the advancement of intelligent and sustainable technologies.

Dr. Nishu Gupta is a Senior Member, IEEE and presently working as a Senior Scientist as the VTT Technical Research Centre of Finland Ltd, located in Oulu, Finland. Before this position, he was a Postdoctoral Fellow in the Smart Wireless Systems research group, Department of Electronic Systems, Faculty of Information Technology and Electrical Engineering at Norwegian University of Science and Technology (NTNU) in Gjøvik, Norway. He has worked as Research Associate in the Smart Living Lab, at the University of Fribourg, Switzerland. Prior to this position, he was a Postdoctoral Fellow in the Department of Information Technology, Faculty of Computing and Information Technology, King Abdulaziz

University in Rabigh, Saudi Arabia. He is also a visiting researcher at the University of Oviedo, Spain, under the research group on Systems for Multimedia and the Internet of Things. He is serving as a Member, Zero Trust Architecture working group of MeitY-C-DAC-STQC project under "e-Governance Standards and Guidelines," Ministry of Electronics and Information Technology (MeitY), Government of India. He is a PhD in Electronics and Communication Engineering Department from Motilal Nehru National Institute of Technology Allahabad, India (an institute of national importance as declared by the Government of India), with specialization in Vehicular Adhoc Networks. His research interests include IoT; 5G/6G technologies; Dynamic Spectrum Sensing, Intelligent Transportation, Edge Computing; etc.

Dr. Polat Goktas is an Assistant Professor at Faculty of Engineering and Natural Sciences, Sabancı University, Türkiye. He previously served as a Senior Researcher at the UCD School of Computer Science and at Ireland's Centre for Applied Artificial Intelligence (CeADAR), University College Dublin, Ireland. His research focuses on AI, machine learning, computer vision, the Internet of Things, and explainable AI, addressing critical challenges in healthcare, security, and sustainability. His PhD work integrated image acquisition, processing, AI, and life sciences to develop label-free, high-throughput cellular analysis methods for in vitro diagnostics. As a Fulbright Doctoral Research Fellow at Harvard Medical School, he deepened his expertise in biomedical imaging and AI-driven healthcare solutions. Throughout his career, Dr. Goktas has received prestigious honors, including the Young Scientist Award at the Lindau Nobel Laureates Meeting, the IEEE AP-S Doctoral Research Grant, the Fulbright Fellowship, the Marie Curie Individual Fellowship, and the METU Serhat Ozyar Young Scientist of the Year Award and among others. He received the 2024 Best Reviewer Award at the *AI, Computer Science and Robotics Technology* (ACRT) journal. He is an Associate Editor for *Fulbright Chronicles* and *PLOS One*, and contributes to the ACRT, Marie Curie Alumni Association (MCAA) Newsletter and Sustainable Futures as an Editorial Board Member. He actively leads initiatives as Chair of the IEEE Young Professionals Innovation Hub Subcommittee, Vice-Chair of the MCAA (European Union) Sustainability Working Group, and Sub-Lead Chairman of the IEEE Engineering in Medicine and Biology Society (EMBS) SAC's Professional Development Portfolio, driving innovation and interdisciplinary collaboration in AI.

Acknowledgments

Sumita Mishra

First and foremost, I express my deepest gratitude to my mother, Smt. Sushila Shukla, whose unconditional love, strength, and lifelong sacrifices have been the cornerstone of my achievements. Her constant prayers have guided and sustained me through every challenge. I am equally grateful to my father, Shri V.K. Shukla, for his blessings.

I am profoundly thankful to my husband, Mr. Gaurav Mishra, whose unwavering belief in me, endless patience, love, and constant encouragement have empowered me to pursue my goals. I am deeply grateful to my daughter, Sanvi Mishra, and my son, Aditya Mishra, whose innocent smiles, unconditional love, and understanding have filled my life with purpose and joy. Their cheerful presence has been a continuous source of strength and motivation. I also extend heartfelt thanks to my other family members for their continued support, encouragement, and understanding throughout this journey.

I am eternally grateful to Dr. Ashok K. Chauhan, Founder President of the Amity Group, and Dr. Aseem Chauhan, Chairman of Amity University, Lucknow Campus, for their visionary leadership and for providing a research environment that nurtures innovation and excellence.

My sincere appreciation goes to Wg. Cdr. Dr. Anil Kumar, Deputy Pro-Vice Chancellor, Amity University, Lucknow Campus, whose motivation and guidance have been instrumental in helping me grow professionally.

I also thank Dr. O.P. Singh, Offg. Head of Amity School of Engineering and Technology, for his continuous encouragement and support.

I deeply acknowledge the collaboration and support of my co-editors, Dr. Nishu Gupta and Dr. Polat Goktas, whose invaluable contributions have been instrumental to the success of this book. I am also sincerely grateful to my friends and colleagues and students for their unwavering motivation, support, and inspiration throughout this journey.

Above all, I thank God for countless blessings, for guiding my path, and for giving me the perseverance to bring this work to completion.

Nishu Gupta

I acknowledge the inspiration and blessings of my mother Smt. Rita Rani Gupta and father Prof. K.M. Gupta, who are the pillars of my strength and personality. Having parents like them is the greatest gift and biggest advantage anyone could ever have given me. It is because of the confidence and values that they have instilled in me that made me who I am today.

I am full of gratitude to my wife Smt. Anamika Gupta, son Ayaansh Gupta, sister Mrs. Nidhi Gupta, brother-in-law CA Ritesh Shankar Gupta, and other family members for the patience shown and encouragement given to complete this venture.

I deeply acknowledge the blessings of my PhD Supervisor Prof. Arun Prakash, Electronics and Communication Engineering Department, MNNIT Allahabad, India, whose guidance has always encouraged me to do my best. I express my profound gratitude to Dr. Anil Gupta at the Centre for Development of Advanced Computing (CDAC), Pune, Maharashtra, India, who has played a very significant role in shaping my research career. I wholeheartedly acknowledge the motivation given to me by my Scientific Advisors: Dr. Jukka Mäkelä, Principal Scientist at VTT Technical Research Center of Finland; Prof. Mohammad Derawi at the Norwegian University of Science and Technology (NTNU), Gjøvik, Norway; Prof. Sara Paiva at the Polytechnic Institute of Viana do Castelo, Portugal; Prof. Manuel J.C.S. Reis at UTAD, Portugal; Prof. Ahmad Hoirul Basori at King Abdulaziz University, Saudi Arabia; and Prof. Ariel Soares Teles at the Federal Institute of Maranhão, Brazil. Thank you for shaping me into a person I am proud to be.

I deeply acknowledge the support and encouragement from my co-editor Dr. Sumita Mishra and Dr. Polat Goktas in shaping this book.

I express my heartfelt thanks to my colleagues, friends, and students for their support and motivation in several ways.

Polat Goktas

I gratefully acknowledge the unconditional love, inspiration, and enduring prayers of my family. Their values, discipline, and deep-rooted belief in education and integrity have profoundly shaped my journey. It is through their unwavering support and countless sacrifices that I have become the person I am today.

My deepest gratitude goes to my beloved wife, Aycan, whose constant encouragement, immense patience, and steadfast understanding have been a source of strength throughout my life.

I wish to express sincere appreciation to my esteemed scientific mentors and collaborators, whose vision, intellectual guidance, and trust have been instrumental in shaping both my academic and professional path. I am especially thankful for the collaborative spirit of my peers and co-authors from diverse disciplines and geographies, whose contributions have enriched this work immensely.

Special thanks to my dedicated colleagues and friends from the IEEE, Fulbright, and Marie Skłodowska-Curie communities for their insightful perspectives, encouragement, and lasting camaraderie.

I also warmly acknowledge the invaluable support of my co-editors, Dr. Sumita Mishra and Dr. Nishu Gupta, whose commitment and collaboration were vital in shaping this book with clarity and coherence.

Finally, I extend my heartfelt thanks to my students and mentees. Their curiosity, enthusiasm, and willingness to explore the unknown continue to inspire me and reaffirm the purpose of this journey.

Last but not least, we, the editors, express our heartfelt gratitude to the publishers and the team behind it for their continued support and cooperation in publishing this book.

About the Companion Website

This book is accompanied by a companion website:

http://wiley.com/go/mishra_aiml

The website includes:

- Multiple Choice Questions for the chapters.
- Case study for Chapters 12 and 13.
- Exercises for Chapter 13.
- One-Page Summary/Teaching Aid for Chapter 13.

Part I

Fundamentals of IoT, AI, and ML for Sustainable Development

1

Introduction to IoT, AI, and ML in Sustainable Communication

Shumukh Alharrani[1,], Amani Ibraheem[1], and Polat Goktas[2,3]*

[1] *College of Computer Science, King Khalid University, Aseer, Abha, Saudi Arabia*
[2] *UCD School of Computer Science, University College Dublin, Dublin, Ireland*
[3] *Faculty of Engineering and Natural Sciences, Sabanci University, Istanbul, Turkey*

*Corresponding author: 445817025@kku.edu.sa

1.1 Introduction

In the contemporary digital era, the convergence of the Internet of Things (IoT), artificial intelligence (AI), and machine learning (ML) has emerged as a pivotal force reshaping communication infrastructures, decision-making paradigms, and sustainable development strategies across critical sectors. These technologies are not only advancing technical capacities but also addressing complex global challenges – ranging from climate change and resource scarcity to healthcare delivery and urban management [1, 2].

IoT technologies provide the foundational infrastructure for real-time data acquisition through a distributed network of sensors and actuators. This continuous stream of data becomes valuable when integrated with AI systems capable of extracting actionable insights via advanced computational techniques, while ML models refine and enhance predictive capabilities through iterative learning from historical and dynamic datasets [3, 4]. The synergy between these technologies plays a central role in enhancing operational efficiency, reducing waste, and promoting sustainability across domains such as agriculture, smart cities, and healthcare [5–8].

AI and ML Techniques in IoT-based Communication: A Path to Sustainable Development Goals, First Edition. Edited by Sumita Mishra, Nishu Gupta and Polat Goktas.

In agriculture, for example, IoT-enabled sensors facilitate precise monitoring of environmental variables such as soil moisture, crop health, and weather patterns. These data streams are analyzed by AI systems to optimize irrigation, fertilization, and pest control practices. Simultaneously, ML models enable predictive analytics for yield estimation, disease forecasting, and adaptive decision support [9, 10]. In urban environments, smart cities leverage IoT technologies to manage infrastructure systems, such as traffic, energy consumption, waste collection, and environmental monitoring, through real-time sensing and control mechanisms [11]. AI algorithms enable real-time analysis of traffic congestion and environmental data to optimize public service delivery, while ML facilitates long-term planning by modeling patterns of population growth, energy demand, and mobility behaviors [12, 13]. In healthcare, wearable IoT devices enable continuous monitoring of physiological parameters, generating data that support early disease detection and remote patient management. AI-based diagnostic tools assist clinicians in interpreting complex datasets, while ML techniques enhance the personalization of treatment plans and predict disease progression trends [14–16].

Despite their transformative potential, the deployment of IoT–AI–ML ecosystems poses considerable challenges. Key issues include the management and integration of large-scale heterogeneous data, ensuring privacy and cybersecurity across interconnected systems, and mitigating the environmental impact of energy-intensive computational processes [17, 18]. To address these barriers, emerging solutions such as edge computing, federated learning (FL) [19], and energy-efficient AI-specific hardware [20] are being increasingly adopted. Thus, the chapter aims to offer a comprehensive examination of the foundational principles, interdependencies, and emerging applications of IoT, AI, and ML in the context of sustainable communication systems. By drawing upon recent academic and industrial advancements [21–23], we highlight the transformative impact of these technologies and discuss their role in enabling scalable, intelligent, and ethically responsible infrastructures that align with global sustainability imperatives.

1.2 Foundational Principles of IoT, AI, and ML

1.2.1 IoT as the Backbone of Intelligent Communication Systems

The IoT signifies a transformative technological paradigm, enabling physical objects – ranging from basic sensors to advanced embedded systems – to connect, interact, and exchange data across network infrastructures in real time. This networked environment empowers systems to make autonomous decisions, adapt to environmental changes, and offer predictive capabilities in diverse domains including agriculture, healthcare, smart cities, logistics, and manufacturing [24, 25].

1.2.1.1 Architectural Overview

IoT systems are typically structured into a three-layer architecture, which underpins the end-to-end flow of data and functionality:

- **Perception Layer**: This foundational layer consists of sensors and actuators that interact directly with the physical environment. These components are responsible for detecting variables such as temperature, pressure, humidity, motion, and biochemical changes [26]. For example, in smart farming, sensors are used to monitor soil moisture and nutrient levels, providing the groundwork for intelligent decision-making [24].
- **Network Layer**: The intermediary layer is responsible for transferring the data acquired at the perception layer to cloud platforms or edge systems. It employs a variety of communication protocols, including Wi-Fi [27], Bluetooth [28], ZigBee [29], and LoRaWAN [30], to ensure reliable data transmission. Recent studies also explore the integration of blockchain and secure protocols to ensure lightweight authentication and confidentiality in medical and industrial IoT applications [31, 32].
- **Application Layer**: At the top of the hierarchy, this layer processes the transmitted data, transforming it into actionable insights for various verticals such as healthcare diagnostics, industrial automation, or intelligent transportation systems [26, 33]. ML models and AI-based inference engines are commonly implemented at this level to drive intelligent decision-making [34].

Figure 1.1 visually captures this tri-layer framework and the seamless interplay of IoT components across it. This structure is essential for enabling scalability, interoperability, and system resilience in complex environments.

The perception layer consists of physical sensors and devices responsible for data collection. The network layer transmits the collected data using communication technologies such as 4G/LTE, ZigBee, Z-Wave, Bluetooth, and Wi-Fi. The application layer processes and utilizes the data for intelligent decision-making in domains such as cloud computing, smart transportation, smart environments, and smart homes.

1.2.1.2 Foundational Capabilities and Communication Technologies

Modern IoT applications demand high scalability and energy efficiency, necessitating the development of lightweight architectures and adaptive routing protocols [35]. Innovative approaches such as multimedia sensing as a service (MSaaS) and cognitive-LPWANs are being introduced to support real-time analytics at the edge while reducing network congestion [36]. Similarly, concepts like Hybrid Energy Harvesting Things (HEHT) propose self-sustainable IoT deployments that leverage renewable energy sources [37].

Security and privacy remain major concerns in IoT ecosystems. Numerous frameworks have been proposed to safeguard data integrity and user anonymity, including Stackelberg game-based physical layer security enhancements [38] and

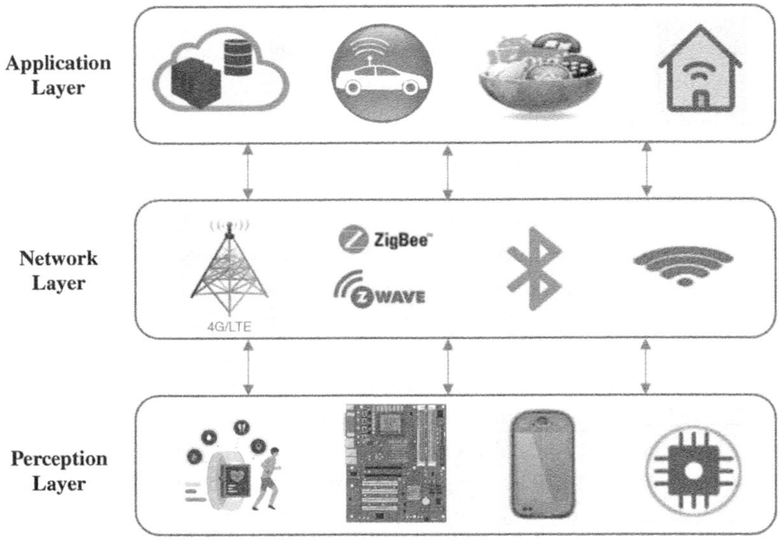

Figure 1.1 Three-layer architecture of the Internet of Things (IoT).

secret key generation mechanisms using turbo codes for vehicular networks [39]. With the proliferation of ultra-dense sensor networks, the integration of semantic communication and intelligent network slicing has also become essential [40].

1.2.1.3 Scalability, Positioning, and Real-time Processing

From a foundational perspective, IoT systems are defined by their ability to ensure interoperability, scalability, low latency, and energy efficiency. Scalability is achieved through modular device deployment, which allows seamless integration of new devices into the network [41]. Interoperability, on the other hand, is addressed through standardization protocols and middleware frameworks that enable devices from different manufacturers to work together [42]. Furthermore, the evolution of IoT has been accelerated by the integration of cloud and edge computing, allowing more efficient and distributed processing of data close to the source. Edge computing particularly enhances the performance of time-sensitive applications by reducing latency and minimizing bandwidth requirements [43].

Recent innovations in IoT have emphasized scalable scheduling mechanisms like PID-based control for 6TiSCH networks [44] and the simulation of interference patterns in LPWANs for optimized network planning [45]. Meanwhile, efforts in PNT (positioning, navigation, and timing) using LEO satellites demonstrate the feasibility of enhancing IoT localization accuracy and temporal synchronization at scale [46]. These developments not only optimize operational efficiency but also address critical concerns regarding data privacy, security, and energy consumption [47].

1.2.2 AI: Foundations, Evolution, and Role in Sustainable Communication

AI is broadly defined as the ability of machines to mimic cognitive functions such as learning, reasoning, and problem-solving, traditionally associated with human intelligence [48]. Over the past few decades, the rapid development of computational power, availability of big data, and advances in ML have collectively propelled AI into virtually every industry. In the context of sustainable communication systems, AI acts as a transformative enabler that supports intelligent decision-making, automation, and predictive analytics across domains such as energy, healthcare, agriculture, and urban planning [49–54].

AI's current landscape reflects a shift from rule-based systems to data-driven paradigms, driven predominantly by ML, deep learning (DL), and natural language processing (NLP). These techniques allow systems to adapt and learn from real-time data, rendering them highly valuable in dynamically changing environments [55]. For example, NLP has become instrumental in automating clinical documentation and extracting actionable insights from unstructured healthcare records [56–59]. In sustainable communication systems, AI is integrated with the IoT to analyze sensor-generated data, identify inefficiencies, and optimize performance without human intervention. This synergy allows for real-time monitoring of environmental variables, infrastructure, and human activity, supporting resource conservation and operational efficiency [60].

Furthermore, AI contributes to energy sustainability by enhancing energy management systems through predictive analytics, enabling demand forecasting, fault detection, and energy-efficient control strategies [60, 61]. In agriculture, AI-driven algorithms are increasingly used to model crop growth patterns, optimize irrigation scheduling, and detect plant diseases from satellite images or drone footage [62]. From an industrial and societal perspective, AI has also made significant strides in improving decision-making in complex environments. For instance, the application of AI in financial forecasting, supply chain optimization, and smart manufacturing demonstrates its role in streamlining processes and reducing wastage [63, 64].

However, alongside these advancements, AI introduces challenges related to interpretability, data privacy, and ethical deployment. There is an increasing call for responsible AI frameworks that promote transparency, fairness, and accountability in automated decision systems, particularly in sensitive applications such as healthcare and human resources [65, 66]. To contextualize the diversity of AI techniques and their applicability across domains, Table 1.1 presents a comparison of widely used AI methods, their foundational concepts, and their relevance to sustainable communication environments.

Table 1.1 Foundational AI techniques and their relevance in sustainable communication.

AI technique	Primary function	Application domain	Strengths	Limitation
Supervised learning	Predict outcomes from labeled data	Energy forecasting, diagnostics	High accuracy, robust prediction	Requires labeled data
Unsupervised learning	Discover patterns in unlabeled data	Customer segmentation, clustering	Insightful for hidden trends	Difficult to validate results
Reinforcement learning	Learn through trial and error interactions	Autonomous systems, robotics	Adaptive, suited for dynamic environments	Slow convergence, exploration risks
Deep learning	Layered neural networks for complex patterns	Image recognition, natural language processing (NLP)	High accuracy, complex feature learning	Opaque, resource-intensive
NLP	Understand and generate human language	Clinical records, chatbot systems	Automates unstructured data analysis	Context understanding remains limited

Note: The content of this table is synthesized from key insights and findings reported in the current literature.

1.2.3 ML: Core Enabler of Intelligent and Sustainable Communication Systems

ML has become a cornerstone of intelligent system design in the era of data-centric innovation. As sustainable communication systems increasingly depend on data analytics, real-time decision-making, and autonomous operations, ML offers powerful tools to process complex data, adapt to dynamic environments, and continuously improve system efficiency. Within the realm of IoT and AI, ML acts as the adaptive layer that enhances system performance, reduces energy usage, and facilitates robust predictive capabilities – thereby supporting the Sustainable Development Goals (SDGs) through smarter technology deployment.

ML encompasses a variety of algorithmic approaches such as supervised learning, unsupervised learning, semi-supervised learning, reinforcement learning, and DL. These techniques are deployed to extract actionable insights from vast datasets in real time, improving applications in domains like edge computing, healthcare diagnostics, environmental monitoring, smart cities, agriculture, and network optimization. Among these, reinforcement learning and FL are gaining momentum due to their ability to personalize learning models at the edge while ensuring data privacy and energy efficiency. For instance, Zhang et al. (2024) proposed deep reinforcement learning (DRL)-enhanced methods for energy-efficient IoT systems, while Lu et al. [67] systematically reviewed the role of ML in intelligent IoT service composition. Ghafari and Mansouri [68] further explored resource optimization through RL in fog computing, addressing scalability and latency in distributed networks.

Furthermore, TinyML, a branch of ML for resource-constrained environments, is being integrated into edge devices, facilitating decentralized learning with minimal energy consumption. Applications include indoor air quality monitoring, smart agriculture, traffic light control, and emotion recognition through physiological signals. These innovations are not only reshaping the technical landscape but also redefining how sustainable, autonomous systems are built and deployed. Table 1.2 highlights prominent applications and trends in ML technologies used in sustainable communication and IoT systems.

Table 1.2 Key ML trends and applications in sustainable communication systems.

ML approach	Application domain	Objective/contribution	Reference(s)
Deep reinforcement learning	Edge computing for IoT	Latency and resource optimization	[69]
Supervised learning	Smart IoT security	Classification of threats and intrusion detection	[70]
Unsupervised learning	Environmental sensor anomaly detection	Real-time fault recognition	[71]
Deep learning	Emotion recognition and medical imaging	Image analysis and diagnosis	[72]
Hybrid learning	Traffic management systems	Dynamic traffic signal control	[73]
TinyML	Audio and keyword spotting	Low-power intelligent processing	[74]
Semi-supervised learning	Smart cities and edge devices	Resource-aware decision-making	[75]
GANs (Generative)	Predictive healthcare and security systems	Enhanced realism and model generalizability	[76, 77]
Reinforcement learning	Routing and task offloading in IoT	Energy-efficient and adaptive scheduling	[76, 78]

Note: This table has been constructed based on a critical synthesis of insights from peer-reviewed literature in the ML and IoT domains.

1.3 Synergistic Integration of IoT, AI, and ML for Sustainable Communication Systems

The convergence of the IoT, AI, and ML represents a pivotal transformation in the design and implementation of sustainable communication systems. Each technology individually contributes to the digitization and automation of processes; however, their integration amplifies the impact, offering scalable, intelligent, and sustainable solutions for global challenges in healthcare, agriculture, infrastructure, smart cities, and industrial sectors [79–83]. This triad enables systems that are not only connected and data-rich but also capable of adaptive, predictive, and autonomous decision-making.

1.3.1 Enabling Architecture and Interoperability

The successful integration of the IoT, AI, and ML into sustainable communication systems hinges on a robust architectural framework and seamless interoperability across diverse components. This interoperability is essential not only for real-time data exchange but also for enabling intelligent decision-making processes that adapt dynamically to changing environmental, infrastructural, and user demands.

1.3.1.1 Middleware and Semantic Interoperability

A key challenge in enabling this multilayered ecosystem is ensuring interoperability, the ability of heterogeneous devices, platforms, and services to communicate and function cohesively. This requires:

- ***Middleware Platforms***: These act as a bridge between hardware and software layers, managing tasks like data formatting, message brokering, security, and device orchestration. Examples include open-source platforms such as FIWARE or proprietary solutions like AWS IoT Core and Google Cloud IoT.
- ***Semantic Interoperability***: Beyond syntax-level compatibility, systems must also understand the meaning of the data exchanged. This is where ontologies and standardized data models (e.g. W3C's Semantic Sensor Network ontology) are used to facilitate meaningful interactions between AI/ML models and IoT systems [81].

1.3.1.2 Distributed Versus Centralized Intelligence

Traditional IoT architectures rely on centralized cloud-based analytics, where vast amounts of data are transmitted to the cloud for processing. However, such models face limitations related to latency, bandwidth, and data privacy. To address these, Edge Computing and Fog Computing architectures are increasingly being adopted:

- *Edge artificial intelligence (Edge AI)* allows AI/ML algorithms to be deployed directly on local devices (e.g. sensors, smartphones, microcontrollers), enabling on-device intelligence with minimal latency and reduced cloud dependency [84, 85].
- *Fog computing* bridges the gap between the cloud and the edge by providing intermediate processing nodes that offer faster response and enhanced scalability.

These distributed paradigms not only improve responsiveness but also reduce energy consumption and support green computing goals, particularly when deployed with efficient ML models such as FL [86].

1.3.1.3 Interoperability Standards and Protocols

Achieving system-wide compatibility requires adherence to standardized communication protocols, data formats, and security frameworks. Commonly used standards include:

- **MQTT and CoAP**: Lightweight messaging protocols suited for constrained devices and networks.
- **JSON, XML**: Common data serialization formats.
- **IEEE 802.15.4/ZigBee/BLE**: Communication standards for low-power wireless personal area networks.

Security and privacy standards such as HIPAA (in healthcare) or General Data Protection Regulation (GDPR) (in Europe) also govern the processing and exchange of sensitive data, especially when AI/ML systems operate on personal information [47, 87–89].

1.3.1.4 Role of Digital Twins and System-level Integration

One emerging trend in integrated architecture is the deployment of digital twins-virtual representations of physical systems. These twins are updated in real time via IoT sensors, allowing AI models to simulate, test, and optimize performance under different scenarios [90]. This is particularly relevant in smart infrastructure, where system-level integration of AI and IoT enhances predictive maintenance, resource allocation, and emergency response planning.

Thus, the integration of IoT, AI, and ML requires a multilayered architectural approach that spans perception, network, and application layers. At the perception layer, IoT devices collect environmental and operational data. The network layer transmits this data to centralized or decentralized platforms using protocols such

Table 1.3 Functional synergies of IoT, AI, and ML across domains.

Domain	Role of IoT	Role of AI	Role of ML
Healthcare	Real-time health monitoring via wearables	Anomaly detection and diagnostic support	Predictive analytics for personalized treatment
Smart cities	Traffic sensors, waste and energy monitoring	Urban planning, dynamic resource allocation Classification of threats and intrusion detection	Demand forecasting, route optimization
Agriculture	Soil and crop sensing	Yield optimization, disease identification	Pattern detection in weather/crop data
Industry 4.0	Equipment and environmental monitoring	Predictive maintenance	Quality control and anomaly detection
Tourism	Visitor movement tracking	Recommendation systems	Customer behavior prediction

Note: This table has been constructed based on a critical synthesis of insights from peer-reviewed literature in the AI, ML, and IoT application domains.

as 5G, ZigBee, or LoRaWAN [91, 92]. AI and ML are embedded in the application layer, where data is processed, analyzed, and interpreted for actionable insights in real time [85]. This architectural interplay enhances the operational intelligence of various systems. For example, in smart agriculture, IoT sensors monitor soil and climate conditions, AI models optimize irrigation schedules, and ML algorithms predict pest outbreaks and crop health [93, 94]. As shown in Table 1.3, the integration of IoT, AI, and ML offers complementary functionalities across multiple sectors, enabling data-driven optimization and real-time responsiveness.

1.3.2 Applications in Key Domains

The integration of IoT, AI, and ML has given rise to intelligent, responsive, and sustainable solutions across various sectors, as outlined in Table 1.4. This synergy not only enhances real-time monitoring and automation but also strengthens system intelligence and adaptability. Below are key domain-specific use cases demonstrating this integration.

1.3.2.1 Healthcare and Internet of Medical Things

The convergence of IoT, AI, and ML in healthcare has laid the foundation for what is now known as the Internet of Medical Things (IoMT). Wearable IoT devices, such as smartwatches, implantable sensors, and portable ECG monitors,

Table 1.4 Comparative strengths of IoT, AI, and ML in sustainable communication systems.

Technology	Core of strengths	Limitations	Potential enhancements via integration
IoT	Real-time data acquisition, device interconnectivity, environmental sensing	Vulnerable to data overload, cybersecurity threats, energy constraints	AI for data filtering and anomaly detection; ML for pattern learning and event prediction
AI	Decision-making, real-time optimization, autonomous system control	High computational demands, data dependency, explainability issues	IoT for real-time contextual inputs; ML for continuous model refinement
ML	Pattern recognition, predictive analytics, anomaly detection	Requires large training datasets, prone to bias, lacks transparency	IoT for diverse data input; AI for intelligent decision-making and interpretability

Note: This table has been developed based on synthesized insights from domain-specific literature across healthcare, urban infrastructure, and agriculture [42, 97, 99–101].

enable continuous collection of vital health data. AI algorithms analyze these data streams to detect abnormalities, flag health risks, and recommend interventions in real time. For instance, AI-powered triage tools use symptom data from mobile apps to prioritize patients during telemedicine consultations.

ML further personalizes this process by identifying disease patterns, predicting hospital readmissions, and supporting clinical decision-making based on longitudinal health records. In chronic disease management, ML models now forecast glucose trends in diabetic patients or predict cardiac arrhythmias based on ECG patterns. Privacy-preserving techniques like FL are becoming essential in healthcare, allowing for decentralized training of ML models across devices while protecting sensitive patient data [95–97]. Moreover, AI-powered robotic process automation is revolutionizing administrative workflows, improving patient throughput, and reducing human error in hospital systems.

1.3.2.2 Sustainable Urban Infrastructure

Smart cities are at the forefront of deploying IoT, AI, and ML to improve urban sustainability and resilience. IoT devices embedded in traffic signals, vehicles, streetlights, and environmental stations enable the real-time collection of data on mobility, air quality, noise levels, and energy consumption. AI dynamically optimizes traffic flow by adjusting signals in response to real-time congestion patterns, reducing idle time and emissions [42].

ML algorithms enhance long-term urban planning by analyzing historical data on public transport usage, pedestrian flows, and accident patterns. These insights inform city planners on infrastructure investments, route adjustments, and emergency preparedness. In waste management, smart bins with IoT sensors alert collection services when full, while AI routes optimize collection and disposal paths, improving efficiency and reducing fuel consumption [81]. Additionally, environmental monitoring systems powered by AI/ML predict pollution surges and guide interventions, such as rerouting traffic or issuing public health alerts. These integrated solutions collectively support data-centric governance for sustainable urban ecosystems.

1.3.2.3 Precision Agriculture and Food Security

In agriculture, the synergy of IoT, AI, and ML underpins precision farming, a transformative approach to maximizing productivity while conserving resources. IoT sensors embedded in fields and on farm equipment gather real-time data on soil moisture, nutrient levels, temperature, and crop health. Drones equipped with IoT modules capture aerial imagery, enabling high-resolution monitoring of crop growth and pest infestations.

AI-driven analytics then evaluate the collected data to recommend optimized irrigation schedules, fertilizer application, and pest control measures. For example, AI can classify crop diseases from leaf images and suggest treatment strategies, minimizing pesticide use. ML models predict yield variations based on weather forecasts, soil profiles, and historical crop data, enabling farmers to make informed planting and harvesting decisions [98, 99]. Furthermore, the use of AI-based robotics in autonomous tractors and harvesters enhances operational efficiency and reduces labor dependency, which is particularly valuable during labor shortages or pandemics. These integrated systems contribute to sustainable food production and bolster food security, especially in regions vulnerable to climate variability.

1.4 Challenges in Implementation

The convergence of IoT, AI, and ML has introduced unprecedented possibilities in advancing sustainable communication systems. However, despite their synergistic potential, the integration and practical implementation of these technologies remain fraught with significant challenges. These obstacles span technical, ethical, legal, and infrastructural dimensions, making their resolution critical for widespread adoption. Furthermore, as summarized in Table 1.5, each technological pillar brings unique strengths and limitations to the integrated system, and

Table 1.5 Key challenges in implementing integrated IoT–AI–ML systems.

Challenge category	Description	Potential mitigation strategies
Data security and privacy	Vulnerability to cyber threats and data breaches	Federated learning, encryption, blockchain
Scalability and interoperability	Lack of standardization, fragmented ecosystems	Open standards, middleware, cross-platform APIs Classification of threats and intrusion detection
Resource constraints	Limited energy, bandwidth, and processing power in edge devices	Requires large training datasets, prone to bias, lacks transparency Energy-efficient hardware, Edge AI, optimized ML models
Ethical and regulatory	Bias in AI, lack of transparency, unclear data governance	Explainable AI, updated data policies, ethics frameworks
Real-time processing	Delays due to centralized computation and network bottlenecks	Edge computing, lightweight AI/ML algorithms
Skills gap	Shortage of qualified professionals across disciplines	Capacity building programs, cross-sector collaboration

Note: This table has been developed based on a synthesis of insights derived from the current literature and expert discussions on implementation bottlenecks across integrated IoT, AI, and ML systems.

understanding these distinctions is essential for effective deployment and optimization in real-world scenarios.

1.4.1 Data Security and Privacy Concerns

A primary concern in the deployment of IoT, AI, and ML systems is data security and privacy. With massive volumes of sensitive data transmitted across networks, systems are vulnerable to cyber threats such as denial-of-service (DoS) attacks, eavesdropping, and data breaches [102–105]. The distributed and heterogeneous nature of IoT devices often leads to inconsistent security protocols, making it difficult to ensure end-to-end data protection [106]. Furthermore, AI-driven analytics can inadvertently expose private user information if models are not adequately anonymized [107]. FL and homomorphic encryption are emerging as privacy-preserving solutions, but they require high computational resources and reliable communication infrastructure, which are often lacking in low-power IoT environments [108, 109].

1.4.2 Scalability and Interoperability

As smart infrastructures grow, ensuring scalability and interoperability becomes essential yet complex. Diverse device manufacturers, proprietary communication protocols, and nonuniform data formats hinder seamless integration [110, 111]. AI models may need to be retrained or adapted for each platform, complicating deployment. Interoperability challenges are especially pronounced in urban infrastructure and healthcare applications where real-time coordination across multiple systems is mandatory [112, 113]. Standardization efforts such as IEEE 1451 for sensor networks and edge computing frameworks provide partial solutions, but widespread adoption remains slow.

1.4.3 Resource Constraints and Infrastructure Gaps

IoT devices are often deployed in environments with limited power, bandwidth, and computational capacity. ML algorithms, particularly DL models, are resource-intensive, requiring robust hardware and software ecosystems [114, 115]. Many developing regions lack the infrastructure to support real-time analytics or secure cloud services, leading to implementation disparities [116, 117]. Furthermore, sustainable communication systems require energy-efficient solutions, yet many IoT deployments contribute to increased carbon footprints due to inefficient architectures or overprovisioned networks [118].

1.4.4 Ethical and Regulatory Issues

The ethical implications of AI decision-making, especially in critical domains such as healthcare and law enforcement, pose substantial challenges. AI models may inherit biases from training data, resulting in unfair or discriminatory outcomes [119]. Additionally, the lack of clear regulatory guidelines for data ownership, accountability, and transparency limits stakeholder trust and hinders large-scale adoption [120]. Frameworks like the EU's GDPR and initiatives for explainable AI (XAI) offer guidance, but more sector-specific regulations are needed for operational clarity [88].

1.4.5 Real-time Processing and Latency

The need for real-time decision-making in critical applications, such as autonomous vehicles, healthcare monitoring, and industrial control systems, demands ultra-low latency and high-speed data processing. Traditional cloud architectures are not sufficient to meet these requirements, and although edge computing presents a solution, its integration with AI/ML is still maturing [121, 122]. Edge devices must be capable of hosting lightweight AI models, managing intermittent connectivity, and coordinating with cloud servers – all of which increase the complexity of deployment.

1.4.6 Human Capital and Skills Gap

Finally, the successful implementation of IoT–AI–ML systems requires a multidisciplinary workforce with expertise in data science, embedded systems, networking, and cybersecurity. However, there exists a significant talent gap, particularly in low-resource settings, limiting the potential for widespread adoption and innovation [103, 107].

1.5 Emerging Trends and Future Perspectives

1.5.1 Edge AI and FL

A key trend in the AI and IoT convergence is the migration of computational intelligence from centralized cloud architectures to the edge. Edge AI enables real-time analytics directly on devices, reducing latency, improving responsiveness, and addressing data privacy concerns [123, 124]. Recent frameworks such as ToEFL [125] and privacy-preserving SVMs [126] demonstrate how edge models can be efficiently trained and deployed with minimal resource consumption. FL, as emphasized in multiple studies [127, 128], complements Edge AI by allowing distributed devices to collaboratively learn global models without exchanging raw data. This preserves privacy while ensuring scalability, making it ideal for sectors like healthcare, finance, and smart cities. Additionally, recent efforts have investigated lightweight FL architectures that can be integrated into resource-constrained IoT systems [129, 130].

1.5.2 Energy-efficient and Green AI

The sustainability implications of massive data processing have led to the rise of energy-efficient AI techniques. Green AI refers to practices that minimize the carbon footprint of model training and inference by optimizing resource usage [131, 132]. This includes hardware-aware model pruning, quantization, and the deployment of TinyML models for edge inference [133]. Moreover, research in lightweight ML deployment has gained traction, with innovations focusing on reducing memory and power consumption while maintaining predictive accuracy [134, 135]. These methods support sustainable applications such as predictive maintenance, precision agriculture, and disaster risk reduction in remote areas.

1.5.3 Integration with Industry 4.0 and Cyber-physical Systems

The transition to Industry 4.0 is inherently dependent on the seamless interaction between digital technologies and physical systems. AI- and IoT-driven cyber-physical systems are now enabling real-time synchronization between digital twins and their physical counterparts, allowing predictive simulations, quality control, and adaptive manufacturing [131, 134]. As highlighted by Anukiruthika and Jayas [137], AI-based automation in agriculture and grain storage can significantly reduce postharvest losses and ensure food security. Similarly, smart logistics using edge-enabled IoT platforms are improving efficiency in transportation and supply chains [138].

1.5.4 Security and Privacy in Intelligent Systems

Security remains a persistent challenge in IoT–AI integration. With devices becoming smarter and more autonomous, vulnerabilities across the network stack require robust solutions. Recent studies have proposed federated intrusion detection systems and blockchain-enhanced IoT platforms as potential defenses [128, 139]. The future of secure AI includes context-aware access control, self-healing systems, and anomaly detection models that operate continuously at the edge. Additionally, privacy-aware data collection methods and GDPR-compliant processing workflows are becoming standard in smart applications [123, 140].

1.5.5 Hybrid and Multimodal Models for Smart Decision-making

An emerging direction in AI research involves hybrid architectures combining symbolic reasoning with DL and multimodal inputs (vision, speech, sensor data). These enable context-aware, human-centric systems with improved explainability [124, 125]. In particular, healthcare, disaster management, and elder care systems benefit from such enriched data-processing capabilities. Table 1.6 summarizes emerging technological directions and their practical applications across various domains.

1.5.6 Future Outlook

Looking forward, the continued evolution of the IoT, AI, and ML ecosystem could be governed by a combination of technological advancements, socio-ethical frameworks, and interdisciplinary collaboration. Among the most transformative trends shaping this trajectory is the rise of XAI, which seeks to demystify the internal workings of complex models and foster trust among stakeholders by offering transparent, interpretable insights [142]. As AI systems increasingly

Table 1.6 Overview of emerging trends in IoT–AI–ML integration and application domains.

Trend	Description	Domains of impact	Reference(s)
Edge AI	On-device inference with reduced latency and privacy risk	Healthcare, transportation, agriculture	[124, 125, 141]
Federated learning	Collaborative learning without centralized data	Finance, smart cities, IoMT	[127, 128]
Tiny ML & green AI	Resource-efficient ML for sustainability	IoT devices, environmental monitoring	[133, 135]
Cyber–physical systems	Real-time coordination between digital and physical components	Industry 4.0, smart grids, warehousing	[131, 134]
AI for food security	Optimized postharvest management and yield prediction	Agriculture, supply chains	[135]
AI-enabled smart logistics	Efficient goods movement with adaptive routing	Transport, logistics, E-commerce	[136]
Blockchain and secure IoT	Ensuring integrity and traceability of data across decentralized networks	Finance, healthcare, industrial IoT	[137]
Hybrid AI systems	Combining symbolic logic with deep neural networks	Emergency response, personalized healthcare	[125, 140]

Note: Table compiled based on synthesized findings from recent literature and technological forecasts.

permeate high-stakes domains, such as healthcare diagnostics, autonomous transportation, and energy management, there is a growing demand for models that not only perform accurately but can also justify their decisions in a manner understandable to humans [123, 125].

In parallel, the emergence of responsible AI frameworks emphasizes the importance of aligning technological development with ethical values, fairness, privacy, and human-centric design principles. This includes mitigating algorithmic biases, ensuring data sovereignty, upholding cybersecurity, and implementing socially accountable governance structures. Regulations such as the EU AI Act and the GDPR exemplify the growing institutional effort to establish legal guardrails around the use of AI and IoT systems. These developments will significantly shape how systems are designed, trained, and deployed in both public and private sectors [88].

Simultaneously, quantum ML (QML) is gaining momentum as an emergent frontier that promises exponential improvements in processing speed and problem-solving complexity. Although still in its infancy, QML could revolutionize data analytics in domains with vast multidimensional datasets, such as climate modeling, genomics, and smart energy systems. As researchers develop hybrid quantum-classical algorithms, the integration of quantum computing with edge and cloud infrastructure is expected to accelerate next-generation decision-making frameworks [133, 140].

Furthermore, interoperability and standardization play a critical role in ensuring seamless integration across platforms and sectors. Without unified data formats, communication protocols, and security policies, the synergistic potential of IoT, AI, and ML cannot be fully realized. Industry alliances, such as the Industrial Internet Consortium (IIC), and initiatives by international standard-setting bodies (e.g. IEEE, ISO) are already contributing to this endeavor. These efforts are crucial for facilitating the scalability and reliability of smart infrastructure across borders and applications. The convergence of academia, industry, and policy-making is also anticipated to become a dominant force in shaping sustainable innovation. Multi-stakeholder collaborations and public–private partnerships will be essential to translate foundational research into scalable, real-world solutions. Academic institutions are expected to play a key role in nurturing transdisciplinary expertise, while industry actors can provide technological infrastructure, and policymakers can establish ethical and legal frameworks to promote responsible innovation [89, 124, 131].

Additionally, the future landscape will likely be characterized by a strong push toward inclusivity and digital equity. Ensuring that AI and IoT technologies benefit all segments of society, including marginalized and underserved populations, will be critical in promoting global sustainability and resilience. This may involve the development of low-cost, localized AI solutions that are tailored to the unique needs of communities in low- and middle-income regions, thereby reducing the digital divide and fostering inclusive development. Technological foresight suggests we are entering an era where AI not only augments human decision-making but also evolves into a strategic enabler for achieving global sustainability objectives. From mitigating climate change to managing pandemics and optimizing food systems, AI-driven solutions, when responsibly implemented, can enhance our capacity to address multifaceted global challenges.

However, as these technologies become increasingly autonomous and self-adaptive, governance and foresight mechanisms must evolve in parallel.

This includes investing in anticipatory regulatory frameworks, dynamic risk assessment models, and participatory platforms that allow citizens to engage in shaping the digital future. Ultimately, a harmonized vision, rooted in human values, ecological responsibility, and technological innovation, will be central to shaping a future where IoT, AI, and ML act as stewards of sustainable progress.

1.6 Conclusion

The integration of the IoT, AI, and ML represents a transformative shift in how modern societies address sustainability, efficiency, and resilience in communication and infrastructure systems. The chapter provided a comprehensive overview of the foundational principles, architectural frameworks, application domains, and synergistic potential of these technologies, underscoring their collective ability to tackle complex global challenges, from healthcare to urban planning, environmental monitoring, and beyond. As demonstrated throughout the discussion, IoT functions as the sensory and data collection backbone, enabling real-time interactions with the physical environment. AI contributes intelligence by enabling systems to analyze, reason, and make decisions, while ML adds adaptability and prediction by learning from patterns within vast data streams. When synergized, these technologies facilitate highly responsive, context-aware, and efficient systems that are essential for sustainable development across critical sectors. However, the pathway toward full-scale deployment is not without obstacles. Challenges including data privacy, interoperability, energy consumption, ethical concerns, and regulatory alignment remain pressing and demand collaborative, multidisciplinary approaches. Addressing these hurdles requires robust governance frameworks, standardized communication protocols, transparent algorithms, and a human-centered design philosophy. Looking ahead, the future of sustainable communication will be increasingly defined by advancements in edge intelligence, FL, quantum computing, and XAI. These emerging technologies offer opportunities to enhance the security, scalability, and transparency of intelligent systems, ensuring that innovation aligns with societal and environmental needs. At the same time, fostering inclusive and equitable access to digital infrastructure will be critical to ensure that the benefits of these technologies are shared globally. In essence, the convergence of IoT, AI, and ML is not merely a technological trend, it is a foundational enabler of a smarter, greener, and more inclusive future. As researchers, engineers, policymakers, and practitioners continue to collaborate across boundaries, the vision of sustainable, intelligent communication systems can be realized in ways that truly serve both people and the planet.

References

1 Gubbi, J., Buyya, R., Marusic, S., and Palaniswami, M. (2013). Internet of Things (IoT): a vision, architectural elements, and future directions. *Future Generation Computer Systems* 29: 1645–1660.
2 Russell, S.J., and Norvig, P. (2016). *Artificial İntelligence: A Modern Approach.* Pearson.
3 Anik, S.M.H., Gao, X., Meng, N., et al. (2022). A cost-effective, scalable, and portable IoT data infrastructure for indoor environment sensing. *Journal of Building Engineering* 49: 104027.
4 Ramya, R. (2024). Analysis and applications finding of wireless sensors and IoT devices with artificial intelligence/machine learning. In: *AIoT and Smart Sensing Technologies for Smart Devices* (pp. 77–102). IGI Global Scientific Publishing.
5 Bodur, G., Turhan, Z., Kucukkaya, A., and Goktas, P. (2024). Assessing the virtual reality perspectives and self-directed learning skills of nursing students: a machine learning-enhanced approach. *Nurse Education in Practice* 75: 103881.
6 Ullah, A., Anwar, S.M., Li, J., et al. (2024). Smart cities: the role of Internet of Things and machine learning in realizing a data-centric smart environment. *Complex & Intelligent Systems* 10: 1607–1637.
7 Kupcu, C., Bodur, G., Kucukkaya, A., and Goktas, P. (2024). Assessing nursing students' e-health literacy and foresights to global health challenges during COVID-19 restrictions: a cross-sectional study with a machine learning approach. *Nurse Education in Practice* 81: 104179.
8 Kucukkuya, A., Bodur, G., Ozata, S.Y., and Goktas, P. (2025). Tapping into awareness: assessing nursing students' water consumption behaviors and sustainability perceptions through a cross-sectional study with machine learning approach. *Public Health Nursing* 42 (1): 444–456.
9 Khan, I.U., Ayub, M.Y., Abdollahi, A., and Dutta, A. (2023). A hybrid deep learning model-based intrusion detection system for emergency planning using IoT-network. *Proceedings of ICT-DM 2023.*
10 Zhang, M., Zhang, H., Yuan, D., and Zhang, M. (2019). Compressive sensing and autoencoder based compressed data aggregation for green IoT networks. In: *2019 IEEE Global Communications Conference (GLOBECOM)* (pp. 1–6). IEEE.
11 Badidi, E., Moumane, K., and Ghazi, F.E. (2023). Opportunities, applications, and challenges of edge-AI enabled video analytics in smart cities: a systematic review. *IEEE Access* 11: 80543.
12 Balasubramanian, S.B., Balaji, P., Munshi, A., et al. (2023). Machine learning based IoT system for secure traffic management and accident detection. *PeerJ Computer Science* 9: 1.

13 Jaiswal, S., Kumar Sharma, D., Jaiswal, T., et al. (2021). Real time analysis of intelligent placing system for vehicles using IoT with deep learning. *Materials Today: Proceedings* 51: 339.

14 Liang, F., Hatcher, W.G., Xu, G., et al. (2019). Towards online deep learning-based energy forecasting. In: *2019 28th International Conference on Computer Communication and Networks (ICCCN)* (pp. 1–9). IEEE.

15 Goktas, P., and Agildere, A.M. (2024). Transforming radiology with artificial intelligence visual chatbot: a balanced perspective. *Journal of the American College of Radiology* 21 (2): 224–225.

16 Goktas, P., and Damadoglu, E. (2024). Future of allergy and immunology: is AI the key in the digital era? *Annals of Allergy, Asthma & Immunology* 134 (4): 396–407.e2.

17 Sagduyu, Y.E., Shi, Y., and Erpek, T. (2019). IoT network security from the perspective of adversarial deep learning. In: *2019 16th Annual IEEE International Conference on Sensing, Communication, and Networking (SECON)* (pp. 1–9). IEEE.

18 He, B., Wang, J., Qi, Q., et al. (2022). Intelligent provisioning of virtualized network functions in cloud of things: a deep reinforcement learning based approach. *IEEE Transactions on Cloud Computing* 10 (2): 1262.

19 Zhang, J., and Li, J. (2022). Mitigating bias and error in machine learning to protect sports data. *Computational Intelligence and Neuroscience* 2022 (1): 4777010.

20 Daghero, F., Pagliari, D.J., and Poncino, M. (2021). Energy-efficient deep learning inference on edge devices. In: *Advances in Computers* (Vol. 122, pp. 247–301). Elseiver.

21 Chen, Z., Tian, P., Liao, W., et al. (2023). Resource-aware knowledge distillation for federated learning. *IEEE Transactions on Emerging Topics in Computing* 11 (3): 706.

22 Liu, J., Wu, J., and Liu, M. (2020). UAV monitoring and forecasting model in intelligent traffic oriented applications. *Computer Communications* 153: 499.

23 Wu, Z., Huang, M., Zhao, A., and Lan, Z. (2021). Traffic prediction based on GCN-LSTM model. *Journal of Physics: Conference Series* 1972 (1): 012107.

24 Triantafyllou, A., Tsouros, D.C., Sarigiannidis, P., and Bibi, S. (2019). An architecture model for smart farming. *Proceedings – 15th Annual International Conference on Distributed Computing in Sensor Systems*. IEEE.

25 Chen, S., Ma, R., Chen, H.-H., et al. (2017). Machine-to-machine communications in ultra-dense networks—a survey. *IEEE Communications Surveys and Tutorials* 19 (3): 1478.

26 Goswami, P., Mukherjee, A., Hazra, R., et al. (2022). AI based energy efficient routing protocol for intelligent transportation system. *IEEE Transactions on Intelligent Transportation Systems* 23 (2): 1670.

27 Gast, M. (2006). *802.11 Wireless Networks: The Definitive Guide*. Southeast University Press.

28 Miller, B.A., and Bisdikian, C. (2001). *Bluetooth Revealed: The Insider's Guide to an Open Specification for Global Wireless Communications*. Prentice Hall.

29 Tennina, S., Koubâa, A., Daidone, R., et al. (2013). *IEEE 802.15.4 and ZigBee as Enabling Technologies for Low-Power Wireless Systems with Quality-of-Service Constraints*. Springer.

30 Jouhari, M., Amhoud, E.M., Saeed, N., and Alouini, M.S. (2022). A survey on scalable LoRaWAN for Massive IoT: recent advances, potentials, and challenges. *arXiv preprint arXiv:2202.11082*.

31 Wang, W., Chen, Q., Yin, Z., et al. (2022). Blockchain and PUF-based lightweight authentication protocol for wireless medical sensor networks. *IEEE Internet of Things Journal* 9 (11): 8883.

32 Du, H., Wang, J., Niyato, D., et al. (2023). Rethinking wireless communication security in semantic internet of things. *IEEE Wireless Communications* 30 (3): 36.

33 Patel, N.R., Kumar, S., and Singh, S.K. (2021). Energy and collision aware WSN routing protocol for sustainable and intelligent IoT applications. *IEEE Sensors Journal* 21 (22): 25282.

34 Aldossari, S.M., and Chen, K.-C. (2019). Machine learning for wireless communication channel modeling: an overview. *Wireless Personal Communications* 106 (1 Special Issue): 41.

35 Cakir, S., Toklu, S., and Yalcin, N. (2020). Rpl attack detection and prevention in the internet of things networks using a GRU based deep learning. *IEEE Access* 8: 183678.

36 Chen, M., Miao, Y., Jian, X., et al. (2019). Cognitive-LPWAN: towards intelligent wireless services in hybrid low power wide area networks. *IEEE Transactions on Green Communications and Networking* 3 (2): 409.

37 Akan, O.B., Cetinkaya, O., Koca, C., and Ozger, M. (2018). Internet of hybrid energy harvesting things. *IEEE Internet of Things Journal* 5 (2): 736.

38 Zhai, L., Zou, Y., Zhu, J., and Li, B. (2022). Improving physical layer security in IRS-aided WPCN multicast systems via Stackelberg game. *IEEE Transactions on Communications* 70 (3): 1957–1970.

39 Epiphaniou, G., Karadimas, P., Ben Ismail, D.K., et al. (2018). Nonreciprocity compensation combined with turbo codes for secret key generation in vehicular ad hoc social IoT networks. *IEEE Internet of Things Journal* 5 (4): 2496.

40 Lu, S., Liu, F., Li, Y., et al. (2024). Integrated sensing and communications: recent advances and ten open challenges. *IEEE Internet of Things Journal* 11 (11): 19094.

41 Sadeghi, A.R., Wachsmann, C., and Waidner, M. (2021). Security and privacy challenges in industrial internet of things. *Proceedings of the IEEE* 105 (1): 96–107.

42 Singh, R., and Rathore, H. (2021). A secure and scalable framework for IoT-based smart cities. *Procedia Computer Science* 185: 539–546.

43 Rahmani, A.M., Gia, T.N., Negash, B., et al. (2021). Exploiting smart e-health gateways at the edge of healthcare Internet-of-Things: a fog computing approach. *Future Generation Computer Systems* 78: 641–658.

44 Domingo-Prieto, M., Chang, T., Vilajosana, X., and Watteyne, T. (2016). Distributed PID-based scheduling for 6TiSCH networks. *IEEE Communications Letters* 20 (5): 1006.

45 Haxhibeqiri, J., Van den Abeele, F., Moerman, I., and Hoebeke, J. (2017). LoRa scalability: a simulation model based on interference measurements. *Sensors (Switzerland)* 17 (6): 1193.

46 Prol, F.S., Ferre, R.M., Saleem, Z., et al. (2022). Position, navigation, and timing (PNT) through low earth orbit (LEO) satellites: a survey on current status, challenges, and opportunities. *IEEE Access* 10: 83971.

47 Salman, T., Zolanvari, M., Erbad, A., et al. (2018). Security services using blockchains: a state of the art survey. *IEEE Communications Surveys & Tutorials* 21 (1): 858–880.

48 Ghahramani, Z. (2015). Probabilistic machine learning and artificial intelligence. *Nature* 521: 452–459.

49 Goktas, P., Rakholia, R., and Carbajo, R.S. (2023). Answering key questions about air pollution dynamics in Ho Chi Minh city: a spatiotemporal analysis using the XAI-SHAP clustering approach. In: *IFIP International Workshop on Artificial Intelligence for Knowledge Management* (pp. 189–203). Cham: Springer Nature Switzerland.

50 Goktas, P., Rakholia, R., and Carbajo, R.S. (2023). Investigating air pollution dynamics in Ho Chi Minh City: a spatiotemporal study leveraging XAI-SHAP clustering methodology. In: *European Conference on Artificial Intelligence* (pp. 201–207). Cham: Springer Nature Switzerland.

51 Goktas, P., Gulseren, D., and Tobin, A.M. (2024). Large language and vision assistant in dermatology: a game changer or just hype? *Clinical and Experimental Dermatology* 49 (8): 783–792.

52 Kucukkaya, A., Arikan, E., and Goktas, P. (2024). Unlocking ChatGPT's potential and challenges in intensive care nursing education and practice: a systematic review with narrative synthesis. *Nursing Outlook* 72 (6): 102287.

53 Goktas, P., and Dirsehan, T. (2025). Using PLS-SEM and XAI for causal-predictive services marketing research. *Journal of Services Marketing* 39 (1): 53–68.

54 Karaçay, P., Goktas, P., Yaşar, Ö., et al. (2025). Investigation of pressure injuries with visual ChatGPT integration: a descriptive cross-sectional study. *Journal of Advanced Nursing*. https://doi.org/10.1111/jan.16905.

55 L'Heureux, A., Grolinger, K., Elyamany, H.F., and Capretz, M.A.M. (2017). Machine learning with big data: challenges and approaches. *IEEE Access* 5: 7776–7797.

56 Demner-Fushman, D., Chapman, W.W., and McDonald, C.J. (2009). What can natural language processing do for clinical decision support? *Journal of Biomedical Informatics* 42 (5): 760–772.

57 Wang, Y., Wang, L., Rastegar-Mojarad, M., et al. (2018). Clinical information extraction applications: a literature review. *Journal of Biomedical Informatics* 77: 34–49.

58 Goktas, P., Kucukkaya, A., and Karacay, P. (2023). Leveraging the efficiency and transparency of artificial intelligence-driven visual Chatbot through smart prompt learning concept. *Skin Research and Technology* 29 (11): e13417.

59 Goktas, P., and Grzybowski, A. (2024). Assessing the impact of ChatGPT in dermatology: a comprehensive rapid review. *Journal of Clinical Medicine* 13 (19): 5909.

60 Ahmad, T., Zhang, D., Huang, C., et al. (2021). Artificial intelligence in sustainable energy systems: a review. *Journal of Cleaner Production* 289: 125449.

61 Li, L., Rong, S., Wang, R., and Yu, S. (2021). Recent advances in artificial intelligence and IoT. *Chemical Engineering Journal* 414: 128802.

62 Mamoshina, P., Ojomoko, L., Yanovich, Y., et al. (2018). Converging blockchain and next-generation artificial intelligence technologies to decentralize and accelerate biomedical research and healthcare. *Oncotarget* 9 (5): 5665–5690.

63 Wamba-Taguimdje, S.-L., Fosso Wamba, S., Kala Kamdjoug, J.R., and Tchatchouang Wanko, C.E. (2020). Influence of artificial intelligence (AI) on firm performance: the business value of AI-based transformation projects. *Business Process Management Journal* 26 (7): 1893–1924.

64 Tambe, P., Cappelli, P., and Yakubovich, V. (2019). Artificial intelligence in human resources management: challenges and a path forward. *California Management Review* 61 (4): 15–42.

65 Chung, S.W., Han, S.S., Lee, J.W., et al. (2018). Automated detection and classification of the proximal humerus fracture by using deep learning algorithm. *Acta Orthopaedica* 89 (4): 468–473.

66 Khan, R.A., Jawaid, M., Khan, A.R., and Sajjad, M. (2023). ChatGPT-reshaping medical education and clinical practice. *Pakistan Journal of Medical Sciences* 39 (1): 301–303.

67 Lu, J., Zhang, W., Hamzei, M., and Jafari, N. (2025). The applications of machine learning mechanisms in the compositions of internet of things services: a systematic study, current progress, and future research agenda. *Engineering Applications of Artificial Intelligence* 147: 110345.

68 Ghafari, R., and Mansouri, N. (2025). Reinforcement learning-based solution for resource management in fog computing: a comprehensive survey. *Expert Systems with Applications* 276: 127214.

69 Majjari, S., Anne, K.R., and George, J. (2023). Deep reinforcement learning (DRL) based data analytics framework for edge based IoT devices latency and resource optimization. *ACCESS 2023 – 3rd International Conference on Advances in Computing, Communication, Embedded and Secure Systems.*

70 Alsharif, M.H., Mosier, W.A., Alomari, O.A., and Yahya, K. (2020). Supervised machine learning for smart data analysis in internet of things environment: an overview. *Journal of Theoretical and Applied Information Technology* 98 (4): 672.

71 Potharaju, S., Tirandasu, R.K., Tambe, S.N., et al. (2025). A two-step machine learning approach for predictive maintenance and anomaly detection in environmental sensor systems. *MethodsX* 14: 103181.

72 Sharma, A., Balasubramanian, V., and Kamruzzaman, J. (2024). A temporal deep Q learning for optimal load balancing in software-defined networks. *Sensors* 24 (4): 1216.

73 Saadi, A., Abghour, N., Chiba, Z., et al. (2024). Traffic light control using reinforcement and deep learning: a survey and comparison. *Lecture Notes in Networks and Systems* 1123: 410.

74 Garai, S., and Samui, S. (2024). Exploring TinyML frameworks for small-footprint keyword spotting: a concise overview. *2024 International Conference on Signal Processing and Communications (SPCOM)*. IEEE.

75 Jiang, Z., Xu, Y., Xu, H., et al. (2024). Semi-supervised decentralized machine learning with device-to-device cooperation. *IEEE Transactions on Mobile Computing* 23 (10): 9757.

76 Zhang, H., Li, M., Yu, H., et al. (2024). Dynamic parameter selection of LoRa edge nodes using reinforcement learning with link prior knowledge. *IEEE Internet of Things Journal* 11 (21): 34420.

77 Sharma, D., Singh, S., Patil, M.R., and Subhan, A. (2024). Medical image processing, disease prediction and report summarization using generative adversarial networks and AIML. In: *2024 2nd World Conference on Communication and Computing (WCONF)* (pp. 1–5). IEEE.

78 Rath, C.K., Mandal, A.K., and Sarkar, A. (2024). Dynamic provisioning of devices in microservices-based IoT applications using context-aware reinforcement learning. *Innovations in Systems and Software Engineering* 1–14. https://doi.org/10.1007/s11334-024-00579-w.

79 Arora, A. (2020). Conceptualising artificial intelligence as a digital healthcare innovation: an introductory review. *Medical Devices: Evidence and Research* 223–230.

80 Duggal, A.S., Singh, R., Gehlot, A., et al. (2021). Infrastructure, mobility and safety 4.0: modernization in road transportation. *Technology in Society* 67: 101791.

81 Ghadi, Y.Y., Mazhar, T., Shah, S.F.A., et al. (2023). Integration of federated learning with IoT for smart cities applications, challenges, and solutions. *PeerJ Computer Science* 9: e1657.

82 Goktas, P., and Grzybowski, A. (2024). Balancing the promises and challenges of artificial intelligence. *Ophthalmology Times Europe* 30–31.

83 Goktas, P., and Grzybowski, A. (2025). Ophthalmology balances the promises and challenges of AI. *Ophthalmology Times* 50.

84 Lee, J. (2020). *Industrial AI. Applications with Sustainable Performance.* Singapore: Springer.

85 Rocha, A., Monteiro, M., Mattos, C., et al. (2024). Edge AI for internet of medical things: a literature review. *Computers and Electrical Engineering* 116: 109202.

86 Qi, Y., and Hossain, M.S. (2024). Harnessing federated generative learning for green and sustainable Internet of Things. *Journal of Network and Computer Applications* 222: 103812.

87 Gopal, G., Suter-Crazzolara, C., Toldo, L., and Eberhardt, W. (2019). Digital transformation in healthcare—architectures of present and future information technologies. *Clinical Chemistry and Laboratory Medicine (CCLM)* 57 (3): 328–335.

88 Goktas, P. (2024). Ethics, transparency, and explainability in generative AI decision-making systems: a comprehensive bibliometric study. *Journal of Decision Systems* 1–29.

89 Goktas, P., and Grzybowski, A. (2025). Shaping the future of healthcare: ethical clinical challenges and pathways to trustworthy AI. *Journal of Clinical Medicine* 14 (5): 1605.

90 Goel, R., and Gupta, P. (2020). *Robotics and Industry 4.0. A Roadmap to Industry 4.0: Smart Production, Sharp Business and Sustainable Development.* Springer Cham.

91 Alotaibi, A., and Barnawi, A. (2023). Securing massive IoT in 6G: recent solutions, architectures, future directions. *Internet of Things* 22: 100715.

92 Gupta, N., Singh, J., Chakraborty, C., et al. (eds.) (2022). *Smart and Secure Internet of Healthcare Things.* CRC Press.

93 Wasatkar, N.N., Chavhan, P.G., and Kolekar, V.K. (2023). IoT and AI-based smart farm: optimizing crop yield and sustainability. In: *Research Trends in Artificial Intelligence: Internet of Things* (pp. 1–17). Bentham Science Publishers.

94 Sarfraz, S., Ali, F., Hameed, A., et al. (2023). Sustainable agriculture through technological innovations. In: *Sustainable Agriculture in the Era of the Omics Revolution* (pp. 223–239). Cham: Springer International Publishing.

95 Bashir, A.K., Victor, N., Bhattacharya, S., et al. (2023). Federated learning for the healthcare metaverse: concepts, applications, challenges, and future directions. *IEEE Internet of Things Journal* 10 (24): 21873–21891.

96 Yang, Y., Wang, H., Jiang, R., et al. (2022). A review of IoT-enabled mobile healthcare: technologies, challenges, and future trends. *IEEE Internet of Things Journal* 9 (12): 9478–9502.

97 Lee, D., and Yoon, S.N. (2021). Application of artificial intelligence-based technologies in the healthcare industry: opportunities and challenges. *International Journal of Environmental Research and Public Health* 18 (1): 271.

98 Kumar, V., Sharma, K.V., Kedam, N., et al. (2024). A comprehensive review on smart and sustainable agriculture using IoT technologies. *Smart Agricultural Technology* 8: 100487.

99 Issa, A.A., Majed, S., Ameer, S.A., and Al-Jawahry, H.M. (2024). Farming in the digital age: smart agriculture with AI and IoT. In: *E3S Web of conferences* (Vol. 477, p. 00081). EDP Sciences.

100 Agrawal, K., Goktas, P., Sahoo, B., et al. (2024). IoT-based service allocation in edge computing using game theory. In: *International Conference on Distributed Computing and Intelligent Technology* (pp. 45–60). Cham: Springer Nature Switzerland.

101 Kalyoncu Atasoy, Z.B., Avery, A., and Goktas, P. (2024). Artificial intelligence-powered nutrition strategies: a focus on vulnerable populations. *Kompass Nutrition & Dietetics* 4 (1): 49–52.

102 Xiao, L., Wan, X., Lu, X., et al. (2018). IoT security techniques based on machine learning: how do IoT devices use AI to enhance security? *IEEE Signal Processing Magazine* 35 (5): 41–49.

103 Bala, B., and Behal, S. (2024). AI techniques for IoT-based DDoS attack detection: taxonomies, comprehensive review and research challenges. *Computer Science Review* 52: 100631.

104 Yagiz, M.A., MohajerAnsari, P., Pesé, M.D., and Goktas, P. (2024). Transforming in-vehicle network intrusion detection: VAE-based knowledge distillation meets explainable AI. In: *Proceedings of the Sixth Workshop on CPS & IoT Security and Privacy* (pp. 93–103).

105 Yagiz, M.A., and Goktas, P. (2025). LENS-XAI: redefining lightweight and explainable network security through knowledge distillation and variational autoencoders for scalable intrusion detection in cybersecurity. *arXiv preprint arXiv:2501.00790.*

106 Bagaa, M., Taleb, T., Bernabe, J.B., and Skarmeta, A. (2020). A machine learning security framework for IoT systems. *IEEE Access* 8: 114066–114077.

107 Ali, M., Naeem, F., Tariq, M., and Kaddoum, G. (2022). Federated learning for privacy preservation in smart healthcare systems: a comprehensive survey. *IEEE Journal of Biomedical and Health Informatics* 27 (2): 778–789.

108 Issa, W., Moustafa, N., Turnbull, B., et al. (2023). Blockchain-based federated learning for securing internet of things: a comprehensive survey. *ACM Computing Surveys* 55 (9): 1–43.

109 Mahmood, N.H., Marchenko, N., Gidlund, M., and Popovski, P. (eds.) (2020). *Wireless Networks and Industrial IoT: Applications, Challenges and Enablers.* Springer Nature.

110 Noor, T.H., Zeadally, S., Alfazi, A., and Sheng, Q.Z. (2018). Mobile cloud computing: challenges and future research directions. *Journal of Network and Computer Applications* 115: 70–85.

111 Ghosh, R.K., Banerjee, A., Aich, P., et al. (2022). Intelligent IoT for automotive Industry 4.0: challenges, opportunities, and future trends. In: *Intelligent Internet of Things for Healthcare and Industry* (pp. 327–352). Cham: Springer.

112 Belliger, A., and Krieger, D.J. (2018). The digital transformation of healthcare. In: *Knowledge Management in Digital Change: New Findings and Practical Cases* (pp. 311–326). Cham: Springer International Publishing.

113 Alshehri, F., and Muhammad, G. (2020). A comprehensive survey of the Internet of Things (IoT) and AI-based smart healthcare. *IEEE Access* 9: 3660–3678.

114 Mohamed, N. (2023). Current trends in AI and ML for cybersecurity: a state-of-the-art survey. *Cogent Engineering* 10 (2): 2272358.

115 Bhandari, G., Lyth, A., Shalaginov, A., and Grønli, T.M. (2023). Distributed deep neural-network-based middleware for cyber-attacks detection in smart IoT ecosystem: a novel framework and performance evaluation approach. *Electronics* 12 (2): 298.

116 Gyamfi, E., Ansere, J.A., Kamal, M., et al. (2022). An adaptive network security system for IoT-enabled maritime transportation. *IEEE Transactions on Intelligent Transportation Systems* 24 (2): 2538–2547.

117 Abdullahi, M., Baashar, Y., Alhussian, H., et al. (2022). Detecting cybersecurity attacks in internet of things using artificial intelligence methods: a systematic literature review. *Electronics* 11 (2): 198.

118 Qu, Q., Hatami, M., Xu, R., et al. (2024). The microverse: a task-oriented edge-scale metaverse. *Future Internet* 16 (2): 60.

119 Moustafa, N., Koroniotis, N., Keshk, M., et al. (2023). Explainable intrusion detection for cyber defences in the internet of things: opportunities and solutions. *IEEE Communications Surveys & Tutorials* 25 (3): 1775–1807.

120 Ashraf, Q.M., Tahir, M., Habaebi, M.H., and Isoaho, J. (2023). Toward autonomic internet of things: recent advances, evaluation criteria, and future research directions. *IEEE Internet of Things Journal* 10 (16): 14725–14748.

121 Alhafnawi, M., Salameh, H.A.B., Masadeh, A.E., et al. (2023). A survey of indoor and outdoor UAV-based target tracking systems: current status, challenges, technologies, and future directions. *IEEE Access* 11: 68324–68339.

122 Rani, S., Kataria, A., Kumar, S., and Tiwari, P. (2023). Federated learning for secure IoMT-applications in smart healthcare systems: a comprehensive review. *Knowledge-Based Systems* 274: 110658.

123 Gill, S.S., Golec, M., Hu, J., et al. (2025). Edge AI: a taxonomy, systematic review and future directions. *Cluster Computing* 28 (1): 1–53.

124 Hemmati, A., Raoufi, P., and Rahmani, A.M. (2024). Edge artificial intelligence for big data: a systematic review. *Neural Computing and Applications* 36 (19): 11461–11494.

125 Mitra, A., Mohanty, S.P., and Kougianos, E. (2024). ToEFL: a novel approach for training on edge in smart agriculture. In: *Proceedings of the Great Lakes Symposium on VLSI 2024* (pp. 657–662).

126 Chen, Y.C., Hsu, S.Y., Xie, X., et al. (2024). Privacy preserving support vector machine based on federated learning for distributed IoT-enabled data analysis. *Computational Intelligence* 40 (2): e12636.

127 Chaves, A.J., Martín, C., and Díaz, M. (2024). Towards flexible data stream collaboration: federated learning in Kafka-ML. *Internet of Things* 25: 101036.

128 Agrawal, S., Sarkar, S., Aouedi, O., et al. (2022). Federated learning for intrusion detection system: concepts, challenges and future directions. *Computer Communications* 195: 346–361.

129 Douaioui, K., Fri, M., Mabroukki, C., and Semma, E.A. (2018). The interaction between Industry 4.0 and smart logistics: concepts and perspectives. In: *2018 International Colloquium on Logistics and Supply Chain Management (LOGISTIQUA)* (pp. 128–132). IEEE.

130 Rane, N.L., Kaya, Ö., and Rane, J. (2024). *Artificial Intelligence, Machine Learning, and Deep Learning for Sustainable Industry 5.0.* Deep Science Publishing.

131 Katari, V., Goyal, S., Nigam, V., et al. (2024). A confluence of emerging technologies like IoT, edge & cloud computing, blockchain, industry 4.0 & 5.0, AI & ML toward the realization of eco-friendly supercapacitors. In: *Eco-Friendly Supercapacitors: Design and Future Perspectives in Sustainable and Green Energy Storage Devices* (pp. 163–204). American Chemical Society.

132 Hussein, D.H., and Ibnkahla, M. (2025). Intelligent and autonomous edge slicing for IoT systems. *IEEE Internet of Things Journal* 12 (2): 21210–21224.

133 Baciu, V.E., Braeken, A., Segers, L., and Silva, B.D. (2025). Secure tiny machine learning on edge devices: a lightweight dual attestation mechanism for machine learning. *Future Internet* 17 (2): 85.

134 Mwase, C., Jin, Y., Westerlund, T., et al. (2022). Communication-efficient distributed AI strategies for the IoT edge. *Future Generation Computer Systems* 131: 292–308.

135 Merenda, M., Porcaro, C., and Iero, D. (2020). Edge machine learning for AI-enabled IoT devices: a review. *Sensors* 20 (9): 2533.

136 Dafflon, B., Moalla, N., and Ouzrout, Y. (2021). The challenges, approaches, and used techniques of CPS for manufacturing in Industry 4.0: a literature review. *The International Journal of Advanced Manufacturing Technology* 113: 2395–2412.

137 Anukiruthika, T., and Jayas, D.S. (2025). AI-driven grain storage solutions: exploring current technologies, applications, and future trends. *Journal of Stored Products Research* 111: 102588.
138 Cichosz, M., Wallenburg, C.M., and Knemeyer, A.M. (2020). Digital transformation at logistics service providers: barriers, success factors and leading practices. *The International Journal of Logistics Management* 31 (2): 209–238.
139 Yogi, M.K., Aiswarya, D., and Mundru, Y. (2023). Security for AI and IoT convergence: novel perspectives. *International Journal of Scientific Research in Network Security and Communication* 11 (5): 1–7.
140 Aboualola, M., Abualsaud, K., Khattab, T., et al. (2023). Edge technologies for disaster management: a survey of social media and artificial intelligence integration. *IEEE Access* 11: 73782–73802.
141 Agrawal, K., Goktas, P., Holtkemper, M., et al. (2025). AI-driven transformation in food manufacturing: a pathway to sustainable efficiency and quality assurance. *Frontiers in Nutrition* 12: 1553942.
142 Panahandeh, A., Rabiei-Dastjerdi, H., Goktas, P., and McArdle, G. (2025). Answering new urban questions: using eXplainable AI-driven analysis to identify determinants of Airbnb price in Dublin. *Expert Systems with Applications* 260: 125360.

2

AI/ML Techniques for Enhancing IoT-based Communication

Kushagra Agrawal[1,], Muhammet Anil Yagiz[2], and Polat Goktas[3,4]*

[1] *School of Computer Engineering, KIIT Deemed to be University, Bhubaneswar, India*
[2] *Department of Computer Engineering, Kirikkale University, Kirikkale, Turkey*
[3] *UCD School of Computer Science, University College Dublin, Dublin, Ireland*
[4] *Faculty of Engineering and Natural Sciences, Sabanci University, Istanbul, Turkey*

*Corresponding author: kushagra.agrawal@ieee.org

2.1 Introduction

The rapid proliferation of the Internet of Things (IoT) has led to a hyper-connected environment where billions of devices, sensors, and actuators constantly exchange data. This massive ecosystem, while enabling unprecedented automation and intelligence, also presents substantial challenges in ensuring secure, efficient, and scalable communication. As IoT systems grow in complexity, traditional rule-based approaches are proving inadequate in meeting modern demands such as low-latency response, real-time anomaly detection, and robust network security. Artificial intelligence (AI) and machine learning (ML) have emerged as transformative solutions to enhance communication across IoT networks. These technologies provide cognitive capabilities to devices and network nodes, enabling them to learn from data, recognize patterns, and make proactive decisions that improve system performance. The application of ML techniques in IoT has spanned a wide array of use cases, including traffic management, anomaly detection, intrusion prevention, and energy-efficient routing [1–4].

The fundamental premise of AI in IoT communication lies in its ability to address core networking issues such as congestion, interference, and dynamic

AI and ML Techniques in IoT-based Communication: A Path to Sustainable Development Goals, First Edition. Edited by Sumita Mishra, Nishu Gupta and Polat Goktas.

topology changes. Deep reinforcement learning (DRL), for instance, has shown promise in offloading decisions and resource allocation in edge computing environments, where latency and bandwidth are constrained [5, 6]. Similarly, fuzzy pattern tree models have been developed for edge-based malware detection to enhance system resilience and communication trust [7]. A critical area where AI/ML plays a pivotal role is in intrusion detection systems (IDS). Traditional IDS often struggle with detecting sophisticated cyberattacks in real time. Recent research has proposed ensemble-based IDS [8] and two-stage hybrid AI architectures [9] that outperform conventional systems in terms of accuracy, false alarm rate, and adaptability. These models process communication flows in real time, ensuring secure transmission and detecting abnormalities early in the communication chain.

The advancement of federated learning (FL) offers additional potential in decentralized IoT environments. By allowing devices to collaboratively train models without transferring raw data, FL reduces communication overhead while preserving privacy – a crucial concern in domains such as smart healthcare, industrial IoT, and intelligent transportation [10–12]. This paradigm shift from centralized to distributed intelligence has significant implications for reducing bottlenecks and enhancing network scalability. AI has also found utility in optimizing quality of service (QoS) and security jointly. For example, reinforcement learning (RL) has been applied to control transmission parameters based on real-time environmental feedback, ensuring that IoT networks remain both efficient and secure under varying conditions [13, 14]. Similarly, intelligent edge mechanisms such as FlowGuard [15] have been designed to provide localized, real-time defense against unauthorized communication attempts, thereby safeguarding data integrity and device trust.

The evolving complexity of IoT ecosystems also demands rich, domain-specific datasets for training robust AI models. Datasets like TON-IoT [16] and ToN_IoT [17] have become valuable benchmarks for evaluating intrusion detection and network performance in heterogeneous IoT environments. These datasets incorporate telemetry, network traffic, and system logs across various domains, aiding researchers in developing generalized and transferable AI models. With the surge in distributed denial-of-service (DDoS) and spoofing attacks, AI/ML techniques have become crucial in proactive defense mechanisms. Several reviews and surveys [18–20] have highlighted the superiority of ML-based solutions over signature-based models, particularly in detecting zero-day attacks. Furthermore, frameworks such as IRAF [21] integrate deep learning (DL) with adaptive routing to mitigate threats dynamically while ensuring uninterrupted communication. The incorporation of intent-based networking and intelligent agents into communication infrastructure is another promising direction. Roopak et al. [22] emphasized the relevance of DL models in enabling context-aware decision-making in

IoT networks. These systems can dynamically adjust routing paths, filter malicious packets, and optimize bandwidth utilization, all without human intervention.

Nevertheless, significant challenges remain in integrating AI with IoT communication protocols. These include the computational overhead of training large models on resource-constrained devices, the need for explainable AI in safety-critical applications, and the risk of adversarial attacks targeting ML components. Techniques such as lightweight models, model pruning, and on-device learning are being explored to address these constraints [23, 24].

Communication demands are also evolving with the integration of CubeSats, autonomous vehicles, and real-time sensing systems into the IoT landscape. Future architectures must support dynamic communication topologies, heterogeneous protocols, and adaptive learning mechanisms [25]. AI/ML holds the key to orchestrating these complex systems, ensuring that data flows remain efficient, secure, and contextually aware. From intrusion detection and malware prevention to traffic optimization and federated model training, intelligent techniques are being embedded at every level of the IoT stack. This chapter aims to explore these advances in depth, providing insights into algorithms, architectures, and datasets that define the future of AI-enhanced IoT communication.

2.2 Background and Fundamentals

2.2.1 IoT Communication Technologies Overview

The IoT ecosystem encompasses a wide range of heterogeneous devices connected through various wireless and wired communication technologies. The effectiveness of an IoT network depends on the communication protocols adopted to enable seamless, scalable, and energy-efficient data exchange. These technologies range from short-range options such as Wi-Fi, Bluetooth Low Energy (BLE), and ZigBee, to long-range, low-power communication systems like LoRa and Narrowband IoT (NB-IoT), and extend to more robust and high-throughput networks enabled by 5G. Unlike traditional IoT systems that rely on reactive decision-making and cloud-based analytics, AI/ML-enhanced IoT systems incorporate intelligent automation, real-time adaptation, and predictive analytics (as illustrated in Figure 2.1).

2.2.1.1 Low-power Wide Area Networks

Low-power wide area networks (LPWAN) technologies, such as LoRa and NB-IoT, have gained popularity due to their long-range coverage and low power consumption-key requirements for massive machine-type communication (mMTC) in smart cities, agriculture, and remote monitoring applications. LoRa,

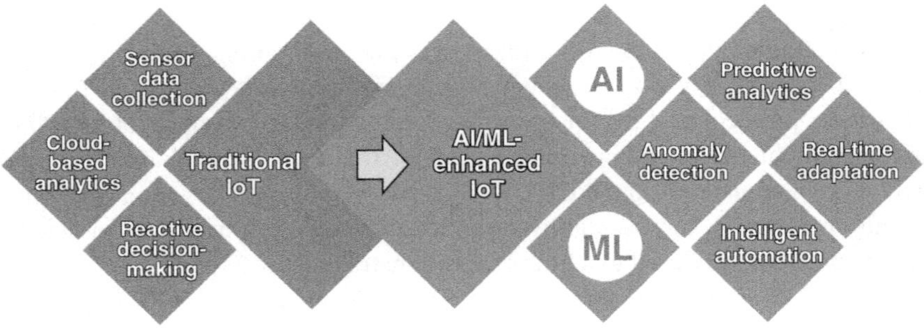

Figure 2.1 Comparison between traditional IoT communication and AI/ML-enhanced IoT communication, emphasizing automation, analytics, and decision-making improvements.

in particular, utilizes chirp spread spectrum modulation and supports adjustable spreading factors to balance data rate and communication range. Zhong et al. [26] proposed an optimized LoRa spreading factor (SF) allocation method based on device grouping and orthogonality metrics, significantly improving communication efficiency and reducing collision rates in dense networks.

Another innovation includes vehicular monitoring systems using LPWAN to ensure reliable and real-time data collection from mobile nodes. Santa et al. [27] introduced a LoRa-based platform for vehicular applications, highlighting its robustness against interference and its capability to operate in energy-constrained environments. In a broader context, Chen et al. [28] presented the concept of Cognitive-LPWAN, integrating AI-driven adaptability into LPWAN technologies to support dynamic spectrum access, optimize coverage, and meet evolving application demands.

2.2.1.2 Short-range Wireless Protocols: Wi-Fi, ZigBee, and BLE

Short-range communication protocols like Wi-Fi, ZigBee, and BLE are commonly used in indoor or high-data-rate applications, including smart homes, industrial monitoring, and healthcare systems. These protocols offer relatively higher bandwidth than LPWAN options, at the cost of shorter range and increased energy consumption. For instance, ZigBee offers low data rates but is highly energy-efficient, making it suitable for sensor-heavy environments. BLE, a derivative of traditional Bluetooth, provides lower power consumption and is widely adopted in wearable and healthcare devices. Meanwhile, Wi-Fi, despite its high energy demand, is utilized in applications requiring significant data transmission and low latency.

Liu et al. [29] addressed the load-balancing challenges in heterogeneous networks that integrate Wi-Fi, ZigBee, and BLE, proposing efficient routing protocols that reduce congestion and packet loss. The need for protocol coexistence was further highlighted in the work of Saavedra et al. [30], who developed a universal testbed for evaluating wireless technologies under realistic conditions, enabling performance benchmarking across different protocol stacks.

2.2.1.3 Cellular Technologies and 5G

NB-IoT, a cellular-based LPWAN technology, leverages existing LTE infrastructure to offer deep indoor coverage and high device density. It is specifically designed for low-throughput and delay-tolerant applications such as smart metering and asset tracking. Buratti et al. [31] provided a comprehensive overview of NB-IoT's role in the broader IoT protocol landscape, noting its strength in licensed spectrum use, interference control, and long battery life.

As IoT applications become more demanding, requiring ultra-reliable low-latency communication (URLLC) and enhanced mobile broadband (eMBB), 5G emerges as the most promising solution. 5G provides higher bandwidth, reduced latency, and supports massive device connectivity, making it suited for Industry 4.0, autonomous vehicles, and real-time analytics. Li et al. [32] explored the role of 5G in enhancing physical layer key generation, contributing to secure communication in both static and mobile IoT scenarios. Furthermore, Salman and Jain [33] examined the evolution of cellular protocols for IoT, emphasizing the importance of backward compatibility and standardization in facilitating a smooth transition from 4G to 5G in connected environments. Their insights on spectrum efficiency and signaling optimization offer a solid foundation for protocol stack enhancements.

2.2.1.4 Protocol Integration and Interoperability

The diversity of IoT applications necessitates interoperability among protocols. Asif et al. [34] proposed a cognitive communication framework that integrates heterogeneous communication technologies, allowing seamless protocol switching based on application requirements and environmental conditions. Similarly, Li et al. [35] introduced a oneM2M protocol conversion platform that bridges gaps between different standards and architectures, facilitating end-to-end interoperability in smart manufacturing systems. To address the interoperability challenges in real-time environments, Khan and Naseer [36] developed a garbage monitoring system based on Wi-Fi and NB-IoT hybrid architecture, demonstrating the effectiveness of multiprotocol systems in smart campus settings. Such integrated models are essential for applications that span multiple domains and require context-aware communication strategies. Last, John et al. [37] provided a foundational overview of IoT communication technologies, emphasizing how protocol selection depends on key factors like data rate, energy constraints, communication range, and environmental interference.

2.2.2 Key Challenges in IoT Communications

As IoT ecosystems get larger and more heterogeneous, the communications infrastructure underpinning such networks experience a set of related and thorny challenges. Such challenges, varying from scalability and latency to power limitations and congestion in networks, can potentially have a broad impact on IoT system performance, dependability, and uptake, especially in mission-critical applications like healthcare, smart cities, and industrial automation. As shown in Table 2.1, communication in IoT networks is constrained by multiple interdependent factors such as device heterogeneity, traffic congestion, and energy limitations.

2.2.2.1 Scalability and Heterogeneity

One of the foremost challenges in IoT communication is scalability. With billions of interconnected devices expected in the coming years, managing seamless communication across this vast landscape requires efficient addressing schemes, routing protocols, and data-processing methods. Zyane et al. [38] proposed a scalable cloud-based collaboration system (IoTScal-SC2) to address these concerns. Their system supports distributed processing while maintaining consistent performance

Table 2.1 Key communication challenges in IoT networks and corresponding AI/protocol-based solutions.

Challenge	Description	Proposed approaches/ solutions	Reference(s)
Scalability	Difficulty handling billions of devices and diverse protocols	Scalable cloud-based architectures	[38, 39]
Latency	Need for real-time response in critical applications	DiffServ-aware middleware; prioritized data flow	[40, 41]
Energy efficiency	Limited battery power and computing in edge devices	Opportunistic routing; energy-aware communication	[42, 43]
Network congestion	Excessive data traffic causing delays and packet loss	Multipath routing; congestion control algorithms	[42]
Protocol overhead	Added complexity from encryption, validation, or consensus mechanisms	Lightweight fuzzy-blockchain communication frameworks	[44]

as device density increases. The heterogeneity of IoT devices – differing in power, protocols, and functions – further complicates network management and service delivery [39].

2.2.2.2 Latency and Real-time Processing

Many IoT applications, especially in fields such as autonomous vehicles and industrial robotics, require ultra-low latency communication to function safely and effectively. Traditional cloud-centric architectures often fall short in meeting these requirements. To address this, Abdellah et al. [40] developed a DiffServ-aware cloud solution that reduces end-to-end delay by prioritizing data flows based on service requirements. Similarly, Bahiri et al. [41] introduced a cloud middleware enhancement framework that streamlines task execution and data exchange, helping to meet real-time constraints.

2.2.2.3 Energy Efficiency and Resource Constraints

Most IoT nodes operate on battery power and possess limited processing capabilities. Consequently, energy efficiency is vital for ensuring system longevity and sustainability. Inefficient communication protocols can lead to rapid energy depletion, reducing the operational lifetime of sensor nodes. Adil [42] proposed an opportunistic multipath routing protocol that not only mitigates congestion but also optimizes energy usage by leveraging dynamic path selection. In another approach, Pavlovian modeling using low-energy wireless modules was introduced for predictive analytics with minimal power consumption [43].

2.2.2.4 Congestion and Data Overload

With continuous data exchange between devices, network congestion is a common challenge in dense IoT environments. High traffic loads can result in increased packet loss, delayed delivery, and reduced overall throughput. Congestion becomes even more problematic when mission-critical data is delayed or lost. Adil [42] addressed this issue through multipath routing mechanisms that balance traffic across alternate paths, reducing bottlenecks in the network.

2.2.2.5 Security and Protocol Overhead

Though not the main focus of this subsection, it is worth noting that security-related overhead can also affect communication performance. Techniques like blockchain and fuzzy logic-based systems are increasingly used to embed lightweight intelligence at the communication layer [44]. However, if not well-optimized, such frameworks can introduce latency and increase power consumption, thereby impacting core communication metrics.

2.3 AI/ML Techniques for Enhancing IoT Communication

2.3.1 Intelligent Routing and Path Optimization

With the ever-increasing magnitude of IoT deployments, there has been an increased need for routing efficiency and reliability in ensuring uninterrupted communication. The dynamic and heterogeneous characteristics of IoT networks pose distinct challenges, such as mobility, limited energy, and varying traffic density, with traditional routing algorithms being unable to address the above complexities. To that end, ML methods – specifically RL, DL, and hybrid intelligent systems – are being used to facilitate intelligent routing and energy-conscious path optimization in real time. Table 2.2 summarizes some of the most frequently applied ML techniques for intelligent routing in IoT systems, showcasing their respective datasets, evaluation metrics, performance outcomes, and typical suitability. AI and ML techniques play a critical role in optimizing IoT communication by enabling adaptive bandwidth allocation, predictive congestion management, and secure authentication (as shown in Figure 2.2).

2.3.1.1 RL and Q-learning in Routing

RL, including Q-learning and Deep Q Networks (DQNs), has shown great promise for adaptive routing in dynamic IoT topologies. These models learn optimal routing

Table 2.2 Comparison of ML algorithms for routing optimization in IoT.

Technique	Dataset/domain	Metrics used	Performance highlight	Suitability
Q-learning (RL)	Synthetic IoT networks	Energy consumption, PDR	Low energy use; adaptive routing	Dynamic, low-power sensor networks [21]
DRL (DQN)	Smart agriculture	Latency, reliability	High reliability in real-time conditions	Agricultural and environmental IoT [46]
Neuro-fuzzy	WSN simulation datasets	Network lifetime, delay	Improved clustering and delay minimization	Dense sensor deployments [45]
Decision tree (IntruDTree)	Bot-IoT	Accuracy, detection rate	High interpretability, good performance	Lightweight edge devices [48]
Random forest	Health IoT datasets	Accuracy, packet delivery	Robust and scalable under varying loads	Healthcare, wearable IoT [49]

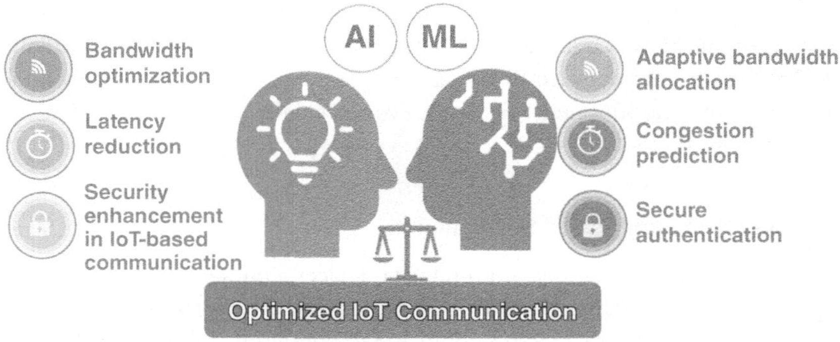

Figure 2.2 AI/ML-enhanced IoT communication improving bandwidth allocation, latency reduction, and security mechanisms.

decisions based on continuous feedback from the network environment, enabling nodes to discover energy-efficient and congestion-free paths autonomously. Chen et al. [14] emphasized the benefits of RL-based routing in 6G-enabled environments, where the scale and complexity of ultra-dense networks make human-designed routing rules ineffective. In energy-constrained environments, such as wireless sensor networks (WSNs), RL-based approaches have helped prolong network lifetime by balancing node workloads. Thangaramya et al. [45] proposed an energy-aware cluster-based routing mechanism using neuro-fuzzy logic that adapts to residual energy and network density, leading to improved packet delivery ratios and reduced transmission delays.

2.3.1.2 DL and Neural Models

DL has contributed significantly to routing by offering multi-hop decision-making, pattern recognition from large-scale telemetry, and context-aware path estimation. Modi et al. [46] provided a comprehensive review of DL in routing, showing its application in predicting traffic congestion and link failures, key inputs for route planning. In smart agriculture, Bu and Wang [47] implemented a DRL system for autonomous route selection among sensor nodes. The model achieved higher reliability and energy efficiency than traditional protocols by dynamically adjusting to soil conditions, moisture levels, and node failures.

2.3.1.3 Hybrid and Ensemble ML Models

Hybrid models, combining decision trees, fuzzy logic, and evolutionary algorithms, have emerged as powerful tools for optimizing path selection in IoT. Sarker et al. [48] introduced IntruDTree, a decision tree-based routing framework that adjusts based on both historical behavior and real-time congestion. Similarly, Kaur et al. [49] demonstrated the effectiveness of random forest models in identifying optimal routing clusters for healthcare monitoring systems. These approaches not only enhance reliability but also reduce packet overhead and computational complexity-key considerations in constrained IoT environments. Ullah and Mahmoud [50] emphasized that ensemble methods provide robustness and adaptability across varying network topologies, especially when integrated with real-time monitoring.

2.3.2 Congestion and Traffic Management

In IoT environments, where tens of millions of devices simultaneously push data on shared, at times bandwidth-constrained, networks, congestion and traffic control are vital in order to enable uninterrupted communication. High-density deployment, varying data rates, and mobility cause cyclic bottlenecks, leading to packet delay, jitter, and loss of mission-critical data. ML and DL algorithms are widely applied nowadays in order to counteract such difficulties and forecast, manage, and reduce network congestion in real time. As summarized in Table 2.3, various ML- and DL-based techniques, such as GCN-LSTM, Deep Q-learning, and FL, have been implemented to manage traffic congestion in IoT networks. These approaches offer adaptive routing, proactive forecasting, and edge-level intelligence for handling network bottlenecks efficiently [51–53].

2.3.2.1 Predictive Congestion Control Using Recurrent Neural Networks and Long Short-term Memory

Time-dependent traffic behavior in IoT networks makes recurrent neural networks (RNNs) and their advanced variant long short-term memory (LSTM) ideal

Table 2.3 ML and DL techniques for traffic management in IoT networks.

Technique	Purpose	Key benefit	Reference(s)
GCN-LSTM	Traffic flow prediction	Accurate spatial-temporal forecasting	[51]
Temporal Deep Q-learning	Load balancing and congestion control	Time-aware adaptive routing	[52]
UAV + predictive modeling	Smart city traffic congestion prediction	Real-time aerial traffic insights	[54]
ECRT edge platform	Edge traffic detection	Low latency congestion response	[56]
FL + distillation	Edge-level traffic prediction	Privacy-preserving distributed learning	[53]
DeepStream-based AI traffic control system	Smart traffic light control	Real-time flow optimization	[58]

for congestion prediction. These models effectively capture temporal dependencies and fluctuations in network load. Wu et al. [51] proposed a GCN-LSTM-based traffic forecasting model, combining graph convolutional networks (GCNs) with LSTM layers to account for both spatial and temporal traffic characteristics. Their approach demonstrated enhanced accuracy in predicting traffic congestion trends, especially in urban IoT applications. In another real-time implementation, Liu et al. [54] integrated UAV-based sensing with predictive modeling to forecast traffic flow and manage congestion in smart cities. This UAV-assisted approach allowed for overhead monitoring and congestion-aware routing in vehicular IoT (V-IoT) systems, significantly improving response latency and traffic coordination.

2.3.2.2 DRL for Adaptive Traffic Control

DRL has emerged as a powerful solution for intelligent, adaptive congestion management. DRL agents can learn optimal traffic shaping and rerouting policies by interacting with dynamic network conditions. Wu et al. [55] reviewed DRL strategies in the context of IoT and edge computing, highlighting their advantages in managing fluctuating traffic volumes while minimizing energy use and maintaining QoS. Sharma et al. [52] proposed a temporal Deep Q-learning framework for real-time load balancing in distributed IoT systems. By using Q-learning in combination with temporal features, their model successfully adapted to time-varying data loads and optimized routing paths to prevent congestion build-up.

2.3.2.3 Edge Intelligence and FL in Traffic Management

To minimize latency and reduce centralized computation overhead, Edge AI is increasingly integrated into IoT communication systems. Zhao et al. [56] demonstrated ECRT, an edge computing platform capable of dynamically allocating computational and communication resources based on predicted congestion levels. This proactive approach enabled faster decisions closer to the data source. Chen et al. [53] introduced a resource-aware knowledge distillation approach for FL environments, enabling smaller edge models to maintain high predictive performance. This framework supports decentralized congestion forecasting without compromising privacy, a critical feature in healthcare and surveillance domains.

2.3.2.4 Hybrid ML Models and Real-world Applications

Hybrid learning systems, combining classical ML with DL models, have also been used for real-time congestion management. Nallaperuma et al. [57] presented an online incremental ML platform that adapts to streaming data for detecting congestion patterns and adjusting resource distribution. Their model showed high adaptability to changing network conditions, especially in fast-evolving smart city infrastructures. Uddin et al. [58] developed an AI-driven traffic control system using NVIDIA DeepStream, capable of real-time congestion detection and vehicle flow optimization at intersections. This system achieved robust performance in both simulated and live environments and offers a scalable solution for urban mobility management.

2.3.3 Resource Allocation and Load Balancing

Efficient resource usage and smart load balancing are critical to ensure performance, energy efficiency, and responsiveness in such distributed settings. Conventional cloud-based solutions are beset with shortcomings in dealing with real-time constraints, data privacy, and power consumption. In this context, FL and Edge AI have emerged as promising paradigms for enabling scalable and intelligent resource management in IoT communication. As summarized in Table 2.4, recent advancements in FL and Edge AI have introduced a diverse set of models and techniques aimed at optimizing resource allocation and load balancing in IoT networks. These approaches demonstrate significant improvements in energy efficiency, fairness, and latency reduction across distributed environments [59–61].

2.3.3.1 FL for Distributed Resource Optimization

FL enables multiple edge devices to collaboratively train ML models without centralizing data, thereby reducing communication overhead while preserving privacy. This makes FL particularly suitable for resource-constrained and

Table 2.4 FL and Edge AI approaches for resource allocation and load balancing in IoT.

Approach/model	Technique used	Benefit achieved	Reference(s)
Deep federated Q-learning (DFQL)	FL + Q-learning	Dynamic network slicing and QoS	[59]
Energy-aware FL	FL + adaptive communication	Reduced energy use in cellular IoT	[61]
FedMCCS	FL + Multi-criteria client selection	Fairness and heterogeneity-aware training	[62]
Edge offloading scheduler	Edge AI + task migration	Adaptive real-time task scheduling	[63]
Digital twin resource allocator	Edge AI + predictive modeling	Latency-aware resource provisioning	[64]
Joint DNN partitioning	DRL + FL	Reduced latency in DNN inference	[66]

data-sensitive IoT environments. Messaoud et al. [59] introduced a deep federated Q-learning-based network slicing algorithm for 5G, demonstrating its potential for dynamic resource allocation across network slices while maintaining task priority and QoS requirements. Similarly, Zhao et al. [61] proposed an energy-efficient FL protocol over cellular IoT systems, which dynamically adjust communication rounds based on local computing resources and bandwidth availability. This approach significantly improved convergence while reducing energy overhead, a critical factor in battery-operated IoT deployments. Another major advancement was presented by Abdulrahman et al. [62] through FedMCCS, a multi-criteria client selection strategy that considers energy levels, computational capacity, and data heterogeneity when scheduling training clients. By selecting clients intelligently, the system achieves improved fairness, reduced straggler effects, and enhanced learning accuracy.

2.3.3.2 Edge AI for Real-time Decision-making

Edge AI brings intelligence closer to the data source, enabling real-time analytics, faster response times, and reduced latency. In IoT contexts, this translates to smarter, localized decisions without relying on cloud infrastructure. Jiang et al. [63] explored edge-intelligence-based offloading mechanisms for dynamic task allocation in mobile IoT applications. Their proposed method balances computation between edge nodes and cloud servers while adapting to fluctuating traffic and node availability. Qiu et al. [64] introduced a digital twin-based intelligent system for smart factories, where Edge AI was used to allocate resources based on predicted operational demand. This system maintained low latency and high reliability for industrial automation, critical in manufacturing 4.0 environments. Edge AI also improves scheduling efficiency. Cui et al. [60] proposed a joint client scheduling and resource management system that optimizes bandwidth and CPU allocation during federated training. Their system used real-time network feedback to dynamically adjust participation rates and resource usage, leading to both better model performance and lower resource waste.

2.3.3.3 Hybrid and RL-enhanced Resource Allocation

Recent studies have combined DRL and FL to jointly optimize task offloading and network resource utilization. Zhang et al. [65] developed a DRL-assisted scheduling policy that reacts to QoS violations and resource fluctuations in industrial IoT. This system demonstrated the capability to balance energy consumption and computation latency while meeting application-specific reliability constraints. Likewise, Dong et al. [66] proposed a hybrid framework that jointly partitions DNN models across devices and allocates bandwidth accordingly. By learning optimal split points and allocation policies, their approach supports real-time inference across collaborative edge nodes, significantly reducing overall computational delay.

2.4 Architectures Integrating AI/ML in IoT Communication

The convergence of ML and AI with IoT communication platforms has transformed data processing, routing, and management. Centralized architecture is being replaced more and more by intelligent, dynamic, and distributed edge and fog computing-based systems, AI-based control planes, and hybrid learning structures. Such architectural advancements are aimed at reversing the escalating complexity of IoT networks while supporting real-time decision-making, energy efficiency, and assured connectivity. A wide range of AI/ML-enabled communication architectures, each tailored to distinct IoT applications, have been proposed in recent literature. These architectures vary significantly in terms of processing location, scalability, and responsiveness. Table 2.5 provides a comparative overview of the major architectures, highlighting their technical configurations, strengths, and limitations [67–69]. The integration of AI/ML in IoT communication is structured across multiple layers, from sensor data acquisition to edge/cloud processing and decision-making modules (as depicted in Figure 2.3).

2.4.1 Edge and Fog Computing Paradigms

Edge and fog computing paradigms are designed to address the latency, bandwidth, and processing limitations of cloud-based IoT systems. By bringing computation closer to the data source, edge intelligence facilitates real-time inference, localized control, and reduced communication overhead. Alhammadi et al. [70] emphasized the importance of AI-enabled 6G architectures, where edge devices execute ML models for immediate responses in latency-critical applications such as autonomous vehicles and remote surgeries. These edge platforms also minimize the need for backhaul communication, significantly reducing response times and enhancing security.

In industrial and smart city environments, fog nodes serve as intermediate processing hubs that balance tasks between cloud and edge. Karjee et al. [68] proposed a split computing framework that utilizes fog infrastructure to divide ML model inference, enabling efficient pose estimation with reduced device-side computation. This architecture is especially valuable in resource-constrained wearable and embedded systems. Firouzi et al. [71] discussed IoT communication support via fog networks, highlighting the value of context-aware fog nodes in maintaining seamless device connectivity in dynamic environments. Their findings reinforce the role of fog systems in enhancing communication robustness and reducing the burden on cloud servers.

Table 2.5 Comparison of AI/ML-enabled IoT communication architectures.

Architecture type	AI technique used	Communication protocol	Use cases	Benefits	Limitations
Edge computing	Lightweight ML	MQTT, CoAP	Smart cities, health monitoring	Low latency, privacy-preserving inference	Limited compute and energy resources
Fog computing	Split inference	HTTP/LoRaWAN	Wearables, real-time analytics	Offloads processing from end devices	Complexity in orchestration
SDN + AI	Reinforcement learning	OpenFlow	Traffic routing, QoS control	Dynamic reconfiguration	Controller bottlenecks, data plane latency
NFV + ML orchestration	Deep learning + automation	REST APIs	Service chaining	Elastic resource scaling	Virtualization overhead
Centralized learning	CNN, DNN	5G/6G	Autonomous vehicles	High accuracy with complete datasets	Privacy risks, high bandwidth consumption
Federated/ distributed FL	Federated averaging	Local aggregation	Edge sensing, federated IoT	Reduced communication load, privacy	Slower convergence, straggler clients

Note: It is based on a synthesis of recent literature examining AI/ML-enabled communication architectures in IoT environments.

Figure 2.3 Illustration of AI/ML-driven IoT communication architecture, incorporating edge/cloud computing and decision-making layers.

2.4.2 AI-enabled Software-defined Networking/Network Function Virtualization Architectures for IoT

Software-defined networking (SDN) and network function virtualization (NFV) architectures provide programmability, modularity, and dynamic resource management for large-scale IoT networks. The integration of AI/ML into these architectures enhances their ability to adapt to traffic patterns, predict anomalies, and optimize routing in real time. Nassef et al. [67] presented a comprehensive survey on distributed ML in 5G and beyond networks, showcasing how AI-driven SDN controllers manage routing and load balancing with high adaptability. These AI-enhanced control planes can react to fluctuating QoS demands and security threats with minimal human intervention.

Pathak et al. [72] highlighted the role of data-driven orchestration in SDN environments, enabling scalable service delivery and efficient spectrum management. They identified dynamic AI policies as critical enablers for cognitive network automation, particularly when deploying resource-heavy services like video analytics or autonomous systems. Wei et al. [69] proposed AccDFL, a decentralized FL protocol integrated into SDN/NFV to improve inference latency while preserving user privacy. Their architecture offers real-time adaptability and balances computation across control and data planes, a vital characteristic in high-density IoT deployments.

2.4.3 Distributed Versus Centralized Learning Models

A central debate in AI-enabled IoT communication architectures revolves around the trade-offs between distributed learning (e.g. FL) and centralized learning approaches. Distributed learning enables decentralized edge nodes to train models locally while exchanging only model updates, thus preserving data privacy and reducing communication costs. Wei et al. [69] demonstrated that their accelerated decentralized FL protocol significantly improves convergence speed without overburdening communication links. In contrast, centralized models require large-scale data aggregation, which can lead to privacy risks, increased latency, and bandwidth congestion. However, centralized learning often yields higher accuracy due to access to comprehensive training data. Saif et al. [73] compared both models and found that distributed schemes like FL are superior in constrained settings, whereas centralized learning performs better in resource-rich environments. Janu et al. [74] explored spectrum sharing strategies using distributed learning, where edge devices make autonomous decisions to avoid interference. Their approach allowed for minimal centralized coordination, reducing control message overhead while maintaining communication quality.

2.5 Challenges, Limitations, and Open Issues

Despite significant advancements in integrating AI and ML into IoT communication systems, several technical, operational, and ethical challenges persist. These limitations hinder large-scale deployment, degrade performance in dynamic environments, and introduce such complexities at the intersection of intelligence and connectivity.

2.5.1 Computational Constraints at the Edge

One of the primary challenges lies in the computational and energy limitations of edge devices, which are often battery-powered and resource-constrained. While lightweight ML models and FL approaches aim to reduce resource usage, many real-time applications, such as those in healthcare or smart manufacturing, still demand higher processing capabilities than current edge infrastructure can sustain [61, 70]. Additionally, performing model updates locally or running inference tasks in situ may introduce thermal and battery drain issues.

2.5.2 Trade-offs Between Accuracy and Efficiency

A recurring trade-off exists between model accuracy and communication/computation efficiency. Centralized learning systems tend to achieve higher performance due to access to comprehensive datasets, but they are bandwidth-intensive and pose significant privacy risks [54]. Conversely, decentralized models such as FL enhance privacy and reduce communication overhead but often suffer from slower convergence, lower accuracy, and challenges with straggler clients and data heterogeneity [62, 69].

2.5.3 Lack of Standardization in Architectures

The diversity of proposed AI/ML-enabled IoT architectures, from edge and fog systems to SDN/NFV-based networks, has resulted in a fragmented landscape lacking unified standards. This makes interoperability across platforms difficult and increases the integration burden, particularly in large-scale and multi-vendor deployments [71]. Moreover, the architectural complexity of hybrid models (e.g. split inference, joint orchestration, and AI-powered control planes) introduces system overhead and configuration challenges. Without clear reference models, reproducibility and deployment at scale remain limited.

2.5.4 Data Quality, Privacy, and Security

The quality, granularity, and labeling of data used to train AI/ML models in IoT systems vary significantly, especially in real-world environments [75]. Inconsistent or imbalanced datasets can lead to biased outcomes or unreliable predictions, particularly harmful in safety-critical applications like autonomous driving or healthcare monitoring [48, 50]. Additionally, despite privacy-preserving techniques like FL, data leakage risks remain due to model inversion, gradient attacks, or poor anonymization techniques. Robust security mechanisms are needed not just for communication protocols but also for AI pipelines themselves.

2.5.5 Explainability and Trust in AI Decisions

AI-powered IoT networks increasingly make autonomous decisions, ranging from routing and load balancing to congestion prediction and anomaly detection. However, most current models, especially DNNs, are considered black boxes, making it difficult to interpret their behavior [9, 76, 77]. This lack of explainability impedes user trust and limits deployment in regulated industries such as finance or healthcare. There is a growing need for explainable AI (XAI) frameworks that can provide transparency, traceability, and confidence in AI-driven communication decisions.

2.5.6 Scalability and Real-time Adaptation

While many AI-based models perform well in controlled or small-scale scenarios, scalability remains a major bottleneck when extended to real-world, high-density IoT environments. AI systems need to adapt to varying traffic loads, device mobility, and fault conditions, all in real time [52, 59]. Achieving such responsiveness requires dynamic, self-optimizing algorithms and fast retraining methods, which are still under active research.

2.5.7 Open Research Directions

Future studies should address the following open questions:

- How can hierarchical FL be structured across edge, fog, and cloud layers?
- What are effective strategies for joint optimization of communication protocols and AI workflows?
- Can lightweight XAI models be developed specifically for constrained IoT environments?
- How do we ensure green AI practices that reduce the energy footprint of continual AI training?

Addressing these open challenges could play a crucial role in realizing scalable, secure, and intelligent IoT communication ecosystems.

2.6 Conclusion

The convergence of AI/ML with IoT communication systems holds transformative potential in addressing the complex demands of modern distributed environments. As evidenced throughout this chapter, AI techniques such as RL, DL, and FL offer significant improvements in routing efficiency, congestion mitigation, resource management, and adaptive spectrum use. Architectures leveraging edge and fog computing further reduce latency and enable near real-time decision-making, while AI-enhanced SDN and NFV frameworks provide scalable and programmable network control. Despite these advancements, several challenges remain unresolved, particularly around model deployment in resource-constrained environments, balancing accuracy and efficiency, ensuring interoperability, and fostering trust through explainability. Additionally, the absence of standardized frameworks and benchmarks limits the replicability and scalability of current solutions. Looking ahead, future research should focus on developing lightweight, energy-efficient models, hierarchical learning frameworks that span edge-to-cloud, and explainable AI mechanisms tailored for IoT. By addressing these open challenges, AI-integrated communication systems can unlock the full potential of next-generation IoT ecosystems – fostering resilience, sustainability, and intelligence at scale.

References

1 Canedo, J., and Skjellum, A. (2016). Using machine learning to secure IoT systems. In: *SoutheastCon 2016* (pp. 1–6). IEEE.

2 Kayode Saheed, Y., Idris Abiodun, A., Misra, S., et al. (2022). A machine learning-based intrusion detection framework for software-defined IoT networks. *Sensors* 22 (9): 3376.

3 Yagiz, M.A., MohajerAnsari, P., Pesé, M.D., and Goktas, P. (2024). Transforming in-vehicle network intrusion detection: VAE-based knowledge distillation meets explainable AI. In: *Proceedings of the Sixth Workshop on CPS&IoT Security and Privacy* (pp. 93–103). New York: Association for Computing Machinery.

4 Yagiz, M.A., and Goktas, P. (2025). LENS-XAI: redefining lightweight and explainable network security through knowledge distillation and variational autoencoders for scalable intrusion detection in cybersecurity. *arXiv preprint arXiv:2501.00790*.

5 Liu, Y., Yu, H., Xie, S., and Zhang, Y. (2019). Deep reinforcement learning for offloading and resource allocation in vehicular edge computing and networks. *IEEE Transactions on Vehicular Technology* 68 (11): 11158–11168.

6 Agrawal, K., Goktas, P., Sahoo, B., et al. (2024). IoT-based service allocation in edge computing using game theory. In: *International Conference on Distributed Computing and Intelligent Technology* (pp. 45–60). Cham: Springer Nature Switzerland.

7 Dovom, E.M., Azmoodeh, A., Dehghantanha, A., et al. (2019). Fuzzy pattern tree for edge malware detection in IoT. *Journal of Computational Science* 34: 72–85.

8 Moustafa, N., Turnbull, B., and Choo, K.-K.R. (2019). An ensemble intrusion detection technique based on proposed statistical flow features for protecting network traffic of IoT and cloud environments. *IEEE Internet of Things Journal* 6 (3): 4815–4830.

9 Li, J., Zhao, Z., Li, R., and Zhang, H. (2019). AI-based two-stage intrusion detection for software-defined IoT networks. *Sensors* 19 (18): 4091.

10 Savazzi, S., Nicoli, M., and Rampa, V. (2020). Federated learning with cooperating devices: a consensus approach for massive IoT networks. *IEEE Internet of Things Journal* 7 (5): 4641–4654.

11 Mothukuri, V., Khare, P., Parizi, R.M., et al. (2022). Federated-learning-based anomaly detection for IoT security attacks: concepts, applications, and future directions. *IEEE Internet of Things Journal* 9 (4): 2545–2562.

12 Ghimire, B., and Rawat, D.B. (2022). Recent advances on federated learning for cyber-security and privacy in IoT: a survey. *IEEE Internet of Things Journal* 9 (1): 422–440.

13 Mao, B., Kawamoto, Y., and Kato, N. (2020). AI-based joint optimization of QoS and security for real-time communication in software-defined vehicular networks. *IEEE Wireless Communications* 27 (6): 96–102.

14 Chen, M., Challita, U., Saad, W., et al. (2019). Artificial neural networks-based machine learning for wireless networks: a tutorial. *IEEE Communications Surveys & Tutorials* 21 (4): 3039–3071.

15 Jia, Y., Zhong, F., Alrawais, A., et al. (2020). FlowGuard: an intelligent edge defense mechanism against IoT DDoS attacks. *IEEE Internet of Things Journal* 7 (10): 9552–9562.

16 Moustafa, N. (2021). A new distributed architecture for evaluating the trust of video surveillance data in smart cities using edge computing and deep learning. *Sustainable Cities and Society* 72: 102994.

17 Booij, T.M., Chiscop, I., Meeuwissen, E., and Moustafa, N. (2022). ToN_IoT: the role of heterogeneity and the need for standardisation in the security of IoT and IIoT systems. *Future Internet* 14 (6): 165.

18 Khraisat, A., and Alazab, A. (2021). A critical review of intrusion detection systems in the Internet of Things. *Journal of Network and Computer Applications* 171: 102806.

19 Vishwakarma, R., and Jain, A.K. (2020). A survey of DDoS attacking techniques and defence mechanisms in the IoT network. *Telecommunication Systems* 73: 3–25.

20 Hussain, F., Hussain, R., Hassan, S.A., and Hossain, E. (2020). Machine learning in IoT security: current solutions and future challenges. *IEEE Communications Surveys & Tutorials* 22 (3): 1686–1721.

21 Chen, J., Chen, S., Wang, Q., et al. (2019). IRAF: a deep reinforcement learning approach for anti-jamming and trajectory control in wireless sensor networks. *IEEE Internet of Things Journal* 6 (2): 3481–3495.

22 Roopak, M., Tian, G.Y., and Chambers, J. (2019). Deep learning models for cybersecurity in IoT networks. *Computer Communications* 148: 213–228.

23 Alsaedi, A., Moustafa, N., Tari, Z., et al. (2020). TON-IoT telemetry dataset: a new generation dataset for IoT and IIoT network intrusion detection and threat intelligence. *Sensors* 20 (20): 5767.

24 Chaabouni, N., Mosbah, M., Zemmari, A., et al. (2019). Network intrusion detection for IoT security based on learning techniques. *IEEE Communications Surveys & Tutorials* 21 (3): 2671–2701.

25 Saeed, N., Elzanaty, A., Almorad, H., et al. (2020). CubeSat communications: recent advances and future challenges. *IEEE Communications Surveys & Tutorials* 22 (3): 1839–1870.

26 Zhong, H., Ning, L., Wang, J., et al. (2022). Optimization of LoRa SF allocation based on deep reinforcement learning. *Wireless Communications and Mobile Computing* 2022 (1): 1690667.

27 Santa, J., Sanchez-Iborra, R., Rodriguez-Rey, P., et al. (2019). LPWAN-based vehicular monitoring platform with a generic IP network interface. *Sensors* 19 (2): 264.

28 Chen, M., Miao, Y., Jian, X., et al. (2018). Cognitive-LPWAN: towards intelligent wireless services in hybrid low power wide area networks. *IEEE Transactions on Green Communications and Networking* 3 (2): 409–417.

29 Liu, Y., Tsang, K.F., Zhu, H., et al. (2022). Efficient load balancing for heterogeneous radio-replication-combined LoRaWAN. *IEEE Transactions on Industrial Informatics* 18 (11): 7400–7411.

30 Saavedra, E., Mascaraque, L., Calderon, G., et al. (2022). A universal testbed for IOT wireless technologies: abstracting latency, error rate and stability from the IOT protocol and hardware platform. *Sensors* 22 (11): 4159.

31 Buratti, C., Ström, E., Feltrin, L., et al. (2021). IoT protocols, architectures, and applications. *Inclusive Radio Communications for 5G and Beyond* (pp. 187–220). Academic Press.

32 Li, G., Sun, C., Zhang, J., et al. (2019). Physical layer key generation in 5G and beyond wireless communications: challenges and opportunities. *Entropy* 21 (5): 497.

33 Salman, T., and Jain, R. (2017). Networking protocols and standards for internet of things. In: *Internet of Things and Data Analytics Handbook* (pp. 215–238). John Wiley & Sons, Inc.

34 Asif, A.R., Zahra, F., and Matin, M.A. (2020). Cognitive solution for IoT communication technologies. *International Journal of Computer Science and Network Security* 20 (2): 1–9.

35 Li, Y., Zheng, C., and Wang, S. (2019). Efficient oneM2M protocol conversion platform based on NB-IoT access. *China Communications* 16 (11): 56–69.

36 Khan, M.N., and Naseer, F. (2020). IoT based university garbage monitoring system for healthy environment for students. In *2020 IEEE 14th International Conference on Semantic Computing (ICSC)* (pp. 354–358). IEEE.

37 Leo John, F., Lakshmi, D., and Kuncharam, M. (2023). Introduction to the internet of things: opportunities, perspectives and challenges. In: *Smart Grids and Internet of Things: An Energy Perspective* (pp. 1–34). Scrivener Publishing LLC.

38 Zyane, A., Bahiri, M.N., and Ghammaz, A. (2022). IoTScal-SC2: two cloud computing systems-based collaboration solution for scalability issue in IoT networks. In: *International Conference on Electronic Engineering and Renewable Energy Systems* (pp. 671–685). Singapore: Springer Nature Singapore.

39 Bahiri, M.N., Zyane, A., and Ghammaz, A. (2020). IoTScal-C: a based cloud computing collaboration solution for scalability issue in IoT networks. In: *International Conference on Electronic Engineering and Renewable Energy* (pp. 123–133). Singapore: Springer Singapore.

40 Abdellah, Z., Nabil, B.M., and Abdelilah, G. (2019). A based cloud computing solution using DiffServ architecture for scalability issue in IoT networks with multiple SLA requirements. In: *Proceedings of the 2nd International Conference on Networking, Information Systems & Security* (pp. 1–6). New York: Association for Computing Machinery.

41 Bahiri, M.N., Zyane, A., and Ghammaz, A. (2018). An enhancement for the autonomic middleware-level scalability management within IoT system using cloud computing. In: *International Conference on Electronic Engineering and Renewable Energy* (pp. 80–88). Singapore: Springer Singapore.

42 Adil, M. (2021). Congestion free opportunistic multipath routing load balancing scheme for Internet of Things (IoT). *Computer Networks* 184: 107707.

43 Lin, J., Nagazawa, R., Nguyen, K., et al. (2023). Pavlovian conditioning modeling using wireless spiking neural network. In: *2023 20th International SoC Design Conference (ISOCC)* (pp. 163–164). IEEE.

44 Yazdinejad, A., Dehghantanha, A., Parizi, R.M., et al. (2023). Secure intelligent fuzzy blockchain framework: effective threat detection in IoT networks. *Computers in Industry* 144: 103801.

45 Thangaramya, K., Kulothungan, K., Logambigai, R., et al. (2019). Energy aware cluster and neuro-fuzzy based routing algorithm for wireless sensor networks in IoT. *Computer Networks* 151: 211–223.

46 Modi, Y., Teli, R., Mehta, A., et al. (2022). A comprehensive review on intelligent traffic management using machine learning algorithms. *Innovative Infrastructure Solutions* 7 (1): 128.

47 Bu, F., and Wang, X. (2019). A smart agriculture IoT system based on deep reinforcement learning. *Future Generation Computer Systems* 99: 500–507.

48 Sarker, I.H., Abushark, Y.B., Alsolami, F., and Khan, A.I. (2020). IntruDTree: a machine learning based cyber security intrusion detection model. *Symmetry* 12 (5): 754.

49 Kaur, P., Kumar, R., and Kumar, M. (2019). A healthcare monitoring system using random forest and internet of things (IoT). *Multimedia Tools and Applications* 78: 19905–19916.

50 Ullah, I., and Mahmoud, Q.H. (2021). Design and development of a deep learning-based model for anomaly detection in IoT networks. *IEEE Access* 9: 103906–103926.

51 Wu, Z., Huang, M., and Zhao, A. (2021). Traffic prediction based on GCN-LSTM model. *Journal of Physics: Conference Series* 1972 (1): 012107.

52 Sharma, A., Balasubramanian, V., and Kamruzzaman, J. (2024). A temporal deep Q learning for optimal load balancing in software-defined networks. *Sensors* 24 (4): 1216.

53 Chen, Z., Tian, P., Liao, W., et al. (2023). Resource-aware knowledge distillation for federated learning. *IEEE Transactions on Emerging Topics in Computing* 11 (3): 706–719.

54 Liu, J., Wu, J., and Liu, M. (2020). UAV monitoring and forecasting model in intelligent traffic oriented applications. *Computer Communications* 153: 499–506.
55 Wu, C., Kim, I., and Ma, Z. (2023). Deep reinforcement learning based traffic signal control: a comparative analysis. *Procedia Computer Science* 220: 275–282.
56 Zhao, Z., Jiang, Z., Ling, N., et al. (2018). ECRT: an edge computing system for real-time image-based object tracking. In: *Proceedings of the 16th ACM Conference on Embedded Networked Sensor Systems* (pp. 394–395). New York: Association for Computing Machinery.
57 Nallaperuma, D., Nawaratne, R., Bandaragoda, T., et al. (2019). Online incremental machine learning platform for big data-driven smart traffic management. *IEEE Transactions on Intelligent Transportation Systems* 20 (12): 4679–4690.
58 Uddin, M.I., Alamgir, M.S., Rahman, M.M., et al. (2021). AI traffic control system based on deepstream and IoT using NVIDIA Jetson nano. In: *2021 2nd International Conference on Robotics, Electrical and Signal Processing Techniques (ICREST)* (pp. 115–119). IEEE.
59 Messaoud, S., Bradai, A., Ahmed, O.B., et al. (2020). Deep federated Q-learning-based network slicing for industrial IoT. *IEEE Transactions on Industrial Informatics* 17 (8): 5572–5582.
60 Cui, Y., Cao, K., Cao, G., et al. (2021). Client scheduling and resource management for efficient training in heterogeneous IoT-edge federated learning. *IEEE Transactions on Computer-Aided Design of Integrated Circuits and Systems* 41 (8): 2407–2420.
61 Zhao, T., Chen, X., Sun, Q., and Zhang, J. (2023). Energy-efficient federated learning over cell-free IoT networks: modeling and optimization. *IEEE Internet of Things Journal* 10 (19): 17436–17449.
62 AbdulRahman, S., Tout, H., Mourad, A., and Talhi, C. (2020). FedMCCS: multicriteria client selection model for optimal IoT federated learning. *IEEE Internet of Things Journal* 8 (6): 4723–4735.
63 Jiang, K., Sun, C., Zhou, H., et al. (2021). Intelligence-empowered mobile edge computing: framework, issues, implementation, and outlook. *IEEE Network* 35 (5): 74–82.
64 Qiu, S., Zhao, J., Lv, Y., et al. (2022). Digital-twin-assisted edge-computing resource allocation based on the whale optimization algorithm. *Sensors* 22 (23): 9546.
65 Zhang, X., Li, S., Tang, J., et al. (2024). DRL-enabled computation offloading for AIGC services in IIoT-assisted edge computing networks. *IEEE Internet of Things Journal* 12 (9): 12829–12844.
66 Dong, C., Hu, S., Chen, X., and Wen, W. (2021). Joint optimization with DNN partitioning and resource allocation in mobile edge computing. *IEEE Transactions on Network and Service Management* 18 (4): 3973–3986.

67 Nassef, O., Sun, W., Purmehdi, H., et al. (2022). A survey: distributed machine learning for 5G and beyond. *Computer Networks* 207: 108820.

68 Karjee, J., Naik, S.P., Anand, K., et al. (2023). Split computing based quantized PoseNet model for distributed AI architecture in IoT-edge. In: *2023 15th International Conference on COMmunication Systems & NETworkS (COMSNETS)* (pp. 192–194). IEEE.

69 Wei, M., Yu, W., and Chen, D. (2024). AccDFL: accelerated decentralized federated learning for healthcare IoT networks. *IEEE Internet of Things Journal* 12 (5): 5329–5345.

70 Alhammadi, A., Shayea, I., El-Saleh, A.A., et al. (2024). Artificial intelligence in 6G wireless networks: opportunities, applications, and challenges. *International Journal of Intelligent Systems* 2024 (1): 8845070.

71 Firouzi, F., Farahani, B., and Marinšek, A. (2022). The convergence and interplay of edge, fog, and cloud in the AI-driven Internet of Things (IoT). *Information Systems* 107: 101840.

72 Pathak, Y., Prashanth, P.V.N., and Tiwari, A. (2023). AI meets SDN: a survey of artificial intelligent techniques applied to software-defined networks. In: *6G Enabled Fog Computing in IoT: Applications and Opportunities* (pp. 395–412). Cham: Springer Nature Switzerland.

73 Saif, M., Hassan, M.Z., and Hossain, M.J. (2024). Decentralized aggregation for energy-efficient federated learning in mmWave aerial-terrestrial integrated networks. In: *IEEE Transactions on Machine Learning in Communications and Networking*. IEEE.

74 Janu, D., Singh, K., and Kumar, S. (2022). Machine learning for cooperative spectrum sensing and sharing: a survey. *Transactions on Emerging Telecommunications Technologies* 33 (1): e4352.

75 Goktas, P. (2024). Ethics, transparency, and explainability in generative AI decision-making systems: a comprehensive bibliometric study. *Journal of Decision Systems* 1–29. https://www.tandfonline.com/doi/full/10.1080/12460125.2024.2410042?src=#

76 Sarker, I.H. (2021). Machine learning: algorithms, real-world applications and research directions. *SN Computer Science* 2 (3): 160.

77 Goktas, P., and Grzybowski, A. (2025). Shaping the future of healthcare: ethical clinical challenges and pathways to trustworthy AI. *Journal of Clinical Medicine* 14 (5): 1605.

3

IoT and AI in Sustainable Agriculture

Pankaj P. Tasgaonkar, Srushti Khomane, Komal Suryawanshi, and Shachi Kokare*

COEP Technological University, Pune, India

*Corresponding author: ppt.extc@coeptech.ac.in

3.1 Introduction

Fruit quality monitoring refers to the process of assessing and tracking various attributes of fruit to ensure that it meets specific quality standards. This monitoring involves the use of different tools and technologies to evaluate factors such as appearance, taste, texture, ripeness, and nutritional content. These assessments help determine whether the fruit is suitable for sale, consumption, or further processing.

IoT-based systems use sensors to collect data continuously of environmental factors, enabling continuous monitoring of crop growth while also to detect fruit conditions throughout the supply chain. This approach helps maintain freshness and safety by detecting ripening stages and potential quality degradation. Recent studies have highlighted the effectiveness of IoT systems, such as methane gas sensors, which assess fruit ripening without harvesting, minimizing damage and loss [1]. Additionally, the integration of cloud computing allows for remote data storage and analysis, offering actionable insights that enhance decision-making. IoT applications also incorporate machine learning (ML) algorithms to predict shelf life and detect diseases, thus reducing food waste and ensuring timely interventions [2]. Furthermore, IoT enhances traceability, providing consumers with transparency regarding the journey of their produce from farm to table.

AI and ML Techniques in IoT-based Communication: A Path to Sustainable Development Goals, First Edition. Edited by Sumita Mishra, Nishu Gupta and Polat Goktas.

This chapter explores the function of IoT in crop monitoring and its underlying architecture. IoT devices, such as sensors and actuators, facilitate real-time data collection on crucial factors like weather conditions, soil moisture, and other variables influencing crop development. These sensors can measure soil pH, moisture levels, temperature, and humidity while also tracking crop growth, pest infestations, and insect populations [3]. Additionally, farmers can use this data to determine which crops are best suited to specific environmental conditions.

Drones are generally associated with military, industrial, and other specialized operations. The most promising industry is agriculture, but it is currently plagued by a number of problems, including a lack of workers in the farming sector, harsh weather, inadequate fertilizers, improper fertilizers, and diseases. Irrigation, crop health monitoring, weather prediction, and livestock monitoring are some of the most common agricultural drone applications. With open-source technology, advanced sensors, increased integration, longer flight durations, and the ability to identify forests and other disaster areas, drones are becoming increasingly intelligent. Through cost control, performance monitoring, and maintenance, the use of IoT and cloud computing in agriculture will improve crop yields, benefiting farmers and the increase Country GDP [4]. This chapter focuses on sustainable Smart Agriculture using the UAV.

3.2 Fruit Quality Monitoring

3.2.1 Evolution of Fruit Quality Monitoring

Early fruit quality monitoring relied on subjective visual inspections and basic chemical tests, such as measuring sugar content, which lacked real-time feedback and were often inefficient. This approach led to significant waste and suboptimal storage conditions. The turning point came in the early 2000s with the Smart Agriculture. IoT-enabled wireless sensor networks (WSNs) allowed for real-time, remote monitoring of critical factors like temperature, humidity, ethylene concentration, and firmness. These sensors improved monitoring accuracy and minimized waste by providing continuous feedback on fruit quality throughout the supply chain.

By the 2010s, the data collected through IoT sensors gave better predictions of ripening stages, shelf life, and optimal storage conditions with ML models. Cloud computing platforms enabled the analysis of vast amounts of sensor data, helping optimize harvest timing, storage, and transportation. Advances included noninvasive sensors like near-infrared (NIR) and hyperspectral sensors for internal fruit quality analysis.

Blockchain technology has greater data security, traceability, and transparency, allowing stakeholders to verify the quality and safety of fruits along the supply chain. IoT capabilities have reached to remote rural areas, enabling monitoring through LoRAWAN, where traditional communication technologies were unavailable to overcome the hurdle.

3.2.2 IoT Technologies for Fruit Quality Monitoring

IoT systems rely on various sensors to measure critical parameters for freshness and safety of fruits, including:

- **Temperature and Humidity Sensors**: Essential for maintaining optimal conditions to prevent spoilage and ensure freshness.
- **Gas Sensors**: Used to detect ethylene levels, which indicate fruit ripening, helping optimize harvest and sale timing.
- **Optical Sensors**: Cameras and spectrometers analyze fruit color and texture to assess ripeness and detect defects.
- **Weight Sensors**: Track moisture loss and decay, important indicators of fruit quality.

Data Collection and Transmission: Sensors continuously collect data on temperature, humidity, gas concentrations, and visual attributes. This data is processed locally through microcontroller then transmitted to cloud platforms through MQTT and HTTP. Cloud storage and advanced analytics tools include ML and deep learning techniques for fruit quality management. By integrating IoT technologies, the accuracy of fruit quality assessments is enhanced, along with traceability [5].

3.2.3 Current Trends and Developments in Fruit Quality Monitoring Using IoT

In fruit quality monitoring, several methodologies are commonly used. These methodologies often integrate various **sensing technologies**, **data analysis techniques**, and **communication protocols** to assess fruit quality in real time and throughout the supply chain. The methodologies typically involve the design of IoT-based systems that combine hardware (sensors) with software (data processing and analytics) to provide actionable insights [6–8]. Recent studies on IoT-based fruit quality monitoring have highlighted key advancements in sensor technologies, ML, communication protocols, and blockchain integration.

Key trends have aimed to improve precision, scalability, and efficiency:

- **Advanced Sensor Technologies**
  Noninvasive sensors like NIR spectroscopy and hyperspectral imaging are increasingly used to assess attributes such as ripeness, sugar content, and moisture. Smart packaging with IoT integration allows real-time monitoring

of conditions, while wearable sensors track fruit handling to prevent damage. Sensors like NIR, hyperspectral imaging, pressure sensors, and gas sensors measure parameters such as ripeness, humidity, and temperature. They are deployed during harvesting, storage, and transport for real-time data collection. Sensor fusion enhances the accuracy of the fruit quality detection. Notable advancements include NIR and hyperspectral sensors for noninvasive internal quality measurement, such as sugar content and ripeness. Gas sensors that detect ethylene are also increasingly used for ripening monitoring, providing valuable data for predicting shelf life.

- **Machine Learning**
  ML models automate decision-making, providing real-time predictions for harvest timing and storage, reducing waste, and optimizing fruit handling. Various models like CNN, SVM, decision trees, and Random Forest predict fruit ripeness and shelf life by analyzing IoT sensor data (e.g. temperature, humidity, ethylene gas). Predictive shelf-life models help manage inventory and reduce waste [9].
- **Edge Computing**
  It enables real-time data processing faster by taking quicker decisions in critical situations such as temperature-sensitive storage on the edge devices for reducing latency. Real-time data visualization platforms enhance monitoring. Edge computing processes data locally for real-time decision-making, while cloud computing supports large-scale data storage and advanced analysis.
- **Blockchain for Traceability and Transparency**
  Blockchain technology ensures data integrity and traceability throughout the supply chain, providing secure records of fruit quality data and enhancing consumer trust. It provides immutable records of fruit quality data from farm to retail. It assures that fruit quality data cannot be tampered with, increasing consumer trust. Smart contracts automate decision-making for optimal fruit handling. Blockchain improves transparency and accountability, ensuring safety and authenticity.
- **Wireless Communication and Low-power IoT**
  LPWAN technologies like LoRaWAN and 5G offer long-range, low-power communication, supporting real-time monitoring in remote areas and improving data handling efficiency. Protocols like LoRaWAN, NB-IoT, ZigBee, and Wi-Fi transmit sensor data over long distances to centralized platforms for analysis, ensuring real-time monitoring. LPWANs and edge computing improve energy efficiency and reliability, making large-scale fruit quality monitoring feasible.
- **Automated Handling and Robotics**
  Robotic systems equipped with sensors can inspect and sort fruits, while drones are used for large-scale monitoring of orchards. These systems automate quality assessments and improve efficiency.

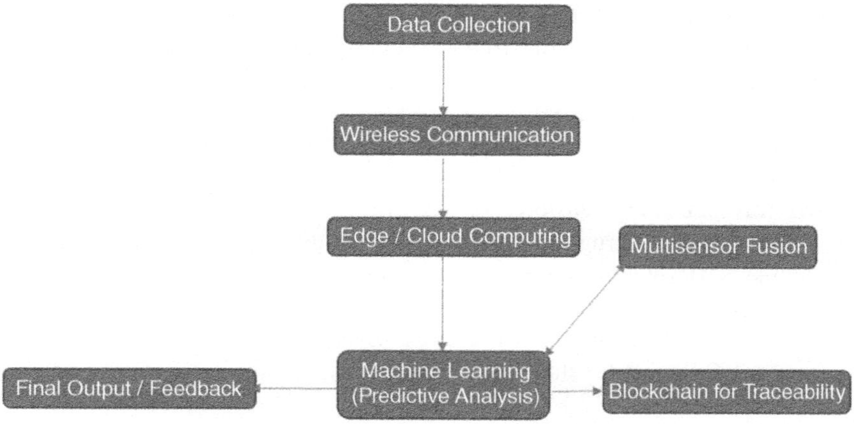

Figure 3.1 Flowchart of IoT-based fruit quality-monitoring system. *Source*: Authors.

- **Sustainability and Waste Reduction**
  IoT helps reduce food waste by improving shelf-life predictions and optimizing resource use such as water and energy in agricultural practices. IoT-based fruit quality monitoring is evolving through advancements in sensors, data analytics, and automation. These innovations enhance quality, reduce waste, and increase transparency, benefiting producers, retailers, and consumers. Figure 3.1 shows the flowchart of IoT-based fruit quality-monitoring system.

3.2.4 Challenges of Fruit Quality Monitoring

Data integration and system scalability remain challenges, especially in large-scale deployments. The integration of diverse data and bulk volume data are complex tasks. Network reliability and data privacy concerns are also significant. While IoT-based systems show promise, addressing data integration, scalability, and network reliability issues is crucial for widespread adoption. Recent studies highlight the potential of IoT to revolutionize fruit quality monitoring through advanced sensors, ML, and blockchain. However, following challenges like data integration, system scalability, and network reliability need to be resolved.

- **Connectivity and Network Issues**
  Poor internet infrastructure in rural areas can delay data transmission, increasing spoilage risks. Reliable connectivity is crucial.
- **Data Integration and Scalability**
  Managing large data volumes and scaling systems for large agricultural operations requires efficient data-processing solutions.

- **Cost of Implementation**
  High costs for sensors, infrastructure, and software limit adoption, particularly for smallholder farmers. Reducing costs is essential.
- **Data Privacy and Security**
  Sensors collect vital data, raising privacy and security concerns. Strong cybersecurity measures are needed for adoption.
- **ML Model Generalization**
  ML models often struggle to generalize across different fruit types and environments. Developing adaptable models is essential.
- **Energy Consumption**
  High energy demands for IoT devices, especially in remote areas, require energy-efficient and sustainable solutions.

3.3 Crop Growth Monitoring Using IoT

Diseases significantly impact crop growth by disrupting essential physiological processes, often leading to reduced yields and poor quality. Common diseases such as late blight (*Phytophthora infestans*), rice blast (*Magnaporthe oryzae*), and powdery mildew (*Erysiphe* spp.) are triggered by pathogens that thrive in specific environmental conditions, affecting crops like potatoes, rice, and wheat [10]. The onset and severity of these diseases can often be linked to nutrient imbalances in the soil. For instance, nitrogen deficiency can weaken plant immunity, making crops more susceptible to fungal infections, while excessive nitrogen may promote diseases like leaf rust. Potassium plays a crucial role in strengthening cell walls and improving plant resistance to pathogens, while phosphorus supports root development and disease resistance. Additionally, micronutrients such as zinc, copper, and manganese are vital for enzyme activation and immune response, helping crops combat bacterial and fungal attacks [11]. Monitoring these nutrients through soil and plant tissue analysis is essential for early disease detection and effective management, ensuring healthier crops and sustainable farming practices.

3.3.1 IoT Architecture for Crop Growth Monitoring

IoT systems for crop growth monitoring typically follow a multilayered architecture:

3.3.1.1 Integration of Sensing Technologies with Other Emerging Technologies

Smart agriculture leverages emerging technologies to modernize agricultural practices through a multilayered architecture comprising sensing, communication, edge/fog, and cloud layers. Each layer integrates specific technologies to

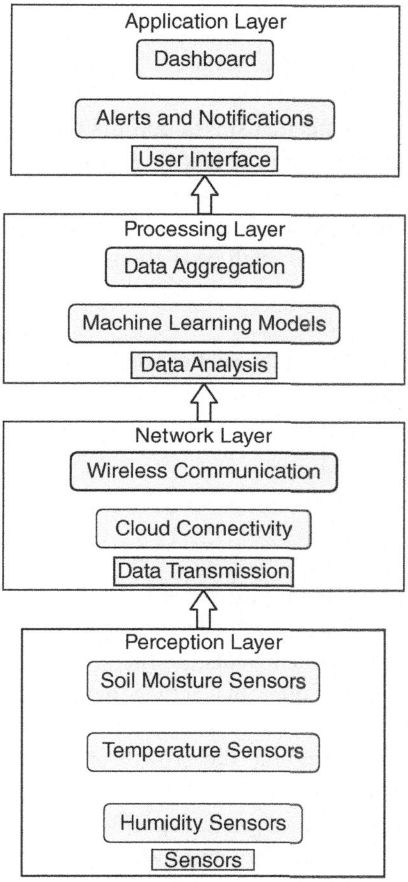

Figure 3.2 Architecture of IoT systems for crop growth monitoring. *Source*: Authors.

enhance efficiency across various agricultural domains. Figure 3.2 displays the architecture of IoT systems for crop health monitoring.

Sensing Layer

The sensing layer employs diverse sensors, such as those for monitoring soil moisture, temperature, humidity, and plant health. These sensors form the backbone of the smart agriculture ecosystem by collecting real-time data essential for decision-making.

Communication Layer

The communication layer connects sensors and IoT devices to the internet, enabling real-time data transmission for tasks like crop monitoring, controlled irrigation, and livestock management. Various wireless communication technologies are employed, including RFID, Bluetooth, ZigBee, Wi-Fi, LoRa, SigFox, WiMAX, NB-IoT, and cellular networks. For example, RFID tags store soil or crop identification data, while LoRa and SigFox enable low-power, long-range communication. Cellular networks, particularly 5G, support large-scale farms by providing high-speed and reliable connectivity.

Edge/Fog Layer

The edge/fog layer processes data close to the network, addressing real-time agricultural needs like precision farming and reducing latency. It supports resource-sharing for cost-effective solutions, especially in underdeveloped areas. Blockchain integration enhances security and traceability, enabling reliable food supply chains and automated payments through smart contracts.

Internet/Cloud Layer

The cloud layer facilitates large-scale data storage and analysis, supporting big data analytics and AI applications in agriculture. This layer enables advanced insights, such as recommending crop protection mechanisms or optimizing irrigation based on weather and crop data. AI models and deep learning techniques are applied for precision farming tasks, such as pest detection, crop monitoring, and predicting optimal harvesting times. For example, image processing using deep learning models can assess crop ripeness and rainfall predictions.

Perception Layer

This layer includes sensors and actuators responsible for data collection. Selecting, deploying, and ensuring the proper functioning of sensors for **crop growth monitoring** requires a thorough understanding of various **technical, environmental, and operational factors** [12].

Every sensor is effortlessly interfaced with the microcontroller device, ready to carry out its assigned function precisely.

- **Temperature and Humidity Sensor:** Temperature directly influences plant growth, photosynthesis, and respiration rates. Extreme temperatures (too high or too low) can stress plants and reduce yields. Soil temperature affects seed germination, root development, and microbial activity in the soil. Similarly relative humidity (RH) affects transpiration rates, water loss, and disease development. Greater humidity can promote the growth of fungal diseases, while less amount can cause water stress. Therefore, sensors like DHT11 are used to maintain ideal environmental ranges for specific crops.
- **pH Sensor:** pH sensor identifies the soil's acidity level and indirectly influences the amount of pesticides needed for the crops.
- **Stem Sensor:** Stem sensors help assess the water status of plants by measuring changes in stem diameter caused by water uptake and loss. By monitoring stem diameter over time, stem sensors provide insights into the plant's growth rate. Abnormal variations in stem diameter indicate stress conditions such as drought, nutrient deficiencies, or disease.
- **Leaf Sensor:** Leaf sensors assess water availability by measuring leaf hydration and turgor pressure. Leaf sensors can measure chlorophyll fluorescence and photosynthetic activity. By monitoring stomatal opening and closing, leaf sensors provide insights into plant transpiration and gas exchange. Changes in leaf temperature, turgor, or hydration often precede visible signs of stress.
- **Raindrop Sensor:** Raindrop sensors sense the presence of rain. Raindrop sensors also identify precipitation, which helps with irrigation scheduling and water conservation.

- **Ultrasonic Sensor:** Ultrasonic sensors measure the distance between the sensor and the plant canopy, providing accurate data on plant height. Ultrasonic sensors detect crop height variations and canopy density, allowing for precise application of fertilizers and pesticides. Lodging (when crops bend or fall due to wind, rain, or nutrient imbalances) can be detected by monitoring changes in crop height.
- **Wireless Image Transmission Sensor:** Drones equipped with camera sensors can monitor crop development by capturing images of the fields. The leaf area index plays a crucial role in assessing the energy utilized in photosynthesis, evapotranspiration, crop growth, nutrient requirements, and variations in water demand. The sensor with camera takes images of the crops and transmits to the base station. The leaf area index is determined based on the occupancy in the pixels.
- **Raspberry Pi Fan:** A 5V DC Relay connection can be used to govern a Raspberry Pi fan to control temperature. The fan kicks on to keep the temperature at a level that is suitable for crops if it gets too hot. It is closely interwoven with the DHT11 sensor, waiting for the signal to initiate when certain temperature thresholds are met.
- **Soil Moisture Sensor:** It measures the moisture level in the soil by inserting probes to detect resistance. The water pump is managed by a 5V DC Relay module, enabling automatic watering when the soil becomes dry.
- **Light Intensity Sensors:** Light intensity sensors measure the range of light (400–700 nm) used by plants for photosynthesis. In controlled environments, light intensity sensors help manage artificial lighting systems. Uneven light distribution in fields or greenhouses can lead to inconsistent crop growth.

Network Layer

This layer enables communication between devices and data storage systems:

- **Communication Protocols:**
  LORAWAN: Among the latest advancements, LoRaWAN is beneficial for its low-power, greater range capabilities, and suits for large fields requiring real-time environmental monitoring.
  Narrowband Internet of Things (NB-IoT) offers wide coverage and deep penetration, supporting extensive sensor networks in remote areas to gather critical agricultural data.
  5G: The adoption of **5G technology** revolutionizes agriculture by providing high-speed, low-latency communication, supporting applications like drone-based monitoring and autonomous machinery for precision farming.
  Wi-Fi 6: For controlled environments like greenhouses, **Wi-Fi 6 (802.11ax)** ensures high throughput and efficient data handling, enabling the integration of multiple sensors.

BLE: Additionally, **Bluetooth Low Energy (BLE) 5.0** provides enhanced range and low power consumption, making it suitable for short-range communication between field sensors and central hubs [9]. Each protocol is selected based on specific factors such as field size, data requirements, and power availability, ensuring efficient, reliable, and scalable communication for modern agricultural systems.
- **Gateways:** Devices that aggregate sensor data and transmit it to cloud servers.

Processing Layer

This layer processes and analyzes the data.

- **Cloud Computing:** Private cloud provides the security while Public cloud Platforms are easily available.
- **Edge Computing:** Reduces latency by performing computations at the device level.

Application Layer

This layer provides user interfaces for visualization and decision-making:

- **Mobile Apps:** Display real-time data for farmers.
- **Decision Support Systems (DSS):** Recommend irrigation schedules, fertilizer applications, and pest control measures.

3.3.2 Technologies Supporting IoT in Crop Monitoring

Several technologies power IoT applications in crop monitoring.

Crop growth monitoring using Deep Learning models leverages various techniques to predict and analyze plant development under different conditions. Convolutional neural networks (CNN) are widely used for tasks such as identifying crop diseases, classifying growth stages, and assessing environmental factors affecting yield. ANN can model complex relationships in crop growth by learning from vast datasets, while CNN excels at image-based analysis, such as interpreting satellite or drone imagery to monitor plant health and detect anomalies. Regression techniques, including linear and polynomial regression, are employed to predict growth patterns, yield estimates, and the impact of environmental variable. Decision trees provide an intuitive way to classify growth outcomes and recommend actions based on feature thresholds like soil quality, temperature, and water levels. Bayesian models integrate probabilistic reasoning, enabling the inclusion of uncertainties in crop predictions, such as variations in weather patterns or pest outbreaks. Lastly, Backpropagation (BP) neural networks optimize the performance of deep learning models by adjusting weights iteratively to minimize prediction errors, making them suitable for precise crop monitoring and management. Together, these technologies form a robust framework for efficient and sustainable agriculture.

3.4 Case Study

This case study focuses on developing an IoT-based solution for effective sugarcane crop management. By continuously monitoring soil nutrients (NPK), pH levels, and temperature, this system aims to enhance crop yield and improve harvest quality.

Sugarcane is classified as a Kharif crop that requires a hot and humid climate with ideal temperature of 27 °C. The crop thrives in areas receiving 75–150 cm of rainfall, but irrigation is necessary in regions with insufficient rainfall. The ideal soil for sugarcane cultivation is deep, rich, and loamy, capable of retaining moisture. Additionally, it maintains neutral pH, as excessive acidity or alkalinity can negatively impact growth, however the phosphorous, calcium and nitrogen level is exceeded with temperature rise.

The proposed IoT-based solution is designed to monitor essential parameters like temperature, soil pH, and NPK levels in real time. Threshold values for these factors are set according to research-backed optimal conditions for sugarcane growth. If sensor readings exceed the defined thresholds, an alert is triggered, notifying users of the current temperature and soil nutrient levels so they can take necessary corrective actions.

For optimal sugarcane cultivation, soil should be loamy with good drainage and a neutral ideal range of 6.5–7.5 pH. The soil should be loose and friable, at least 45 cm deep, free from harmful salts, and rich in essential nutrients. The best geographical conditions for growing sugarcane include an average annual rainfall of 150 cm. An ideal NPK input for maximum yield is 150–80–80 kg ha^{-1}.

3.5 Drones in Agriculture

Unmanned aerial vehicles (UAVs) can be utilized in a variety of applications related to crops management that captures high spatial and temporal resolution images. These technologies are very faster for data analysis that can increase the crop productivity if monitored regularly. UAV technology has applications in precision agriculture, like irrigation, fertilization, pest control, weed management, livestock monitoring, plant growth monitoring, crop disease management, and field-level phenotyping. ML and deep learning methods can be used for analysis of the parameters of farming that can yield the farming.

3.5.1 Drone Applications in Agriculture

Drones are revolutionizing various branches of agriculture by providing precise and efficient tools for monitoring, analysis, and intervention. In **crop production (agronomy)**, drones are used to monitor crop health with multispectral or hyperspectral imaging, perform precision spraying of fertilizers and pesticides, map soil conditions, and assess moisture levels [13]. Similarly, in **horticulture**, they facilitate monitoring the health and growth of fruits, vegetables, and flowers, detecting pests and diseases, optimizing irrigation, and estimating yields in orchards and vineyards, often using multispectral cameras, LiDAR, and thermal imaging.

In **livestock management**, drones equipped with thermal cameras and GPS are used to track livestock movement, monitor grazing patterns, identify sick or injured animals, and inspect water sources or fences. For **forestry**, drones help monitor forest health, detect illegal logging and forest fires, and map canopy cover and biomass using multispectral sensors, LiDAR, and infrared cameras. In **soil science**, they map soil types and conditions, monitor erosion, assess drainage, and measure soil moisture and nutrient levels, often complemented by multispectral sensors and ground-penetrating radar.

Drones also play a vital role in **irrigation management**, helping optimize water distribution through moisture mapping, detecting leaks, and identifying water stress using thermal cameras and soil moisture sensors. In **aquaculture**, drones are used to monitor fish populations, water quality, and aquatic vegetation with tools such as thermal cameras, water quality sensors, and RGB imaging. For **agroforestry**, drones assess interactions between crops and trees, monitor soil and crop health, and evaluate biodiversity and carbon sequestration, utilizing multispectral sensors and LiDAR.

In **plant pathology**, drones detect early signs of diseases, monitor pest spread, and evaluate treatment effectiveness using hyperspectral sensors, RGB cameras, and fluorescence imaging. Common sensors in agricultural drones include RGB cameras for high-resolution imagery, multispectral and hyperspectral sensors for analyzing crop and soil health, thermal cameras for detecting water stress and livestock health, LiDAR for 3D mapping, and gas sensors for measuring emissions like methane. These innovations are transforming agriculture, enabling sustainable and precise farming practices across multiple domains.

UAVs, though initially limited in their agricultural applications, are gaining prominence due to advancements in communication technologies and the widespread adoption of the IoT. Precision farming with UAVs holds significant potential, offering the ability to operate in diverse weather conditions with low operational costs and minimal human intervention. Equipped with advanced sensors, UAVs provide high-resolution images and valuable sensing data, facilitating more efficient and sustainable farming practices [14].

The versatility of UAVs makes them indispensable in modern agriculture. They are employed in various tasks, including irrigation, planting, crop spraying, crop inspection, and soil analysis. Their ability to cover large areas and capture data at specific times and frequencies, unlike satellites, offers farmers greater flexibility and precision in data collection. UAVs are outfitted with advanced sensors, such as 3D cameras, thermal cameras, and multispectral and optical imaging systems, enabling comprehensive monitoring of agricultural activities [15].

Through UAV data, key physical factors like height, canopy coverage, vegetation density, flowering stages, pesticide needs, plant health indicators, fertilization, and crop yield can be effectively tracked. This information aids in predicting the performance of different crop types, allowing farmers to make informed decisions. By integrating UAVs into agriculture, precision and efficiency are significantly enhanced, leading to optimized resource use and improved agricultural outcomes [16].

IoT and UAV are integrated for testing the environmental parameters on the agricultural area of Medenine, Tunisia, during 1 year till March 2021. The sensors are deployed on the field as well as above the field. The data is transmitted periodically every hour while it is analyzed every 12 hours on the cloud. The UAV consists of communication module and camera [17]. The IoT technology comprises control station, gateway, soil moisture sensors, and environment-monitoring sensors that detect temperature, humidity, precipitation, and solar intensity. The obtained results predicted that data was useful for increasing crop productivity.

Research study was carried on the farms in Switzerland and Germany using the UAV for crop and weed classification. The JAI-dataset captured 97 images of the sugar beet farm for one and half months every week [18]. The system comprised vegetation detection, feature extraction, and classification. Here, Random Forest was used for classification. Based on the individual decision tree, class labels were assigned and spatial distribution in the adjacent area were recognized.

Researchers proposed a method for identifying and classifying the image of the fruit. RNN was used over CNN for finding the adjacent features and optimizing the classification. From the Kaggle dataset pictures were selected for training and testing was done on Orchards of Khalsa College Amritsar [19]. LSTM was applied for classification of the fruit images. The achieved results were better in terms of accuracy, root mean square, and coefficient of correlation analysis.

Experiments were carried out on the detection of apples defect and estimation of banana ripeness using the ML algorithms [20]. The model obtained the accuracy equal to 1% or less below the best-performing CNN. The banana ripeness dataset was collected from Fayoum University that contains 273 single fruit images of bananas. The apple fruits contain 5858 apple images with or without worm defects. With the small dataset they have achieved higher accuracy without processing on GPU.

3.6 Benefits of ICT in Agriculture

The integration of Information and Communication Technology (ICT) in agriculture has revolutionized the sector by introducing innovative technologies such as IoT, AI, blockchain, and UAVs. ICT-driven technologies have revolutionized agriculture by improving efficiency, productivity, and sustainability while enabling data-driven decision-making and resource optimization.

These sensors can sense their surroundings to gather vital data, which can then be transmitted online to a server for storage or directly to a user's device, such as a smartphone. These tools enable farmers to monitor their fields remotely, anytime and from anywhere, significantly enhancing productivity. Compared to traditional farming techniques like manual planting, reaping, and tilling, IoT in agriculture generates a vast amount of data, giving rise to the concept of big data. This influx of information presents challenges for IoT systems, including data storage, recording, analysis, and search capabilities. It promises increased efficiency, sustainability, and better management of agricultural resources. At the physical layer, sensors, cameras, and drones collect essential data, including meteorological conditions, soil moisture, crop images, and animal activity. AI empowers farmers to take proactive measures and improve farm management. AI has diverse applications in agriculture, starting with efficient crop management. AI models optimize yield prediction, resource allocation, and crop protection, while also aiding in soil health monitoring, genetic modifications, and disease diagnosis. Advanced ML techniques further enable plant growth estimation and yield forecasting. In irrigation, AI-driven "smart irrigation" systems use sensors to detect plant water needs, ensuring precise water delivery, enhancing soil health. By combining IoT devices, drones, and AI, agriculture is becoming more efficient and sustainable. These technologies address critical challenges across crop production, irrigation, weather forecasting, and livestock management, modernizing farming practices and ensuring better productivity and profitability.

3.7 Blockchain in Agriculture

Blockchain technology offers a revolutionary approach to improving transparency, security, and efficiency in various sectors, including agriculture. As a distributed database maintained across multiple computers, blockchain ensures data integrity by preventing unauthorized changes to previously recorded information [21]. Entities in the network routinely verify the accuracy of the database, creating a reliable and immutable system. Below is a detailed exploration of how blockchain can transform the agricultural sector:

3.7.1 Secure System

Blockchain provides robust data protection, addressing concerns of farmers and stakeholders regarding the confidentiality of their information. The system's architecture stores data in blocks that are tamper-proof and inaccessible without proper authorization. This secure structure ensures that agricultural data, such as crop yields, supply chain logistics, and pricing information, is protected against unauthorized access or manipulation, fostering trust among users.

3.7.2 Enhanced Transparency

Transparency is vital in agriculture due to its central role in human survival. Blockchain facilitates openness in operations by enabling precise and reliable tracking of products throughout the agri-food supply chain. This traceability begins at production and extends to final delivery to consumers.

Blockchain integrates sensors and smart contracts to offer verifiable information, ensuring customers can access immutable data about products. For example, QR code labels on packages can provide detailed insights into an animal's living conditions, transportation history, and storage times, empowering consumers with knowledge about product origins and quality. This transparency supports manufacturing efficiency, sustainable resource management, and security in the supply chain.

3.7.3 Streamlined Communication and Payment Processes

Blockchain accelerates the pace of communication and transactions among agricultural stakeholders, simplifying the exchange of goods and payment processes. Traditional delays in payments – weeks or months – are eliminated, allowing farmers to receive compensation immediately after product delivery.

Digital currencies tailored for the agricultural sector, such as PavoCoin, provide a secure and efficient payment system. These cryptocurrencies reduce transaction times, improve user experience, and ensure safe financial exchanges between farmers and buyers. The integration of blockchain-enabled payment systems supports a seamless agricultural economy by reducing costs and enhancing trust.

Blockchain technology holds immense potential for revolutionizing agriculture by addressing critical challenges like data security, transparency, and inefficiencies in communication and payment systems. By enabling tamper-proof data storage, precise product tracking, and fast transactions, blockchain fosters trust, sustainability, and growth within the agricultural industry. Its integration promises to improve food supply chains and ensure a more transparent and efficient agricultural ecosystem.

3.8 Summary

The use of IoT for fruit quality monitoring, crop growth analysis, and drone-assisted farming is transforming modern agriculture by leveraging sensors, data analytics, and automated systems to enhance efficiency, precision, and sustainability. Real-time monitoring and predictive analytics help reduce spoilage, extend shelf life, and improve consumer satisfaction while optimizing economic outcomes for producers and retailers. In crop growth monitoring, IoT devices such as sensors, drones, and automated systems provide farmers with precise insights into soil moisture, temperature, humidity, and nutrient levels, enabling better resource management and early detection of potential issues like pest infestations or diseases. Drones further enhance agricultural efficiency by integrating AI, IoT, and cloud computing, supporting precision irrigation, crop health monitoring, and livestock management. Despite challenges like regulatory constraints and scalability concerns, continuous advancements in ML, big data analytics, and cloud computing are driving smarter, more resilient farming practices. As these technologies evolve, they hold immense potential to address global agricultural challenges, ensuring sustainability and food security in the face of climate change and a growing population. UAVs can be utilized in a variety of applications related to crops management, which captures high spatial and temporal resolution images. These technologies are very faster for data analysis that can increase the crop productivity if monitored regularly.

References

1 Mishra, S., Khatri, S., and Johri, P. (2019). IoT based automated quality assessment for fruits and vegetables using infrared. In: *2019 4th International Conference on Information Systems and Computer Networks (ISCON)*, pp. 134–138. Mathura, India. https://doi.org/10.1109/ISCON47742.2019.9036165.

2 Saha, A., Ali, L., Rahman, R., et al. (2023). IoT based fruit quality inspection and lifespan detection system. In: *2023 7th International Conference On Computing, Communication, Control and Automation (ICCUBEA)*, pp. 1–6. Pune, India. https://doi.org/10.1109/ICCUBEA58933.2023.10392254.

3 Yadav, R., Saroj, R., Verma, A.K., and Mishra, A.K. (2023). A survey of IoT and machine learning-based monitoring of the growth of crops using blockchain technology. In: *2023 International Conference on IoT, Communication and Automation Technology (ICICAT)*. Gorakhpur, India: IEEE.

4 Namani, S., and Gonen, B. (2020). Smart agriculture based on IoT and cloud computing. In: *2020 3rd International Conference on Information and Computer Technologies (ICICT)*. San Jose, CA: IEEE.

5 Islam, N., Ray, B., and Pasandideh, F. (2020). IoT based smart farming: are the LPWAN technologies suitable for remote communication? In: *2020 IEEE International Conference on Smart Internet of Things (SmartIoT)*, pp. 270–276. Beijing, China: IEEE. https://doi.org/10.1109/SmartIoT49966.2020.00048.

6 Satyaraj, D., Prabakaran, T., Blessy, Y.M., et al. (2024). Enhancing fruit and vegetable preservation with support vector machine and IoT connectivity. In: *2024 International Conference on Automation and Computation (AUTOCOM)*, pp. 237–242. Dehradun, India: IEEE. https://doi.org/10.1109/AUTOCOM60220.2024.1048614.

7 Mowla, M.N., Mowla, N., Shah, A.F.M.S., et al. (2023). Internet of things and wireless sensor networks for smart agriculture applications: a survey. *IEEE Access* 11: 145813–145852.

8 Yang, X., Li, M., Yu, H., et al. (2021). A trusted blockchain-based traceability system for fruit and vegetable agricultural products. *IEEE Access* 9: 36282–36293.

9 Pandey, V., Patil, Y., Kanagaraj, R., et al. (2024). CHMPCNN: to implement the IoT-based crops health monitoring and prediction using deep learning. In: *2024 International Conference on Advances in Computing Research on Science, Engineering and Technology (ACROSET)*, pp. 1–7. Indore, India: IEEE.

10 Manoharan, G., Ali, S.D.N.H., Sathe, M., et al. (2024). IoT-based crop disease detection and management system using machine learning algorithms. In: *2024 International Conference on Science, Technology, Engineering and Management (ICSTEM)*, pp. 1–5. Coimbatore, India: IEEE.

11 Sundari, S.L.K., Rana, M., Ahmed, S.T., and Anitha, K. (2021). Real-time IoT-based temperature and NPK monitoring system for sugarcane crop yield improvement. In: *2021 Innovations in Power and Advanced Computing Technologies (i-PACT)*. Kuala Lumpur, Malaysia: IEEE.

12 Rozario, G.S., and Vasanthi, V. (2021). Crop growth monitoring and leaf area index estimation using wireless sensor network and CNN. In: *2021 Third International Conference on Inventive Research in Computing Applications (ICIRCA)*. Coimbatore, India: IEEE.

13 Jasim, A.N., and Fourati, L.C. (2023). Agriculture 4.0 from IoT, artificial intelligence, drone, & blockchain perspectives. In: *2023 15th International Conference on Developments in eSystems Engineering (DeSE)*. Baghdad & Anbar, Iraq: IEEE.

14 Singh, D.P. (2022). Evaluation of drone systems in precision agriculture. *International Journal of Food and Nutritional Sciences (IJFANS)* 11: 1984–1990.

15 Al-Shareeda, M.A., Manickam, S., and Saare, M.A. (2022). Intelligent drone-based IoT technology for smart agriculture system. In: *International Conference on Data Science and Intelligent Computing (ICDSIC)*. Karbala, Iraq: Kerbala University.

16 Raj, K.M., Balaji, N., Vairavel, K.S., et al. (2022). IoT-based agricultural drones for pest control. In: *2022 1st International Conference on Computational Science and Technology (ICCST)*. Chennai, India: IEEE.

17 Almalki, F.A., Soufiene, B.O., Alsamhi, S.H., and Sakli, H. (2021). A low-cost platform for environmental smart farming monitoring system based on IoT and UAVs. *Sustainability* 13 (11): 5908. https://doi.org/10.3390/su13115908.

18 Lottes, P., Khanna, R., Pfeifer, L., et al. (2017). UAV-based crop and weed classification for smart farming. In: *International Conference on Robotics and Automation (ICRA)*. arXiv:1709.06764. Singapore: IEEE.

19 Gill, H., and Khehra, B. (2022). An integrated approach using CNN-RNN-LSTM for classification of fruit images. *Materials Today: Proceedings* 51: 591–595. https://doi.org/10.1016/j.matpr.2021.06.016.

20 Knott, M., Perez-Cruz, F., and Defraeye, T. (2023). Facilitated machine learning for image-based fruit quality assessment. *Journal of Food Engineering* 345: 111401. https://doi.org/10.1016/j.jfoodeng.2022.111401.

21 Pranav, A.H., Senthilmurugan, M., Rahul, K.P., and Chinnaiyan, R. (2021). IoT and machine learning-based peer-to-peer platform for crop growth and disease monitoring system using blockchain. In: *2021 International Conference on Computer Communication and Informatics (ICCCI)*. Coimbatore, India: IEEE.

4

AI and IoT-based Robust and Resilient Healthcare Services

Shilpi[1,*], *Sachin Kumar*[1], *Sudhanshu Tyagi*[2], *and Sumita Mishra*[1]

[1] *Department of ECE, Amity School of Engineering and Technology, Lucknow Campus, Uttar Pradesh, India*
[2] *Thapar Institute of Engineering and Technology, Deemed to be University, Patiala, Punjab, India*

*Corresponding author: shilpi.scriet@gmail.com

4.1 Introduction

Both artificial intelligence (AI) and Internet of Things (IoT) are transforming healthcare by creating robust and resilient services. These services are capable to address many challenges such as resource constraints, personalized care, real-time monitoring, disaster preparedness, and many more. The two technologies are able to enhance healthcare services. In healthcare, AI is used in predictive analytics, decision support systems, and automation. This technique can predict resource needs, patient deterioration, and disease outbreaks. Also, by using medical imaging or patient data, AI-powered tools can identify diseases such as cancer, heart issues, etc., with good accuracy. AI can also work as virtual health assistant. AI assistant provides preliminary consultations, appointment scheduling, and medication reminders. AI approaches drug development by analyzing vast datasets and simulation outcomes. AI prepares treatment plans based on patient genetics, lifestyle, and health data.

IoT connects devices and systems, ensuring seamless data flow and improved decision-making. IoT devices are competent to connect and monitor patients, healthcare providers, and medical systems. IoT devices are used for remote patient monitoring (RPM). An example of RPM is wearable device, which is operated

AI and ML Techniques in IoT-based Communication: A Path to Sustainable Development Goals, First Edition. Edited by Sumita Mishra, Nishu Gupta and Polat Goktas.

to track vital signs such as heart rate, blood pressure, and glucose levels and to send real-time data to healthcare providers. IoT-enabled devices are also utilized to manage equipment, monitor patient flow, and ensure safety in smart hospitals. IoT pill dispensers remind patients to take their medicine and notify caregivers if doses are missed. In emergency situation like falls or cardiac events, IoT sensors trigger alerts for immediate response [1].

These two techniques, AI and IoT, together provide efficient and scalable solutions to healthcare problems. The two techniques work in various ways. AI analyzes IoT data to predict potential health risks and notify providers for disease management. For real-time decision-making, AI processes the collected data through IoT for real-time guidance. Remote consultations and diagnostics are also enhanced by AI-powered analytics in combination with IoT tools. System made up with the combination of AI and IoT provides the assurance of uninterrupted healthcare delivery during natural disasters or pandemics by monitoring supplies, staff availability, and patient needs. The outcomes of elderly and chronic disease patients are improved by continuous monitoring with IoT devices and AI driven alerts. By the use of the combination of AI and IoT, early detection is possible and thus these phenomena can reduce complications in healthcare [2]. This automation lowers the costs and boosts the efficiency. These technological solutions can serve a large number of people without proportional change of resource expansion. This combination of technique can bridge the gap in remote or underserved areas. By integrating AI and IoT, healthcare systems can become more adaptive, reliable, and patient centered, paving the way for a more resilient future.

4.1.1 General Architecture

AI and IoT-based framework for healthcare services integrates advanced technologies to provide efficient, adaptive, and reliable healthcare solutions. The general architecture is made up of four layers: IoT layer, data-processing layer, AI-based analytics layer, and application layer shown in Figure 4.1 with the help of block diagram. 5G and LPWAN are the network technologies used for reliable and fast communication. Overview of such architecture is given below.

4.1.1.1 IoT Layer (Data Collection and Communication)

IoT layer comprises IoT devices, IoT sensors, edge devices, and communication protocol. Smart devices, for example, wearable, implantable devices, environmental sensors, gather real-time data of patient. Data may be heart rate, temperature, glucose levels, ambient conditions, etc. Edge computing nodes preprocess the data to reduce latency and bandwidth usage, ensuring near-real-time insights. Secure and energy-efficient communication protocols, for example, Bluetooth, ZigBee, MQTT, and 5G, are used to enable device to device and device to cloud

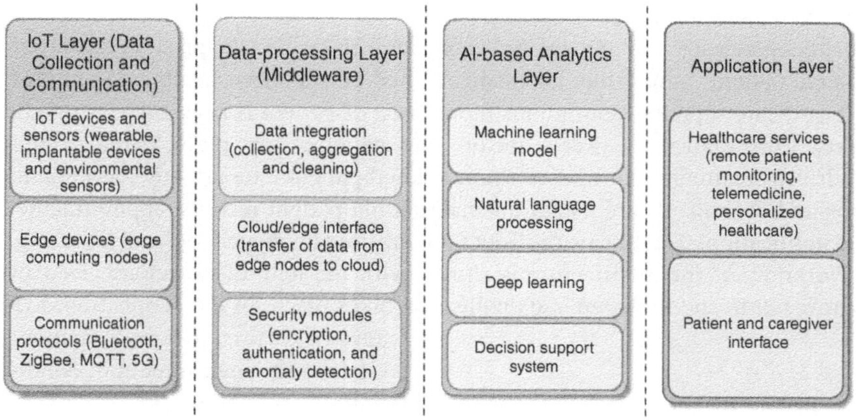

Figure 4.1 General architecture.

communication. AWS IoT Core, Google Cloud IoT, and Azure IoT Hub are the technologies that make healthcare services robust and resilient. Nvidia Jetson, Raspberry Pi, and Intel Movidius are the edge computing techniques used for real-time analysis.

4.1.1.2 Data-processing Layer (Middleware)

In data-processing layer, integration of data, cloud, and/or edge interfacing and security modules takes place. Collection, aggregation, and cleaning of data from diverse IoT devices for consistency occur in this layer. Data from nodes to cloud transfers for advanced analytics when necessary. Security modules incorporate encryption, authentication, and anomaly detection mechanisms to ensure secure data transmission and storage. Blockchain for secure, decentralized data handling are the techniques used for data security.

4.1.1.3 AI-based Analytics Layer

Machine learning models, natural language processing, deep learning, and decision support systems are involved in this analytical layer. Algorithms used in models analyze data for predictive insights. Processing technique enables AI systems to interpret unstructured data like clinical notes and patient queries. Learning methods support complex tasks like medical image analysis and diagnostic support. Support system recommends interventions based on data-driven insights, assisting clinicians and caregivers. TensorFlow, PyTorch, and Scikit-learn are the technologies that make healthcare services robust and resilient [3].

4.1.1.4 Application Layer

Healthcare services and patient and caregiver interfacing take place in this layer. Remote patient monitoring, telemedicine, and personalized healthcare are part of healthcare services. Continuous monitoring of patient is necessary for timely alerts. Consultations between doctors and patients take place virtually, and AI-driven treatment plans based on individual patient data are tailored. Mobile apps and dashboards are useful for maintaining patient records, appointments, and notifications. Healthcare provider tools are used for diagnostics and reporting.

Workflow of the architecture is shown with the help of flowchart given in Figure 4.2. In the first step, data collection takes place. In data collection, IoT devices are used to monitor patient vitals and environmental conditions in real time [4]. After data collection, preprocessing of collected data occurs. In preprocessing, edge devices filter, compress, and preprocess the collected data locally. After processing data is transmitted securely to cloud platforms or local servers for deeper analysis. This step is named data transmission. AI analytics identifies anomalies or potential health issues and predicts patient outcomes using historical data and AI models. On the basis of prediction, decision is to be made. And alerts go to caregivers or automated interventions are triggered. The intervention provides personalized treatment recommendations. This flow produces alerts that are sent to patients, caregivers, and healthcare providers. A continuous feedback loop works to improve system performance and patient outcomes. This architecture can significantly advance modern healthcare by making it smarter, more accessible and resilient to challenges.

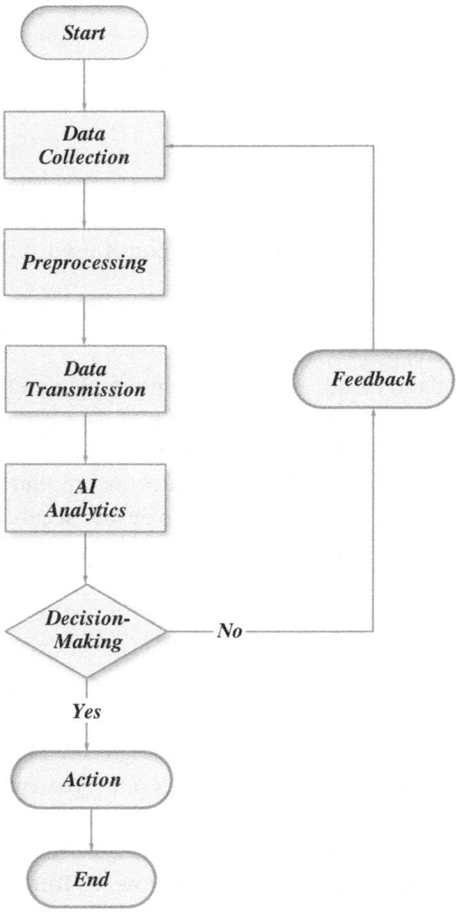

Figure 4.2 Workflow of architecture.

4.2 Role of IoT in Predictive Healthcare

The IoT plays a pivotal role in predictive healthcare. This technique allows healthcare providers to predict medical issues, improve outcomes, and optimize resource allocation. By manipulating connected devices and real-time data, IoT gives the facilities of proactive healthcare solutions rather than reactive care [5]. Some of the primary roles of IoT in predictive healthcare are shown in Figure 4.3. The description of all are given as follows:

4.2.1 Continuous Monitoring and Early Detection

IoT devices such as wearable, implants, and sensors collect real-time data on vital signs and behavior patterns. Signs may be heart rate, blood pressure, glucose levels, etc. Wearable can be used for tracking of sleep and fitness. They can predict potential sleep disorders or lifestyle-induced risks. Risks may be in terms of obesity and stress. Other IoT devices are used to detect anomalies in a patient's health data such as irregular heart rhythms. They also allow early intervention. For the

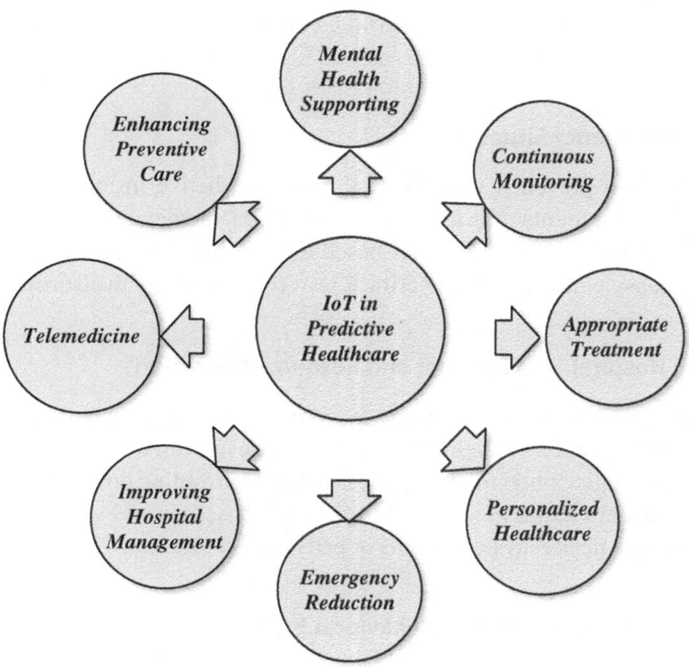

Figure 4.3 IoT in predictive healthcare.

special condition like diabetes, IoT device monitors glucose levels and predicts risks like hypoglycemia or complications.

4.2.2 Data-driven Predictive Analytics

IoT generates vast amounts of health data. When this data is processed with predictive analytics, it can identify the trends and envision of various potential issues. These health trends analyzed by IoT systems are also used to predict the progression of chronic diseases. This aggregated IoT data can predict disease outbreaks or patterns. This can give assistance for public health planning. Also, IoT devices can analyze the impact of medication by tracking how medication affects the patients and predicts adverse reactions.

4.2.3 Personalized Healthcare

IoT is able to do personalized healthcare. It is used to adapt the predictions for individual patients by analyzing their specific health data. This technique is very much useful for risk assessment. The predictive model based on IoT-collected metrics is used to assess an individual's risk for conditions like heart attacks or strokes. This makes an alert for the patients. And based on real-time data trends, patients will receive health warnings.

4.2.4 Reducing Emergency Situations

IoT helps anticipate and prevent emergencies through predictive insights. IoT devices monitor elderly patients and identify fall based on movement patterns and give some prevention. An example of IoT-based EEG monitors is used to predict seizures for epilepsy patients, offering critical lead time for safety measures.

4.2.5 Improving Hospital Management and Resource Allocation

IoT systems contribute to predictive healthcare by optimizing hospital operations. Maintenance is possible now a days by the use of IoT, because IoT can track medical equipment usage and predict when maintenance is required, avoiding downtimes. IoT can also predict patient admission and resource needs based on real-time and historical data. Resource need may be in terms of beds, ventilators, etc.

4.2.6 Supporting Mental Health and Behavioral Insights

IoT devices monitor behavioral patterns, sleep cycles, and stress levels to predict mental health crises. This technique is very useful to prevent mental health crises.

To monitor stress, wearable devices based on IoT are used. These devices detect stress markers like heart rate variability and predict burnout or anxiety episodes. Behavioral risks induced by lifestyle such as depression or addiction can also be read by IoT devices with the help of available data.

4.2.7 Telemedicine and Remote Monitoring

IoT supports telemedicine by allowing predictive care even in remote locations. Continuous IoT monitoring predicts health issues for patients in underserved areas. In emergency situation, IoT alerts healthcare providers of impending emergencies for remotely monitored patients.

4.2.8 Enhancing Preventive Care

IoT systems enhance potential health issues by prediction and therefore promote prevention. IoT data influenced by lifestyle is used to predict risks from unhealthy habits, and suggestions are also advised for changes in habits or lifestyle. IoT is used to predict disease outbreaks, enabling timely vaccination efforts.

4.3 Improved Decision-making and Treatment Planning

Improved decision-making and treatment planning in healthcare are critical outcomes of integrating advanced technologies like IoT and AI. These technologies enable healthcare professionals to make data-driven, timely, and accurate decisions, leading to better patient outcomes and efficient care delivery.

4.3.1 Role of IoT in Decision-making and Treatment Planning

IoT devices are designed to collect real-time health data from patients continuously to provide valuable insights for personalized and effective healthcare. This real-time data is used for decision support system. IoT devices like wearable and medical implants provide real-time data on vitals such as heart rate, oxygen levels, and blood pressure. This helps doctors to assess patients' conditions instantly [6]. IoT systems also predict and alert providers to impending health crises (e.g. cardiac arrest or respiratory failure), enabling rapid intervention. For example, for diseases like diabetes or hypertension, IoT devices track and analyze trends, aiding treatment adjustments.

IoT sensors gather comprehensive data that can help clinicians diagnose conditions remotely, especially in underserved areas for disease risk prediction. IoT data is integrated with predictive analytics to identify risk factors early and inform preventive strategies.

IoT systems allow shared data access. It means multiple healthcare professionals can access patient's data simultaneously, fostering collaboration in complex cases. IoT devices enable patients to track their own health metrics, empowering them to make informed decisions alongside their healthcare providers.

4.3.2 Role of AI in Decision-making and Treatment Planning

AI complements IoT by analyzing vast datasets and delivering actionable insights for decision-making and treatment planning. AI supports disease progression forecasting by analyzing patient history, and IoT collects data to predict disease progression, enabling timely interventions. This AI models simulate various treatment scenarios, helping doctors choose the most effective options.

AI-powered clinical decision support systems provide tailored treatment suggestions based on patient data, improving the precision of care plans. AI is also used for risk assessment. AI evaluates potential complications or side effects of treatments, guiding safer decision-making [7].

AI-integrated IoT systems provide surgical assistance. Systems, such as robotic surgery platforms, aid in planning and executing complex procedures. AI suggests personalized medication plans based on patient responses and IoT-monitored data trends.

4.4 Personalized Healthcare

Personalized healthcare tailors medical treatment and prevention strategies to individual patients based on their unique biological, environmental, and lifestyle characteristics. This approach contrasts with traditional "one-size-fits-all" medicine, emphasizing patient specific care to improve outcomes and reduce side effects [8]. Technologies like AI, IoT, genomics, and wearable devices play a significant role in enabling personalized healthcare.

4.4.1 Key Points of Personalized Healthcare

Key points of personalized healthcare are as follows:

4.4.1.1 Precision Medicine

Precision medicine is one of the main components of personalized healthcare. It focuses on understanding a patient's genetics, environment, and lifestyle to customize treatments, for example. Genetic testing identifies biomarkers to determine the most effective cancer therapies or the likelihood of drug side effects.

4.4.1.2 Real-time Monitoring with IoT

IoT devices such as wearable and smart sensors are used to continuously monitor vital signs like heart rate, glucose levels, and sleep patterns of the patients [9]. These devices provide data that healthcare providers can use to tailor treatments to an individual's needs, for example, a wearable insulin pump adjusts dosage based on real-time blood sugar readings.

4.4.1.3 AI-driven Analytics

AI analyzes vast datasets from patient records, genetic tests, and IoT devices to uncover patterns and recommend personalized treatments, for example. AI algorithms predict which patients are likely to benefit from specific medications based on past data and real-time metrics [10].

4.4.1.4 Genomics and Molecular Profiling

Genomic sequencing identifies specific mutations or genetic predispositions that guide treatment planning, for example. Pharmacogenomics tailors medication dosages based on how an individual metabolizes drugs while reducing adverse reactions.

4.4.1.5 Personalized Prevention

By analyzing lifestyle and environmental factors, healthcare providers can offer tailored advice on diet, exercise, and preventive measures, for example. IoT wearable track physical activity and provide personalized fitness recommendations to prevent chronic diseases.

4.4.1.6 Telemedicine and Virtual Care

Remote consultations those are powered by IoT and AI allow healthcare providers to deliver tailored care regardless of location, for example. Virtual health platforms use patient-reported data to offer specific advice for managing chronic conditions.

4.5 Remote Diagnostics and Efficient Resource Allocation

Remote diagnostics and efficient resource allocation are two critical advancements enabled by technologies like IoT, AI, and telemedicine. These innovations improve healthcare access, optimize resource use, and enhance patient outcomes, particularly in underserved or resource-constrained settings.

4.5.1 Remote Diagnostics

Remote diagnostics involves using technology to identify, monitor, and assess medical conditions without the patient needing to visit a healthcare facility [11].

Remote cardiology, diabetes management, and virtual dermatology are the examples of remote diagnostics. IoT-enabled ECG devices detect arrhythmias or other heart issues and alert healthcare providers. Continuous glucose monitors track blood sugar levels and provide feedback to both patients and doctors. AI-powered applications analyze skin images for conditions like melanoma, eczema, or rashes.

4.5.1.1 Key Technologies Enabling Remote Diagnostics

1. **IoT Devices:**
   IoT devices, for example, wearable devices such as smart watches, fitness trackers are used to monitor vital signs such as heart rate, blood pressure, glucose levels, and oxygen saturation in real time. Home-based diagnostic kits and connected devices also enable remote monitoring of chronic diseases.
2. **AI and Machine Learning:**
   AI analyzes data collected by IoT devices to detect abnormalities, predict disease progression, and suggest next steps. AI-powered algorithms interpret chest X-rays or ECG readings remotely.
3. **Telemedicine Platform:**
   Video consultations allow doctors to assess symptoms and diagnose conditions remotely. Connected diagnostic tools such as stethoscopes can transmit data directly to healthcare providers.
4. **Mobile Health Applications:**
   Apps enable patients to input symptoms, track health trends, and receive preliminary assessments.

4.5.2 Efficient Resource Allocation

Efficient resource allocation involves optimizing the use of healthcare resources such as staff, equipment, and facilities to deliver timely and effective care. Smart hospitals, ambulance routing, and vaccination campaigns are the examples of efficient resource allocation. IoT systems track patient flow, bed occupancy, and equipment availability in real time. AI optimizes ambulance routes based on traffic and proximity to the nearest equipped hospital. Predictive models identify regions with low vaccination rates and allocate doses accordingly [12–14].

4.5.2.1 Key Technologies Enabling Efficient Resource Allocation

1. **Predictive Analytics:**
   AI analyzes historical and real-time data to forecast patient demand, equipment needs, and staffing requirements. Prediction of hospital admissions during flu season to allocate resources effectively is the common example.

2. **IoT in Hospital Management:**
   Smart beds and IoT-enabled equipment track availability of needs along with ensuring optimal utilization of resources. Sensors are used to monitor medical inventory, for example, oxygen tanks, PPE, etc., to prevent shortages.
3. **Telemedicine:**
   Telemedicine is used to reduce the burden on physical facilities by enabling virtual consultations and follow-ups. It also ensures that specialists are accessible to patients without requiring travel.
4. **Electronic Health Records:**
   A centralized patient record in electronic form is used for streamline care coordination. It is also used to reduce redundant tests.

4.6 Benefits

There are several benefits of IoT and AI-based robust and resilient healthcare services [15]. These technologies are very helpful for the future of advance healthcare services. The flowchart in Figure 4.4 shows the different types of benefits in healthcare services.

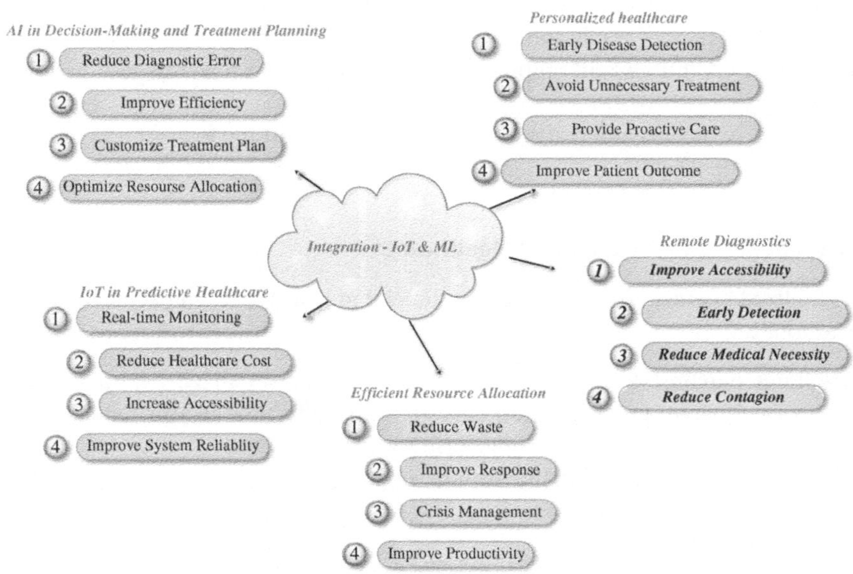

Figure 4.4 Integration of AI and IoT in healthcare.

4.6.1 Benefits of IoT in Predictive Healthcare

The following are some benefits of the use of IoT in predictive healthcare:

1. IoT is used to enhance patient outcomes with continuous real-time monitoring and predictive insights.
2. Healthcare costs through early detection and remote management is reduced with the use of IoT [16].
3. The accessibility to healthcare services in remote and underserved areas is increased by using IoT.
4. In predictive healthcare system, reliability is increased using fault tolerant and self-healing mechanisms through the use of IoT.

4.6.2 Benefits of AI in Decision-making and Treatment Planning

The following are some benefits of the use of IoT and AI in decision-making and treatment planning:

1. Data-driven insights reduce diagnostic errors and optimize treatment decisions. Thus, in turn, accuracy is improved.
2. The efficiency of overall work is enhanced due to automatic routine tasks. Data collection and analysis frees up clinicians to focus on patient care.
3. AI and IoT combined enable highly customized treatment plans tailored to individual patient needs and conditions [17].
4. Early interventions and optimized resource allocation lower overall healthcare costs.
5. Predictive models and real real-time monitoring shift the focus from reactive to preventive care.

4.6.3 Benefits of Personalized Healthcare

Personalized healthcare is very much beneficial in healthcare, and some of them are as follows:

1. Patient outcomes are improved by the use of personalized healthcare. In this healthcare, tailored treatments are more effective. It also reduces the likelihood of complications or adverse reactions [18].
2. Disease can be detected in its early stage with the help of personalized healthcare. Continuous monitoring identifies health risks before they develop into severe conditions.
3. Personalized healthcare avoids unnecessary treatments and hospitalizations through precise interventions.

4. In personalized healthcare, patient's engagement is improved. Here, patients actively participate in their care through access to personalized insights and recommendations.
5. Personalized healthcare provides proactive care. It moves healthcare from reactive to preventive. It also reduces the burden on healthcare systems.

4.6.4 Benefits of Remote Diagnostics

1. Accessibility is improved with the help of remote diagnostics. Patients in rural or remote areas can access expert diagnostics without traveling long distances. Remote diagnostics also reduces barriers for patients with mobility issues.
2. Early detection is possible with continuous monitoring of patient health. IoT devices continuously detect health problems before symptoms become severe, enabling early interventions.
3. Remote diagnostics reduces the need for hospital visits and tests by facilitating remote assessments.
4. Remote diagnostics minimizes in-person interactions during infectious disease outbreaks. Thus, it reduces the spread of illnesses.

4.6.5 Benefits of Efficient Resource Allocation

1. Due to efficient resource allocation, waste is reduced. Use of equipment, medications, and staff are also optimized.
2. Timely access of resources ensures that patients receive the right care at the right time. This in turn improves the patient outcome.
3. During emergencies, predictive systems allocate resources to high need areas. Thus, it is useful in crisis management.
4. Staff scheduling and workload distribution tools prevent burnout and ensure adequate coverage. So, it also improves overall efficiency.

4.7 Applications in Specific Areas

4.7.1 Oncology

IoT devices monitor treatment in real time. AI analyzes imaging data for accurate cancer diagnosis. AI identifies genetic mutations in tumors, which helps to determine the most effective chemotherapy or immunotherapy. Targeted therapies are used to treat cancer.

4.7.2 Cardiology

IoT-enabled devices like ECG monitors and AI algorithms are helpful to predict heart attacks. They are used to enable timely interventions [19]. Smart ECG trackers detect heart issues and recommend lifestyle changes. Based on continuous blood pressure monitoring, blood pressure medication can be adjusted.

4.7.3 Mental Health

IoT wearable track stress and sleep patterns, while AI identifies behavioral patterns for early intervention in mental health conditions. AI-driven platforms provide tailored mental health recommendations based on mood. Cognitive behavioral therapy apps personalize exercises based on user interactions.

4.7.4 Surgical Planning

AI assists in preoperative planning using IoT data to minimize risks and improve outcomes.

4.7.5 Diabetes Management

IoT devices monitor glucose levels, while AI analyzes trends to optimize insulin delivery. Closed-loop insulin delivery system is the example of it. It is also known as artificial pancreas.

4.7.6 Rare Diseases

Some rare diseases can also be predicted by the help of AI and IoT-based healthcare services. Genomic sequencing identifies the root cause of rare conditions, enabling targeted treatments [20]. CRISPR-based therapies for genetic disorders are the example.

4.8 Challenges and Opportunities

4.8.1 Challenges in Implementing Healthcare Services

Various challenges exist in the implementation of AI and IoT-based healthcare services. Some of them are as follows:

4.8.1.1 Data Privacy and Security

Data privacy and security is the primary challenge in implementing healthcare services. Sensitive patient health data from IoT devices, genomics, and telemedicine platforms must be safe and protected from breaches [21]. To protect health data, security protocols should be enhanced. Data encryption, data sharing, and robust data governance protocols can be implemented for this. Authentication for all devices and platforms should be secure.

4.8.1.2 High Cost

Cost is the big challenge in the implementation. Initial investment for AI and/or IoT devices, infrastructure, and technologies becomes very high. Advanced diagnostic tools and therapies are also expensive. Managing and analyzing vast volumes of data effectively carries high cost [22]. Government funding and value-based care models can improve affordability of the healthcare system. A collaborative partnership between tech and healthcare providers can work innovatively for it.

4.8.1.3 Data Integration and Digital Divide

Integration of data from diverse sources like IoT, EHRs, genomics, and systems is a complex task and becomes a big challenge in the implementation of healthcare services. Seamless integration of data across diverse systems should be ensured for better service. Integration of data should be unified into a system from diverse IoT devices. For this, interoperable systems and standardize data formats should be developed [23]. The universal protocols for device communication for IoT and EHR interoperability should be adopted. One major issue of limited access of internet or technology is in underserved regions. For this, government and private sector initiatives are helpful to provide low-cost devices and connectivity.

4.8.1.4 Ethical and Regulatory Issues

Ethical issues are related to the misuse of genetic data and the fairness of AI algorithms. To manage AI decision-making, transparency and patient privacy is the primary issue. Ethical guidelines should be established for these issues, and with the help of these guidelines, transparency in AI decision-making must be ensured. For continuous monitoring, robust governance frameworks for ethics must also be implemented [24].

4.8.2 The Future of Healthcare Services

1. In the future of healthcare services, AI integration will become more advanced, more sophisticated, predicting health trends and personalizing care with minimal human input.
2. Affordable and rapid genomic testing will become widespread and it will make personalized care accessible to all in future.
3. IoT will enable seamless communication among all devices to create a unified and patient-centered care platform to enhance the IoT ecosystem.

4. Telemedicine, home-based diagnostics, and wearable technologies will empower patients to manage their health independently focusing on individual needs to improve outcomes and quality of life in future.
5. With advancements in technology, the approach is set to redefine the future of medicine and healthcare services.

4.9 Conclusion

In healthcare, AI is used in predictive analytics, decision support systems, and automation. IoT is revolutionizing predictive healthcare by enabling early detection, personalized care, and proactive resource management. Its role is crucial in shifting healthcare from a reactive to a preventative model, ultimately improving patient outcomes and healthcare system efficiency. IoT and AI are transforming healthcare. Together, they enable proactive, accurate, and efficient care, benefiting patients and overall healthcare systems. The combination is trying to reshape healthcare by improving access, reducing costs, and ensuring optimal use of resources. These innovations make healthcare more resilient and patient-centered, especially in the face of growing demand and limited resources.

References

1 Abdulmalek, S., Nasir, A., Jabbar, W.A., et al. (2022). IoT-based healthcare-monitoring system towards improving quality of life: a review. *Healthcare* 10 (10): 1993.
2 Li, C., Wang, J., Wang, S., and Zhang, Y. (2024). A review of IoT applications in healthcare. *Neurocomputing* 565: 127017. ISSN 0925–2312.
3 Olatinwo, S.O., and Joubert, T.-H. (2022). Deep learning for resource management in internet of things networks: a bibliometric analysis and comprehensive review. *IEEE Access* 10: 94691–94717.
4 Younas, M.I., Iqbal, M.J., Aziz, A., and Sodhro, A.H. (2023). Toward QoS monitoring in IoT edge devices driven healthcare—a systematic literature review. *Sensors* 23 (21): 8885.
5 Taimoor, N., and Rehman, S. (2022). Reliable and resilient AI and IoT-based personalised healthcare services: a survey. *IEEE Access* 10: 535–563.
6 Islam, M.M., Hasan, M.K., Islam, S., et al. (2024). Enabling pandemic-resilient healthcare: narrowband internet of things and edge intelligence for real-time monitoring. *CAAI Transactions on Intelligence Technology* 2024: 1–18.
7 Pamulaparthyvenkata, S., Murugesan, P., Gottipalli, D., and Palanisamy, P. (2024). AI-enabled distributed healthcare framework for secure and resilient remote

patient monitoring. In: *2024 5th International Conference on Smart Electronics and Communication (ICOSEC)*, pp. 2034–2041. Trichy, India: IEEE.

8 Avdan, G., and Onal, S. (2024). Resilient healthcare 5.0: advancing human-centric and sustainable practices in smart healthcare systems. In: *IISE Annual Conference. Proceedings. Institute of Industrial and Systems Engineers (IISE)*, pp. 1–6.

9 Rane, N., Choudhary, S., and Rane, J. (2024). Artificial intelligence for enhancing resilience. *Journal of Applied Artificial Intelligence* 5 (2): 1–33.

10 Maguraushe, K., Ndayizigamiye, P., and Bokaba, T. (2024). A systematic review on the use of AI-powered cloud computing for healthcare resilience. In: *International Conference on Emerging Technologies for Developing Countries*, pp. 126–141. Springer, Cham.

11 Aminizadeh, S., Heidari, A., Dehghan, M., et al. (2024). Opportunities and challenges of artificial intelligence and distributed systems to improve the quality of healthcare service. *Artificial Intelligence in Medicine* 149: 102779.

12 Junior, A.A., Ndayizigamiye, P., and Bokaba, T. Resilient healthcare systems through AI-powered cloud computing: perspectives on resource allocation. *Hawaii International Conference on System Sciences*. https://doi.org/10.24251/HICSS.2025.399.

13 Alshamrani, M. (2022). IoT and artificial intelligence implementations for remote healthcare monitoring systems: a survey. *Journal of King Saud University-Computer and Information Sciences* 34 (8): 4687–4701.

14 Moskalenko, V., and Kharchenko, V. (2024). Resilience-aware MLOps for AI-based medical diagnostic system. *Frontiers in Public Health* 12: 1342937.

15 Moskalenko, V., Kharchenko, V., Moskalenko, A., and Kuzikov, B. (2023). Resilience and resilient systems of artificial intelligence: taxonomy, models and methods. *Algorithms* 16 (3): 165.

16 Vani, G., Naveenkumar, R., Singha, R., et al. (2024). Advancing predictive data analytics in IoT and AI leveraging real time data for proactive operations and system resilience. *Nanotechnology Perceptions* 20 (16): 568–582.

17 Akinsola, A., and Ejiofor, O. (2024). Securing the future of healthcare: building a resilient defense system for patient data protection. Available at *SSRN 4902351*.

18 Bhambri, P., and Tripathy, B. (2024). Role of AI and IoT based medical diagnostics smart health care system for post-Covid-19 world. In: *Smart Healthcare Systems*, pp. 135–145. CRC Press.

19 Subhan, F., Mirza, A., Su'ud, M.B., et al. (2023). AI-enabled wearable medical internet of things in healthcare system: a survey. *Applied Sciences* 13 (3): 1394.

20 Bala, I., Pindoo, I., Mijwil, M.M., et al. (2024). Ensuring security and privacy in healthcare systems: a review exploring challenges, solutions, future trends, and the practical applications of artificial intelligence. *Jordan Medical Journal* 58 (3): 250–270.

21 Hamood Alsamhi, S., Hawbani, A., Shvetsov, A.V., and Kumar, S. (2023). Advancing pandemic preparedness in healthcare 5.0: a survey of federated learning applications. *Advances in Human-Computer Interaction* 2023 (1): 9992393.

22 Vettriselvan, R. (2025). Harnessing innovation and digital marketing in the era of industry 5.0: resilient healthcare SMEs. In: *The Future of Small Business in Industry 5.0*, pp. 163–186. IGI Global Scientific Publishing.

23 Achour, N. (2024). Resilient hospital structures, systems and services. In: *Oxford Research Encyclopedia of Natural Hazard Science*. Oxford Research Encyclopedias of Natural Hazard Science, Oxford University Press.

24 Pradhan, B., Das, S., Roy, D.S., et al. (2023). An AI-assisted smart healthcare system using 5G communication. *IEEE Access* 11: 108339–108355.

Part II

Practical Applications of IoT and AI in Key Sectors Including Smart Infrastructure and Industrial Applications

5

Applications and Impact of Artificial Intelligence in the Field of Agriculture, Education, Healthcare, and Administration

Virender Kumar[1,*] *and Mohammad Derawi*[2]

[1] *Department of Electronics and Communication Engineering, Chandigarh University, Mohali, Punjab, India*
[2] *Department of Electronic Systems, Faculty of Information Technology and Electrical Engineering, Norwegian University of Science and Technology, Gjøvik, Norway*

*Corresponding author: virendergangotra@gmail.com

5.1 Introduction

In this chapter, we explore the impact and future potential of artificial intelligence (AI) in various aspects of human life. AI pertains to creating computer systems with the ability to carry out tasks that traditionally necessitate human-like intelligence. These systems aim to replicate human cognitive abilities such as speech recognition, decision-making, and visual perception by employing algorithms and vast amounts of data. AI technologies enable information analysis, prediction, and adaptive responses to different situations. AI is a transformative technology with applications across a wide range of industries. It is reshaping our lives, work, and interactions, going beyond simple automation and efficiency improvements. Industries such as education, agriculture, healthcare, administration, governance, business, commerce, and customer service are experiencing the positive effects of AI. In the area of education, AI holds tremendous capability to improve learning opportunities and increase accessibility. Personalized learning paths can be developed using AI to cater to each student's unique needs and preferences. Data

AI and ML Techniques in IoT-based Communication: A Path to Sustainable Development Goals, First Edition. Edited by Sumita Mishra, Nishu Gupta and Polat Goktas.

analytics can uncover valuable insights and improve teaching strategies, while intelligent tutoring systems offer targeted feedback and guidance. AI-powered virtual assistants can also assist with administrative duties, liberating teachers' time to concentrate on fostering student engagement and creativity. In agriculture, AI is revolutionizing farming practices by enabling data-driven decision-making for improved crop yield, pest management, and resource optimization. AI algorithms can scrutinize satellite imagery, weather patterns, and soil data to provide precision agricultural solutions. This allows farmers to maximize productivity while minimizing environmental impact. AI-enabled robots and drones can automate labor-intensive tasks like planting, harvesting, and crop monitoring, thereby enhancing production efficiency and sustainability. The healthcare industry is benefiting from AI's capabilities in patient care, diagnosis, and treatment. Machine learning (ML) algorithms can analyze medical data and imaging images to help doctors identify diseases earlier and develop personalized treatment plans. AI-powered virtual nurses and chatbots provide round-the-clock assistance to patients by answering questions, reminding them to take medication, and remotely monitoring vital signs. Moreover, AI-driven drug discovery and research have the potential to accelerate the development of innovative medicines and precision medicine approaches. In the realm of governance and administration, AI technologies are enhancing decision-making and streamlining processes. Intelligent automation and natural language processing (NLP) enable efficient data analysis, facilitating policy creation, resource allocation, and public service provision. These advancements contribute to more effective governance and administration. Overall, AI is revolutionizing various aspects of human life, including education, agriculture, healthcare, and governance. Its transformative impact is evident in improved outcomes, enhanced efficiency, and increased innovation across multiple industries.

AI-driven virtual assistants and chatbots have the capability to address citizen inquiries, resulting in faster response times and improved service quality. By leveraging predictive analytics, proactive measures can be taken to address issues like crime prevention and disaster management by identifying potential areas of concern. AI, ML, and deep learning (DL) have immense capability to positively change our lives. Nevertheless, it is essential to tackle concerns regarding privacy in data and ensure ethical usage of AI technologies. To initiate this chapter, we opted for a research topic and conducted an extensive exploration of existing research papers. Our examination encompassed a detailed analysis of their methodologies, findings, and conclusions. Furthermore, we took into account the most recent developments in AI, Internet of Things (IoT), and other relevant ongoing research domains. Building upon this comprehensive analysis, we present a comprehensive overview of the present applications of AI in diverse facets of human existence, including education, agriculture, healthcare, and administration. Lastly, based on the gathered information, we offer our own conclusion. Figure 5.1 indicates AI, ML, and DL difference and hierarchy.

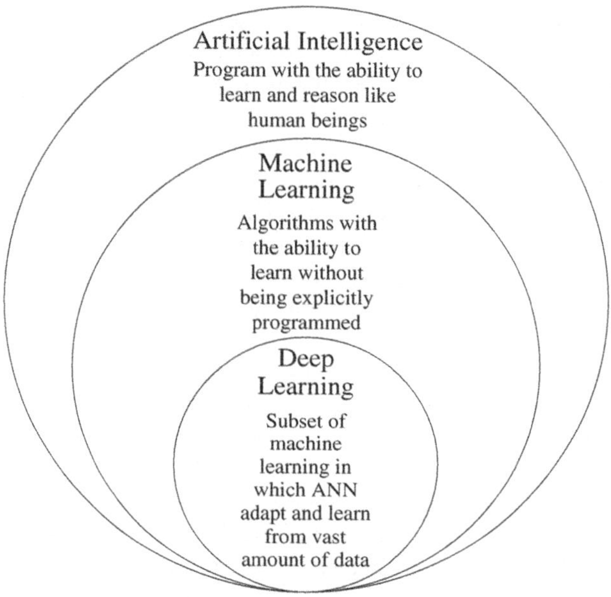

Figure 5.1 AI, ML, and deep learning difference and hierarchy.

5.2 Application of AI in Several Fields of Agriculture

Various AI-based applications in the area of agriculture are given below. These applications also show how it impacts the agriculture field in India.

5.2.1 AI in Agriculture

Undoubtedly, AI has had a profound impact in numerous aspects of life, particularly within the agricultural domain. It has brought about significant advancements, ranging from accurate weather forecasting and soil nutrient analysis to the selection of crop seeds tailored to specific regions. AI-powered robots have also revolutionized farming operations, while efficient food management and distribution systems have transformed the agricultural industry as a whole. Weather prediction plays a crucial role for farmers as it enables them to make well-informed decisions regarding planting, irrigation, and pest control. Through the analysis of extensive weather data, AI algorithms provide precise forecasts, empowering farmers to plan their activities accordingly and minimize potential risks. AI also contributes significantly to soil analysis and nutrient management. By analyzing soil samples and historical data, AI systems offer recommendations on optimal nutrient levels required for specific crops. This valuable information assists farmers in optimizing fertilization techniques that result in enhanced crop well-being and higher yields. AI has also revolutionized seed selection in agriculture. By considering factors such as climate, soil conditions, and historical data, AI algorithms can suggest the most suitable crop varieties for specific regions. This enables farmers to maximize productivity and adjust to evolving environmental circumstances effectively. Additionally, the use of AI-powered robots in farming operations has been a significant advancement. These robots are capable of performing precise and efficient tasks such as planting, harvesting, and weed control. By automating labor-intensive processes, AI-powered robots reduce the reliance on manual labor, increase productivity, and optimize resource utilization. Furthermore, AI has transformed the management and distribution of food produced. AI-powered systems assist in inventory management, quality control, and supply chain optimization. These systems ensure the efficient delivery of food to consumers, reducing waste and improving overall food security.

In summary, AI holds immense potential to revolutionize the agriculture sector by enabling precise decision-making, automation of tasks, and optimal resource utilization. With its applications in weather prediction, soil analysis, seed selection, robotic farming, and food management, AI is reshaping agriculture and giving the path for a more sustainable and effective future in food production.

5.2.1.1 Effective Decision-making

AI possesses the capability to serve as an invaluable instrument in offering farmers with access to valuable information and empowering them to make data-based decisions. Through the integration of AI algorithms, satellite imagery, and IoT devices, farmers can receive up-to-date data regarding soil moisture levels, crop well-being, and fertilizer requirements. This information enables precise allocation of resources, such as targeted fertilization, optimized irrigation, and effective pest control. By leveraging AI, precision agriculture aids in enhancing farmers' productivity while minimizing environmental impact through waste reduction and maximizing crop yields. Numerous ML models have been developed using diverse agricultural data and crop requirements, enabling accurate predictions of soil health and local weather conditions. Consequently, farmers are equipped with optimized options and solutions, promoting increased crop yield, enhanced efficiency, and improved time management in cultivation. Ultimately, these advancements significantly reduce the risk of crop failure [1].

5.2.1.2 Crop's Nourishment and Disease Detection

AI-driven ML algorithms and computer vision systems have enabled rapid and accurate crop monitoring in agriculture. These systems utilize drones, satellites, or ground-based cameras to capture images of fields, which are then analyzed by AI systems for signs of diseases, nutrient deficiencies, or insect infestations. Within seconds, AI can identify patterns and anomalies, alerting farmers to potential issues before they escalate. This early warning system allows prompt action to be taken, preventing crop losses and reducing the need for excessive pesticide and fertilizer use. By training AI models to scan crop images and analyze data on pests and soil nutrition, it becomes easier to predict crop diseases and provide recommendations for necessary measures, such as the appropriate amounts of fertilizers and pesticides to be used. If there is a deficiency in nutrients like nitrogen or others, the AI system can alert the farmer and provide suitable suggestions. Weed control is another common challenge for farmers. In crops like dried beans, maize, and wheat, weed invasion can lead to significant production losses. Weeds compete with crops for resources such as water, nutrients, and sunlight, affecting their growth and output. While herbicides are often used to combat weed growth, excessive usage can have adverse effects on the environment and public health. To address this, AI weed identification systems have been developed to determine the optimal amount of herbicide required and precisely target its application. Implementing AI in weed control reduces costs and minimizes the risk of crop damage [2, 3].

5.2.1.3 Use of Robotic Farming

The utilization of robots and automation in various agricultural practices has always captivated researchers. Autonomous robots equipped with AI have the capability to perform a range of operations, including planting, harvesting, crop spraying, and weeding. These robots are equipped with sensors and cameras that can perceive and analyze information about crops, soil conditions, and weather patterns. They can operate autonomously and carry out specific tasks as programmed. Robotic farming has the potential to increase productivity, reduce labor costs, and enhance agricultural yields. For example, the "Rowbot" robot has been developed to navigate accurately between crop rows and apply fertilizer and weed control measures. Another noteworthy example is the "Harvest CROO" robot, which can pick strawberries at a remarkable rate of eight plants per minute, surpassing human labor in this task. However, the widespread adoption of robots in agriculture faces a significant obstacle: their high cost. These robots possess precise field navigation capabilities and employ computer vision to identify and address the specific needs of plants. While they offer substantial savings in labor and expenses, along with improved overall efficiency due to their tireless and precise work, their expensive price tags currently limit their extensive use in agriculture [4]. Figure 5.2 shows the flowchart of robotic arms in fruit harvesting.

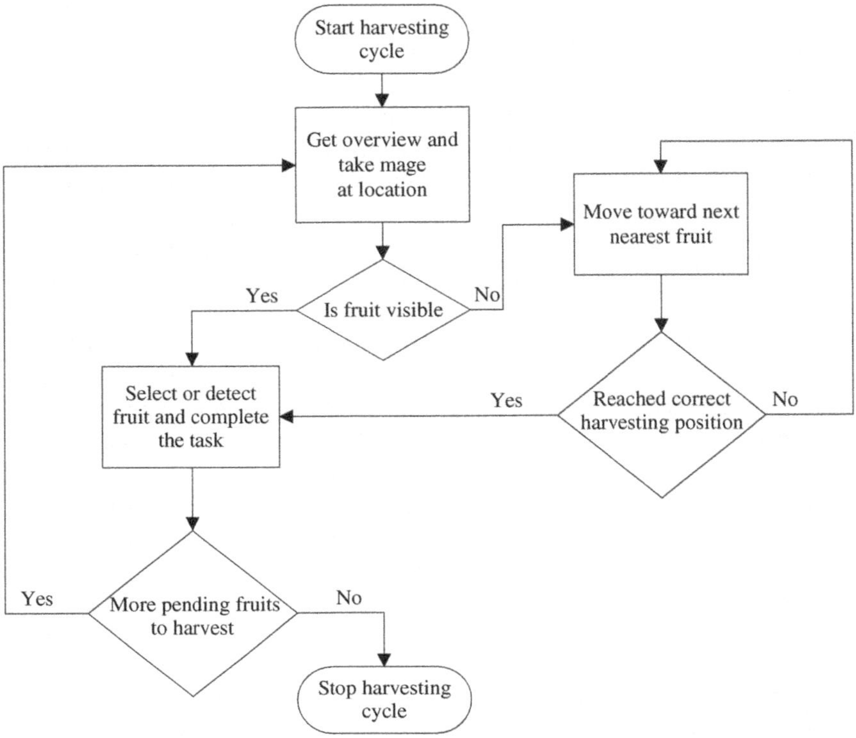

Figure 5.2 Flowchart of robotic arms in fruit harvesting.

5.2.1.4 Sustainable Farming and Crop Management

AI plays a pivotal function in advancing sustainable agricultural approaches and alleviating the negative environmental impacts associated with agriculture. By leveraging AI-driven systems, it becomes possible to optimize resource allocation, reduce the reliance on pesticides and fertilizers, and preserve soil, water, and biodiversity. AI's ability to analyze data from multiple sources enables the adoption of regenerative agricultural techniques and the implementation of climate-smart policies by farmers. In terms of food preservation, AI can be utilized in cold storage facilities to monitor the health of grains, vegetables, and fruits regularly. It can detect any signs of crop infections or spoilage and promptly alert farmers, allowing for timely actions to be taken to minimize losses. Furthermore, AI can also be employed in the transportation and supply chain management of crops, ensuring efficient and effective distribution from one location to another. By harnessing the power of AI in agriculture, sustainable practices can be fostered, leading to improved environmental stewardship and the promotion of a more resilient and efficient food system [2, 4].

5.2.1.5 Future Scope of AI in the Agriculture

By 2028, it is projected that the AI market in agriculture will reach approximately USD$4.7 billion, and it is expected to continue growing at a rate of 20% or more. AI will play a crucial role in providing valuable insights into farming practices, including monitoring weather conditions, soil health, and crop health and disease detection. As AI and ML continue to advance, new applications of these technologies will emerge, such as monitoring cultivation patterns and implementing innovative farming techniques driven by AI. The potential for AI in agriculture is vast, and ongoing advancements will further revolutionize the industry, enhancing productivity and sustainability.

5.3 AI in Education

AI has significant potential and applications in the field of education, offering opportunities to enhance learning methods and stimulate students' interest. By utilizing ML algorithms, AI systems can analyze student data and behavior, providing personalized recommendations and adaptive learning paths. Intelligent tutoring programs can offer immediate feedback and guidance, facilitating better comprehension of concepts. AI-powered chatbots and virtual assistants enable students to access information for independent study. AI algorithms can assist teachers in grading assignments, providing timely feedback, and identifying areas where students may require additional support. Moreover, AI technology allows for the analysis of vast amounts of educational data to identify trends, improve curriculum development, and predict student outcomes. By incorporating AI in education, institutions can enhance learning outcomes, increase student engagement, and create more inclusive and personalized learning environments. A comprehensive examination of empirical studies conducted between 1993 and 2020 in the field of AI in education highlights various AI technologies and applications, along with their proven and potential educational benefits. The chapter also presents real-world examples and serves as a source of inspiration for both AIEd technology developers and educators leading AI innovations in education. Furthermore, the study discusses the limited educational perspectives in AIEd research, development, and implementation, emphasizing the importance of multidisciplinary and transdisciplinary collaboration in long-term research and development projects. The ethical and privacy concerns associated with AI require significant attention in the development of AIEd. The chapter concludes that AI technology is rapidly advancing and its utilization in education is expected to grow rapidly in the near future [5]. Figure 5.3 describes the basic structure of AI system in education.

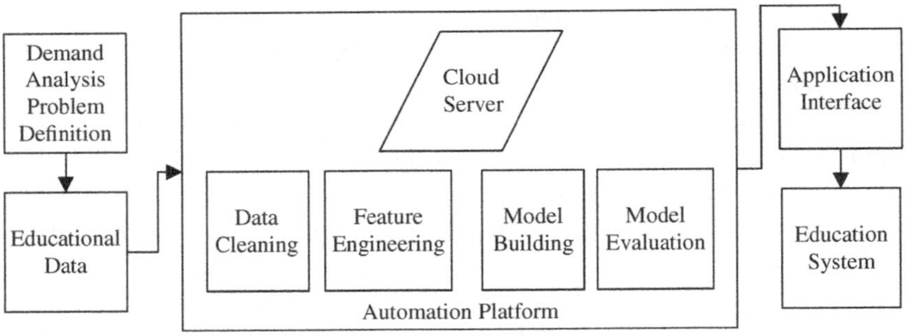

Figure 5.3 Basic structure of AI system in education.

5.3.1 Student and Virtual Teacher or Mentor Interaction

AI-driven intelligent tutoring systems have the potential to revolutionize education by acting as virtual mentors for students. These systems utilize ML and AI to provide personalized assistance and guidance tailored to each learner's needs. One key aspect of computerized tutoring systems is their ability to understand students' strengths, weaknesses, and learning preferences. Through the examination of data gathered from diverse sources like tests, quizzes, and past interactions, the system creates a profile of the student's knowledge gaps and areas of competence. For example, if a student consistently struggles with mathematical formulas but excels in geometry, the system can identify these patterns and adjust the content accordingly. This technology can greatly support teachers in managing large classrooms and delivering instruction. AI also enhances the interaction between teachers and students, as students can access knowledge from anywhere and at any time through AI-driven augmented reality (AR) and virtual reality (VR) demonstration models. Teachers can customize these models to align with their teaching methods and easily explain complex concepts in subjects like science or other fields [6, 7].

5.3.2 Improvised Supervised Learning System

AI-powered intelligent teaching systems act as virtual mentors, providing students with personalized guidance and support. These systems adapt the content and pace of lessons based on an understanding of students' strengths, weaknesses, and preferred learning styles. By offering dynamic and engaging instruction, these intelligent tutors can clarify concepts, address questions, and provide step-by-step instructions. Additionally, they continuously monitor students' understanding levels and provide relevant recommendations to enhance their learning effectively [8].

5.3.3 Automatic Assessment Model and Feedback System

AI, particularly NLP algorithms, can streamline the evaluation process in education. AI-based assessment technologies automate the evaluation procedure, saving teachers considerable time. Instead of manually reading and grading each student's written work, NLP algorithms can analyze the structure and content of the text to generate unbiased evaluations. These AI technologies provide students with prompt feedback, allowing them to receive input on their work immediately rather than waiting for manual grading. This aspect of AI assessment is advantageous as it enables students to identify and correct their mistakes promptly, facilitating faster application of their learning. AI-driven assessment systems offer comprehensive insights into students' performance. By analyzing written assignments, the algorithms can identify areas where students excel or struggle. This data allows teachers to gain a deeper understanding of each student's strengths and weaknesses, enabling more personalized education. With the information provided by AI, teachers can adapt their teaching strategies and focus on specific areas where students need further development [6, 8].

5.3.4 Behavior Analysis and Skill Development

AI plays a crucial role in behavior analysis and skill improvement, utilizing advanced algorithms and information processing capabilities. In behavior analysis, AI can analyze large amounts of data, such as video recordings or sensor inputs, to identify patterns and understand human behavior. Professionals in fields like psychological research, therapy, and special education can leverage this data to analyze behavior, identify potential triggers or causes, and develop targeted interventions. By automating the process, AI enhances the efficiency and accuracy of behavior analysis, enabling data-driven decision-making and providing valuable insights. AI also contributes to skill improvement by offering personalized and flexible learning experiences. By analyzing learner data, including performance, progress, and educational preferences, AI algorithms can adapt educational materials and exercises to meet the unique needs and abilities of each individual. Customization ensures that information is presented at the appropriate level of difficulty, offers focused practice opportunities, and provides real-time feedback, thereby promoting more effective learning. AI-based systems can also monitor and track progress over time, identifying areas of strength and weakness, and adjusting course content accordingly. This adaptive approach to skill development optimizes learning outcomes, increases engagement, and enhances the likelihood of successful skill acquisition [5].

5.3.5 Future Scope of AI in Education

The advancements in AI and the IoT offer significant potential to revolutionize the education process, particularly in countries like India. Personalized education is gaining importance, as it holds tremendous promise for the future. AI algorithms can leverage student data analysis to generate customized learning paths that cater to different learning styles and abilities. By utilizing data from educational systems, AI can create adaptive learning platforms and evaluation tools, providing personalized educational experiences to students worldwide. One crucial area where AI can have a significant global impact is expanding access to high-quality education. Delivering quality education in remote or disadvantaged areas is a challenge, but AI-powered solutions can bridge this gap. Through educational platforms, distance learning programs, and automated tutoring systems facilitated by AI, education can be democratized, ensuring fair access to educational resources and professional assistance across the globe. Moreover, AI can streamline administrative tasks in educational institutions. By analyzing data on enrollment, attendance, and performance, governments can make informed policy decisions, enhance system efficiency, and improve infrastructure development. AI's data-driven approach enables education systems to optimize resource allocation and enhance overall performance. In summary, the advancements in AI and IoT hold immense potential for transforming education globally, including countries like India. Personalized learning, expanded access to quality education, and streamlined administrative processes are among the key areas where AI can have a profound and positive influence.

5.4 AI in the Field of Healthcare

AI plays a crucial role in the healthcare sector, bringing about significant changes in the identification, treatment, and management of illnesses. Through its unparalleled data analysis capabilities, AI is revolutionizing healthcare, enabling healthcare practitioners to enhance patient care and make informed decisions. AI algorithms, leveraging patient data such as genetic information and medical records, can generate personalized treatment plans that maximize effectiveness and minimize side effects. Furthermore, AI's predictive analytics capabilities enable healthcare professionals to identify early warning signs, anticipate disease progression, and foresee potential outcomes. By utilizing real-time data, AI algorithms facilitate proactive interventions, leading to improved patient outcomes and disease management. AI also accelerates medical research through the analysis of extensive datasets, identification of potential therapeutic targets, and optimization of drug-discovery processes. This expedites the pace of innovation, reduces the time required for novel therapies to reach the market, and enhances our understanding of medicine.

5.4.1 Disease Detection Using Image Scanning

AI algorithms have emerged as powerful tools in the field of medical imaging, including X-rays, CT scans, and MRIs, offering remarkable accuracy and speed

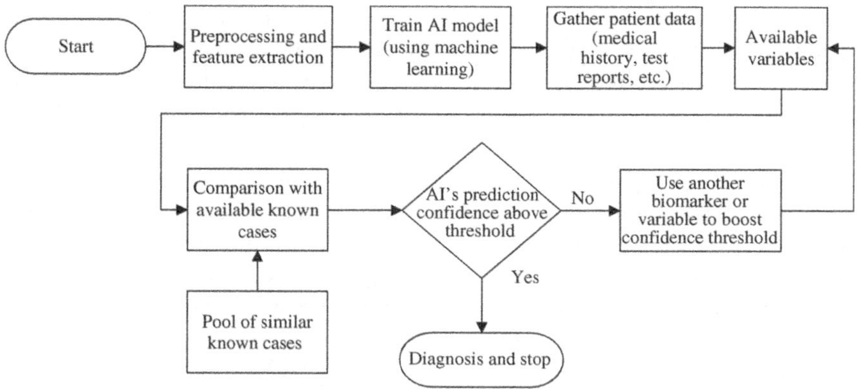

Figure 5.4 Flowchart of using AI to diagnose any disease.

in diagnosing illnesses. Through ML techniques, AI systems can learn from vast amounts of medical image data, enabling them to identify patterns, anomalies, and subtle disease symptoms that may be challenging for human observers to detect. By training AI models, diseases can be recognized with high accuracy, potentially saving lives. Whether it is predicting a brain tumor from scanned brain images or identifying skin conditions, AI can provide valuable assistance. While radiologists play a crucial role in interpreting medical images, the process can consume a significant amount of time and be susceptible to human errors. AI-powered image recognition technology can assist radiologists by swiftly analyzing images and highlighting any abnormalities, simplifying the diagnostic process and making it faster and more accurate. AI's speed and precision are particularly advantageous in time-sensitive scenarios, such as identifying malignant tumors or acute diseases. Early detection of illnesses often leads to improved treatment outcomes. AI algorithms in medical imaging can detect early indications of various conditions, including cancer, cardiovascular diseases, and neurological issues. This enables prompt interventions and treatment planning, ultimately improving patient outcomes and increasing survival rates. AI can also reduce the risk of false positives or negatives, minimize unnecessary procedures, and decrease patient anxiety. Figure 5.4 gives an example of flowchart of using AI to diagnose any disease.

5.4.2 Remote Personalized Monitoring and Treatment

AI technology in healthcare enables the creation of personalized treatment plans tailored to individual patients. By analyzing patient data such as genetic profiles, medical histories, and lifestyle preferences, AI systems can identify the most effective treatment programs and recommend suitable medications. This approach, known as precision medicine, maximizes treatment effectiveness while minimizing adverse effects and improving patient outcomes. AI-powered remote patient-monitoring devices allow for continuous and real-time tracking of patient health data. ML algorithms can analyze this data, detect anomalies, and track chronic conditions. Smart technologies and sensors enable timely alerts to be sent to medical specialists, facilitating rapid detection and proactive treatment of health issues. This enhances overall patient well-being and enables remote access to high-quality healthcare through virtual consultations. In regions with limited healthcare resources or during emergencies, AI-integrated telemedicine systems eliminate the need for in-person visits by providing medical advice and treatment remotely. Furthermore, AI-enabled systems improve patient record management, facilitating quicker tracking and access to patient data for healthcare professionals. This results in more effective and coordinated treatment. The integration of AI technology in healthcare holds the capability to revolutionize the administration, monitoring, and personalization of therapies. By utilizing ML algorithms and remote monitoring technology, healthcare providers can improve patient outcomes, expand access to care, and enhance delivery efficiency. However, it is crucial to prioritize patient confidentiality and data security while using AI in healthcare. Ethical and responsible use of AI must be ensured.

5.4.3 Medicine and Health System Making

AI is highly proficient at analyzing vast amounts of data and identifying patterns, making it a valuable asset in accelerating the drug discovery process, improving efficiency, and enhancing success rates. One key application of AI in drug research is in the early stages of target selection and validation. By analyzing extensive datasets that encompass genetic information, protein structures, and biological processes, AI systems can identify prospective therapeutic targets. This enables researchers to focus their attention on the most promising possibilities, saving time and resources. AI also plays a significant role in virtual screening and drug discovery. Through computer models and ML techniques, AI can predict how medications will interact with their targets. This capability allows for the design and optimization of medication candidates with a higher likelihood of success. AI algorithms can also generate and explore novel chemical compounds with desired characteristics, expanding the pool of potential therapeutic candidates. Additionally, AI plays a crucial role in the development of health monitoring equipment. By providing insights into the human body and analyzing patient data, AI contributes to the creation of advanced health monitoring devices. These devices can offer valuable information for medical professionals, aiding in diagnosis, treatment planning, and patient care. Overall, AI's ability to analyze data, predict outcomes, and generate novel solutions greatly accelerates drug discovery, improves efficiency, and enhances the development of health monitoring equipment.

5.4.4 AI Robots in Healthcare

AI robots are making significant contributions to the medical industry by revolutionizing healthcare delivery and improving patient outcomes. These advanced robotic devices, equipped with AI capabilities, are designed to perform various tasks and support medical professionals in different aspects of patient care. One prominent application of AI robots in medicine is in surgical robotics. With AI algorithms and real-time data analysis, these robots assist surgeons in performing intricate operations with enhanced precision and dexterity. They can manipulate delicate anatomical structures, reduce the risk of human error, and enable less invasive procedures. This leads to faster patient recovery and fewer complications, ultimately improving surgical outcomes. AI robots are also utilized for patient care and monitoring. Equipped with AI sensors and algorithms, these robots can track changes in patients' conditions, continuously monitor vital signs, and alert medical staff to potential issues. They can also perform basic tasks such as delivering supplies, assisting with mobility, and reminding patients to take medications. This alleviates some of the workload on healthcare professionals and enhances overall patient care. In the field of rehabilitation, AI robots play a vital role. These robots employ AI algorithms to develop personalized therapy plans for individuals undergoing rehabilitation or recovering from injuries. They can provide specific exercises, track progress, and adjust therapy schedules based on real-time feedback. This improves the efficiency of rehabilitation programs and enhances functional outcomes for patients. Overall, AI robots in healthcare are transforming the delivery of care and contributing to improved patient outcomes. From surgical assistance to patient monitoring and rehabilitation, these AI-powered robots offer valuable support to medical professionals, enhancing efficiency and overall quality of care [9–11]. Figure 5.5 shows the flowchart of the use of AI robotic arm in the surgery.

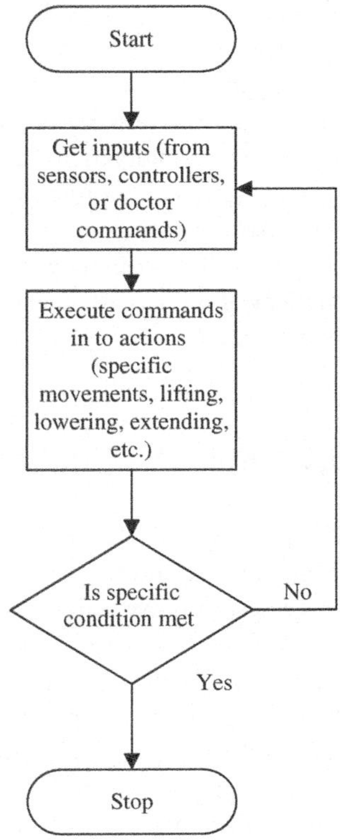

Figure 5.5 Flowchart of the use of AI robotic arm in the surgery.

5.4.5 Future Scope of AI in Healthcare Field

The utilization of ML and AI in the healthcare sector holds great promise for improving healthcare services and expanding access to disadvantaged groups, particularly in remote areas. Through telemedicine and remote surveillance technologies powered by AI, healthcare professionals can efficiently reach patients in distant locations.

Automation of certain processes through AI also enhances the efficiency of medication development and accelerates the search for new drugs. AI robots have the potential to undertake complex procedures that are currently challenging for human surgeons. These robots can enhance accuracy, reduce invasiveness, and improve surgical outcomes. Additionally, medical support robots can improve the quality of life for individuals with disabilities and help restore movement in their limbs. AI advancements in disease detection models continuously surpass human capabilities in terms of effectiveness and precision. This progress can lead to earlier and more accurate diagnoses, enhancing the effectiveness of treatments and potentially saving lives. AI's application in pharmaceutical development is also expected to yield benefits, as computers can swiftly analyze vast amounts of data to identify prospective drug candidates. In the field of cancer treatment, AI enables researchers to analyze complex biological data, identify biomarkers, and develop targeted therapies for different types of cancer. The future of AI in healthcare appears promising, offering potential advancements in medical treatments, improved disease management, and overall better patient outcomes. However, challenges such as ethical concerns and safeguarding patient privacy still need to be addressed.

5.5 AI in the Field of Administration and Governance

The potential of AI to revolutionize government and administration is receiving significant attention. AI offers new opportunities to enhance administrative processes, improve efficiency, and strengthen decision-making in the public sector. With its ability to analyze vast amounts of data, identify patterns, and draw intelligent conclusions, AI has the potential to transform how governments and public organizations operate. One area where AI can make a significant impact is in automating repetitive tasks, freeing up time for administrators to focus on more complex and strategic activities. AI chatbots, for example, can provide instant responses to common citizen inquiries and offer guidance, improving the overall citizen experience. Furthermore, AI algorithms can analyze massive datasets, extract insights, and generate reports, enabling administrators to make data-driven decisions and optimize resource allocation. By leveraging AI technology, administrative procedures can become more efficient, cost-effective, and responsive to public needs. Data-driven policy-making is another area where AI can contribute. By analyzing diverse datasets and identifying trends, AI algorithms can provide valuable insights to support evidence-based policy development. This can lead to more informed decision-making and the formulation of policies that better address societal challenges. Overall, the integration of AI in government and administration has the potential to bring about improved efficiency, responsiveness, and effectiveness. By automating tasks, leveraging data-driven insights, and enhancing decision-making processes, AI can help create more efficient and citizen-centric governance.

5.5.1 Effective Decision-making and Data Analysis

AI-based solutions in governance and administration have the capability to effectively analyze vast amounts of data, providing valuable information for decision-makers and administrators. These technologies utilize ML algorithms to analyze extensive datasets and recognize patterns, and generate prediction models. This data-driven approach enhances decision-making in several ways. One significant benefit is the ability to support policymakers in designing effective policies. AI-generated insights can assist in ensuring that decisions are based on comprehensive and accurate information. By analyzing data, policymakers can identify areas of high demand or inefficiency, allowing for optimized resource allocation and more effective policy implementation. AI technology can also contribute to improving service delivery by enabling governments to address issues and provide tailored solutions. By identifying patterns and trends in public input, AI can help governments understand the needs and preferences of citizens, leading to more responsive and personalized services. Overall, the integration of AI into governance and administration enables more efficient and informed decision-making processes. By harnessing the power of data analysis and prediction models, AI enhances governance practices and delivers better outcomes for citizens [12–15].

5.5.2 Participation and Engagement of Citizens in Governance

Virtual assistants and chatbots with AI capabilities are increasingly valuable tools for enhancing citizen engagement and delivering services in the public sector. These advanced technologies offer a range of benefits as they automate support and assistance for individuals. One key advantage is their round-the-clock availability, providing residents with access to information and assistance beyond regular business hours. This convenience and responsiveness of 24/7 access contribute to overall citizen satisfaction. Virtual assistants and chatbots powered by AI are designed with extensive knowledge bases, enabling them to provide prompt and accurate responses to frequently asked questions. Users can interact with these systems through various channels such as websites, smartphone applications, or chat platforms, receiving immediate answers to their inquiries. AI technologies analyze user data and personalize solutions based on individual needs and preferences, offering tailored assistance. This personalized approach enhances the citizen experience, fostering a sense of inclusion and understanding. By providing relevant advice and assistance tailored to specific circumstances, AI technologies improve citizens' interactions with government services. Governments promote transparency and accessibility in governance by leveraging AI-powered chatbots or virtual assistants. These systems ensure standardized and consistent information, ensuring that citizens receive precise and up-to-date guidance. Moreover, AI technology can automate repetitive tasks, freeing up human resources to focus on more complex and valuable responsibilities. This streamlines operations and enables government personnel to concentrate on tasks that require human judgment and creativity. In summary, the utilization of virtual assistants and chatbots with AI capabilities in the public sector enhances citizen engagement and service delivery. The 24/7 availability, personalized assistance, and automation of routine tasks contribute to an improved citizen experience, increased accessibility, and more efficient governance [12, 13].

5.5.3 Situation Management and Policy Implementation

The management of situations and the application of policies are critical aspects of administration and governance, and AI technology has the potential to significantly enhance these processes. By analyzing vast amounts of real-time data from various sources such as social media platforms, sensors, and government databases, AI can assist in crisis management. Through the use of ML algorithms, AI systems can quickly recognize patterns, detect anomalies, and generate insights that aid decision-making in emergency scenarios or complex situations. This empowers policymakers and administrators to take timely, effective decisions that lead to efficient resource allocation and response plans. AI technology can also expedite policy implementation by automating repetitive administrative tasks, reducing human error, and increasing productivity. AI-powered systems can analyze policy frameworks, rules, and guidelines to generate recommendations and ensure compliance. Additionally, AI algorithms can assess key performance metrics and provide feedback for future improvements while monitoring and evaluating the impacts of established policies. Furthermore, AI can contribute to evidence-based policy-making by analyzing extensive data and creating prediction models. AI algorithms can identify patterns, evaluate potential outcomes, and assist in policy formulation by leveraging historical data. This data-driven approach enhances the precision and effectiveness of policy decisions, increasing the likelihood of achieving positive social impacts. In summary, AI technology has the potential to greatly enhance crisis management, automate administrative processes, and improve policy-making in governance and administration. By analyzing large volumes of data and generating meaningful insights, AI enables good decision-making, efficient resource allocation, and evidence-based policy development [13, 14].

5.5.4 Crime Management and Law Enforcement

The integration of AI technology holds potential solutions for improving crime prevention, investigation, and overall public safety. Crime prevention is a complex and challenging area for law enforcement, and AI can greatly enhance crime management in several ways. AI systems can analyze vast amounts of data, including criminal histories, incident reports, and online activity, to identify trends and potential risks. By utilizing ML algorithms, AI systems can forecast criminal activity, identify crime hotspots, and assist law enforcement agencies in efficiently allocating resources. This proactive approach reduces the likelihood of criminal incidents by enabling the implementation of preventive measures. AI-powered video surveillance systems equipped with facial recognition capabilities can play a crucial role in identifying suspects and detecting suspicious activities in real time. These systems can automatically analyze surveillance footage, compare faces against criminal databases, and alert authorities when potential threats or persons of interest are identified. This technology expedites the investigation process, enhances situational awareness, and aids in apprehending offenders. Moreover, AI can assist in analyzing large volumes of digital evidence, such as emails, text messages, and online activity, to support investigations. NLP algorithms can process text data and extract relevant information, enabling law enforcement agencies to uncover critical evidence and build stronger cases. AI technologies also contribute to predictive policing by analyzing historical crime data and identifying patterns to anticipate future criminal activity. This data-driven approach helps law enforcement agencies strategically allocate resources, focus preventive measures on high-risk areas, and proactively address potential criminal threats. Overall, the incorporation of AI in crime prevention and law enforcement enhances investigation capabilities, improves situational awareness, and enables more proactive and targeted approaches to maintaining public safety. By analyzing data, detecting suspicious activities, and assisting in investigations, AI technology supports law enforcement agencies in their efforts to prevent crime and ensure the security of communities [15].

5.5.5 Future Scope of AI in Administration and Governance

The potential for the revolutionary impact of AI in governance is indeed promising. Through data analysis and predictive modeling, AI has the capacity to transform decision-making, policy formulation, and service provision. Governments can extract valuable insights from massive volumes of data, anticipate trends, and develop evidence-based policies. AI-powered chatbots and virtual assistants enhance citizen engagement by providing personalized assistance, reducing response times, and increasing satisfaction. These intelligent systems enable governments to interact with citizens more effectively and efficiently. By automating administrative tasks, AI solutions improve efficiency and streamline operations in government agencies. This allows for the consideration of factors that would be challenging to manage without the support of these technologies. Governments can operate more smoothly and manage large populations more effectively with the aid of AI. Overall, the integration of AI in governance has the potential to bring about significant improvements in decision-making processes, citizen engagement, and administrative efficiency. By harnessing the power of data analysis and automation, governments can enhance their ability to serve and manage their populations.

5.6 Results and Discussion

The study examined the use and impact of AI in various sectors such as administration, healthcare, agriculture, and education. The findings revealed that AI technologies have the potential to transform these industries, leading to improved outcomes and increased efficiency.

In education, personalized learning platforms and evaluation technologies driven by AI have contributed to better academic performance and increased student engagement. AI systems in agriculture have enhanced production and sustainability by optimizing resource utilization, monitoring crop health, and predicting harvests. In the healthcare industry, AI has successfully analyzed medical images, aiding in early disease diagnosis and improving diagnostics, thus enhancing medical services and accessibility. AI has also played a significant role in drug discovery, study design, and remote patient monitoring. Moreover, AI has improved decision-making processes, resource allocation, and public participation in government through data analysis and predictive modeling. However, in order to fully realize the potential of AI in these fields, it is crucial to address ethical concerns, infrastructure limitations, and data access issues. The study concluded that further research, ethical considerations, and training are necessary for the successful adoption of AI in these sectors. Despite these positive aspects, there are challenges and concerns that hinder the widespread adoption of AI in these fields:

1. Complexity of AI technologies: AI is a complex technology, making it difficult for new users to understand and utilize effectively. Adequate education and training on AI are necessary, which can be challenging in countries with limited resources.
2. Costly availability of AI technologies: AI requires high-powered machines, satellites, advanced computer systems, and various sensors, making it an expensive investment. Limited financial resources can hinder the widespread availability and adoption of AI.
3. Accountability in various fields: The widespread use of AI raises questions about responsibility and potential risks to human life. Biased or inaccurate AI algorithms in the medical field can result in unsafe treatments and incorrect diagnoses. AI decision-making in autonomous vehicles can have life-or-death consequences in collisions. AI algorithms used in criminal justice systems may exhibit biases, leading to unfair outcomes. Additionally, relying on AI for critical infrastructure introduces the risk of malfunctions or cyberattacks that endanger human lives. Ensuring accountability requires legislation, ethical frameworks, and continuous monitoring to address these issues and prioritize human safety and well-being.

In conclusion, while AI offers numerous benefits in various sectors, challenges such as complexity, cost, and accountability must be addressed for its widespread adoption and ethical use.

5.7 Conclusion

The applications of AI in administration, healthcare, agriculture, and education offer significant opportunities for improving efficiency and outcomes. Personalized learning tools enhance student performance, sustainable practices benefit agriculture, and AI enables precise diagnosis and remote monitoring in healthcare. Data-driven decision-making and citizen participation are advantageous for administration. However, challenges such as privacy concerns, biases, and ethical considerations must be addressed. To fully leverage the benefits of AI while mitigating risks, responsible deployment, ongoing research, and collaboration are crucial. AI has the potential to revolutionize various fields, positively impacting people's lives and advancing society. With careful and ethical use, AI can provide solutions to complex societal issues. Although concerns exist, the adoption of AI is expected to increase in various domains as we move forward.

References

1 Mourtzinis, S., Esker, P.D., Specht, J.E., and Conley, S.P. (2021). Advancing agricultural research using machine learning algorithms. *Scientific Reports* 11 (1): 17879.

2 Zha, J. (2020). Artificial intelligence in agriculture. *Journal of Physics: Conference Series* 1693 (1): 012058.

3 Talaviya, T., Shah, D., Patel, N., et al. (2020). Implementation of artificial intelligence in agriculture for optimisation of irrigation and application of pesticides and herbicides. *Artificial Intelligence in Agriculture* 4: 58–73.

4 Jha, K., Doshi, A., Patel, P., and Shah, M. (2019). A comprehensive review on automation in agriculture using artificial intelligence. *Artificial Intelligence in Agriculture* 2: 1–12.

5 Popenici, S.A., and Kerr, S. (2017). Exploring the impact of artificial intelligence on teaching and learning in higher education. *Research and Practice in Technology Enhanced Learning* 12 (1): 1–13.

6 Zhang, K., and Aslan, A.B. (2021). AI technologies for education: recent research & future directions. *Computers and Education: Artificial Intelligence* 2: 100025.

7 Guan, C., Mou, J., and Jiang, Z. (2020). Artificial intelligence innovation in education: a twenty-year data-driven historical analysis. *International Journal of Innovation Studies* 4 (4): 134–147.

8 Seo, K., Tang, J., Roll, I., et al. (2021). The impact of artificial intelligence on learner–instructor interaction in online learning. *International Journal of Educational Technology in Higher Education* 18 (1): 1–23.

9 Secinaro, S., Calandra, D., Secinaro, A., et al. (2021). The role of artificial intelligence in healthcare: a structured literature review. *BMC Medical Informatics and Decision Making* 21: 1–23.

10 Sunarti, S., Rahman, F.F., Naufal, M., et al. (2021). Artificial intelligence in healthcare: opportunities and risk for future. *Gaceta Sanitaria* 35: S67–S70.

11 Rong, G., Mendez, A., Assi, E.B., et al. (2020). Artificial intelligence in healthcare: review and prediction case studies. *Engineering* 6 (3): 291–301.

12 Zuiderwijk, A., Chen, Y.-C., and Salem, F. (2021). Implications of the use of artificial intelligence in public governance: a systematic literature review and a research agenda. *Government Information Quarterly* 38 (3): 101577. https://doi.org/10.1016/j.giq.2021.101577.

13 Reis, J., Santo, P.E., and Melão, N. (2019). Impacts of artificial intelligence on public administration: a systematic literature review. In: *2019 14th Iberian Conference on Information Systems and Technologies (CISTI)*, pp. 1–7. IEEE.

14 van Noordt, C., and Misuraca, G. (2020). Evaluating the impact of artificial intelligence technologies in public services: towards an assessment framework. In: *Proceedings of the 13th International Conference on Theory and Practice of Electronic Governance*, pp. 8–16. Elsevier.

15 Sharma, G.D., Yadav, A., and Chopra, R. (2020). Artificial intelligence and effective governance: a review, critique and research agenda. *Sustainable Futures* 2: 100004.

6

IoT-based Communication in Smart Cities for SDGs: Enhancing Urban Resilience Through Human-centric Hypernetwork Models

Yuhan Zheng[1,*] *and Zeyu Zhao*[2]

[1] *Geography Department, University of Galway, Galway, Ireland*
[2] *Department of Construction Management, Tsinghua University, Beijing, China*

*Corresponding author: y.zheng3@universityofgalway.ie

6.1 Introduction

The Anthropocene epoch underscores the urgency of redefining urban resilience through human–environment synergies. While cities increasingly adopt IoT-enabled systems for infrastructure monitoring, the integration of community-driven data into resilience planning remains underutilized. The connotation of *urban resilience in the Anthropocene* entails more than mere environmental or infrastructural restoration. It also encompasses how humans perceive and interact with the changing environment in response to diverse natural and anthropogenic disasters. The decision-making and implementation processes for resilient solutions to existing urban infrastructures increasingly incorporate amenities that promote a higher quality of urban life (QOUL) [1]. How to ensure QOUL and well-being is prioritized by both the resilience research and international community [2, 3]. Recent studies have been grappling with human-oriented approaches to interrogate the resilience capabilities across critical urban infrastructure systems, which are colloquially known as the urban lifeline systems [4].

Both authors contributed equally.

AI and ML Techniques in IoT-based Communication: A Path to Sustainable Development Goals, First Edition. Edited by Sumita Mishra, Nishu Gupta and Polat Goktas.

Resilience models like the performance optimization models [5], the social ecological conceptual framework of resilience [6], resilience system dynamics models [7], and input–output models [8] primarily connect built environment with human needs in a top-down manner. To date, no comprehensive model captures the dynamic relationship between human needs, public perceptions, the built environment, and the resulting interplay between human behavior and urban systems during disaster recovery process.

This chapter aligns with the book's theme by focusing on IoT-based communication as a catalyst for sustainable development goals (SDGs). We address two gaps: (i) the disconnect between dynamic human needs and infrastructure recovery processes, and (ii) the limited application of IoT data in modeling resilience. Which critical urban functions are significantly affected during the process of recovery?

The chapter is organized as shown in Figure 6.1. By probing the above questions, the joint effects of *urban resilience in the Anthropocene* are unfolded by systematically investigating the interrelationships among QOUL, the built environment, and human needs. Section 6.2 conducts a concise literature review on urban resilience, human agency, and hypernetwork. Section 6.3 explains the methodology employed herein. Section 6.4 presents the results. Section 6.5 further discusses contributions and further directions. Section 6.6 summarizes the key findings and concludes the study.

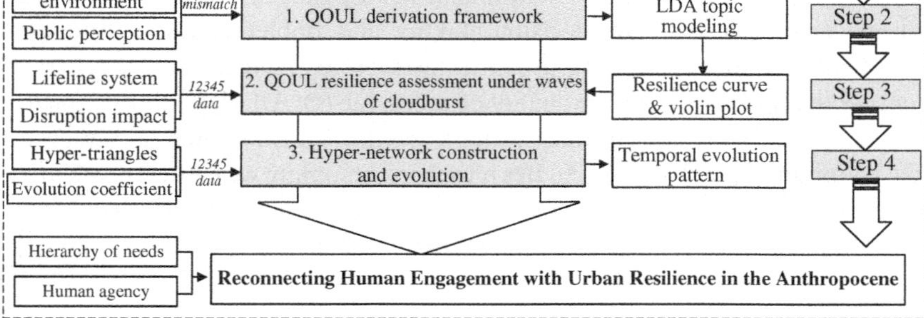

Figure 6.1 Workflow of this research.

6.2 Literature Review

6.2.1 Urban Resilience in the Anthropocene

The Anthropocene, marked by humanity's dominance as a geological force [9], underscores the urgency of redefining urban resilience. Human-driven activities – deforestation, fossil fuel use, and land-use changes – have destabilized Holocene-era stability [10], intensifying climate risks and infrastructure vulnerabilities [11]. Cities, as hubs of socio-technical systems [12], face cascading failures during extreme weather, exposing latent weaknesses in "stable" infrastructure like roads and power grids [13]. Resilience in this epoch shifts toward dynamic human–environment interactions [14], yet gaps persist in addressing urban inhabitants' evolving needs [15]. While studies prioritize livelihood frameworks [16], systemic integration of human needs into infrastructure planning remains nascent. This demands models that treat infrastructure as dynamic enablers of human agency, not passive artifacts [17].

6.2.2 Human Agency and QOUL

Human agency drives planetary transformations [18], shaping urban resilience through risk perception, feedback, and adaptive governance [19]. The Anthropocene necessitates frameworks where QOUL – residents' well-being tied to urban services and environments [20–22] – aligns with infrastructure performance. QOUL and resilience are mutually reinforcing: resilient cities sustain service access during disasters, while high QOUL populations recover faster [23, 24]. Crowdsourced data (e.g. hotline complaints) act as "human sensors," revealing mismatches between built environments and perceived needs [25, 26]. For instance, Zhao et al. [5] link postdisaster needs to infrastructure performance states, while Pan et al. [27] classify QOUL drivers (survival, safety, social, spiritual). However, quantitative synthesis of built environments, urban performance, and QOUL remains underexplored [28], necessitating models that bridge objective conditions with subjective evaluations.

6.2.3 Hypernetwork Emerging in the QOUL Layers

Traditional networks, limited to pairwise connections, fail to model multidimensional urban resilience [29, 30]. Hypernetworks address this by mapping higher order relationships among heterogeneous nodes [31], such as infrastructure assets, human needs, and SDG-aligned outcomes. For example:

- **Infrastructure Networks**: Jayaram and Srinivasan [32] optimize water distribution reliability via topology analysis.

- **Social–environmental Links**: Campbell et al. [33] connect objective built environments to subjective well-being.
- **Dynamic Resilience**: Yuan et al. [34] use hyperedges to model interdependencies between QOUL and lifeline systems.

Hypernetworks overcome limitations of traditional methods [35] by integrating social, technical, and environmental layers [30], enabling granular analysis of critical linkages (e.g. flooded mobility disruptions). This approach prioritizes resource allocation to vulnerable nodes (e.g. suburban drainage systems) and fostering participatory governance through citizen-generated data.

6.3 Methodology

The adoption of both qualitative and quantitative methods is deemed necessary in constructing the hypernetwork model. The most critical aspect of this study lies in determining the QOUL model, which addresses the linkage between QOUL and built environment to evaluate urban resilience and construct the multilayer hypernetwork.

6.3.1 Derivation of QOUL Model

This chapter proposes a human needs-driven model to evaluate urban resilience in the Anthropocene, focusing on cloudburst impacts. By integrating subjective perceptions (e.g. safety, accessibility) with objective infrastructure metrics (e.g. drainage efficiency), the framework holistically assesses QOUL. This dual approach addresses gaps in adaptive planning, ensuring cities strengthen capacity to withstand and recover from disruptions [20, 33]. The study highlights how the mismatch between residents' subjective perceptions and objective built environment generates evaluation of needs satisfaction degree. The smaller the mismatch between people's subjective perception and the objective environment, the higher their QOUL level. Based on the logical framework shown in Figure 6.2, we further conduct the topic system of QOUL that allows for a more precise and quantitative assessment of urban life quality.

As illustrated in Figure 6.2, the diagram provides insight into which physical lifeline attributes (i.e. built environment) and postdisturbance urban topics have an impact on public perceptions and how mismatches in these areas can further affect the QOUL. In complex urban environments, the urban lifeline infrastructure system plays a central role in determining QOUL [36]. Subjective needs, shaped by lifeline systems, critically influence residents' QOUL. This linkage between perceived urban responses and objective infrastructure attributes is modeled

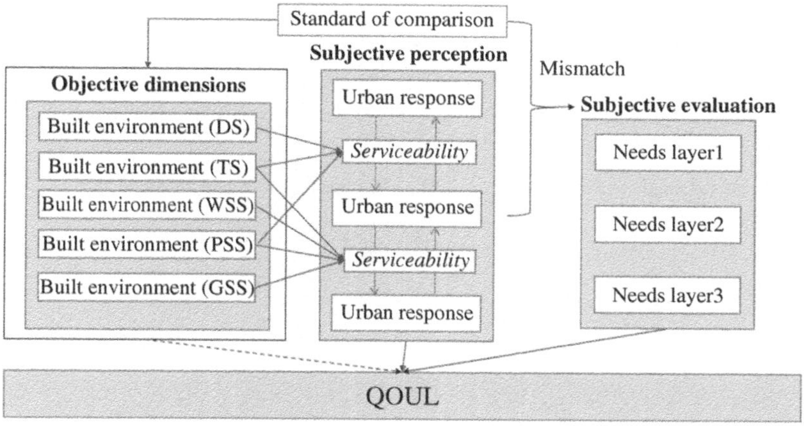

Figure 6.2 Logical framework of QOUL.

in Figure 6.2. Based on Maslow's motivation theory [37], the QOUL includes both the right to survival and development, and often presents a specific hierarchy following the governance of resilience mode, rather than being solely coordinated in a multidimensional manner [2]. In light of previous research [27], the subjective evaluation realm with various urban topics can often be divided into three to four layers. Furthermore, this study employs hyperedges to link the three heterogeneous layers, existing between objective dimensions and subjective perceptions. These hyperedges are instrumental in classifying groups in accordance with the fluctuations in their needs and QOUL over time.

6.3.2 Resilience Assessment Under Waves of Cloudburst

Climate change amplifies extreme weather such as recurrent cloudburst "waves" during flood seasons. Urban resilience now compels cities to rethink adaptive management, as systems often fail to fully recover between successive waves, compounding disruptions. Previous resilience assessment methods, such as indicator-based method [38] and the integral approach [27] failed to quantify the evolution of system performance and the nuanced condition beyond bounce back in waves [39].

In this study, we employed the integrated multiwaves framework to quantify human activity resilience, which was originally developed by Nan and Sansavini [40] and improved through decomposing the system performance curve [41]. Figure 6.3 illustrates QOUL performance curves under successive cloudburst waves, demonstrating infrastructure impacts. Independent shocks (Figure 6.3b)

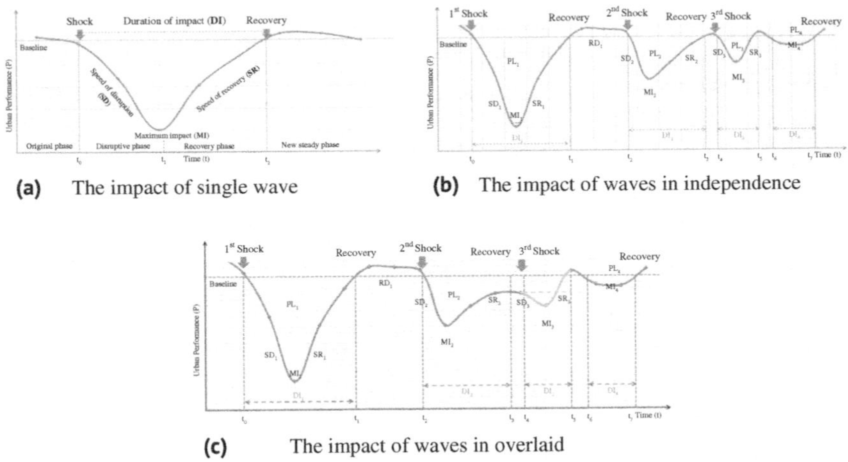

(a) The impact of single wave

(b) The impact of waves in independence

(c) The impact of waves in overlaid

Figure 6.3 Resilience curve of QOUL activity due to waves of shock under rainstorms.

enable direct resilience assessment, while overlapping shocks (Figure 6.3c) disrupt recovery, necessitating curve decomposition. Resilience curves, aligned with QOUL framework, use a baseline (normal conditions) to quantify granular deviations during disruptions [42, 43]. A disrupted level of QOUL of 5% compared to the baseline change threshold was assumed. To enhance the evaluation of QOUL contribution in lifeline systems, the disrupted value was utilized to construct the QOUL resilience curve compared to its corresponding baseline value.

The final objective maps resilience curves (QOUL [y-axis] vs. time [x-axis]) and computes overall and lifeline-specific resilience values. Each QOUL metric reflects evolving human needs during cloudbursts, shaping distinct resilience trajectories. Generally, study calculates the maximum impact (MI), duration of impact (DI), recovery degree (RD), slope ratio (SR_1/SD_1), time average performance loss (PL) for each category, which contains the time-slot of original stage (t_0), disruptive phase ($t_0 \leq Time \leq t_1$), recovery phase ($t_1 \leq Time \leq t_2$), and new steady phase ($Time > t_2$) [44, 45]. Additionally, this chapter modified and employed the GR metric in cloudburst waves under flood season as follows:

a. **The impact of two waves of rainstorms in independence:**

$$R_{QOUL1} = \frac{\int_{t_0}^{t_1} (Baseline - P(t))\,dt}{t_1 - t_0} \times SR_1/SD_1 \times RD_1 \tag{6.1}$$

$$R_{QOUL2} = \frac{\int_{t_2}^{t_3} (Baseline - P(t))\,dt}{t_3 - t_2} \times SR_2/SD_2 \times RD_2 \tag{6.2}$$

$$R_{QOUL3} = \frac{\int_{t_4}^{t_5} (Baseline - P(t))dt}{t_5 - t_4} \times SR_3/SD_3 \times RD_3 \quad (6.3)$$

b. The impact of two waves of rainstorms in overlaid:

$$QOUL_{1※} = \frac{\int_{t_0}^{t_1} (Baseline_1 - P(t))dt}{t_1 - t_0} \times SR_{1※}/SD_{1※} \times RD_{1※} \quad (6.4)$$

$$QOUL_{2※} = \frac{\int_{t_2}^{t_3} (Baseline_2 - P(t))dt}{t_3 - t_2} \times SR_{2※}/SD_{2※} \times RD_{2※} \quad (6.5)$$

$$QOUL_{3※} = \frac{\int_{t_4}^{t_5} (Baseline_3 - P(t))dt}{t_5 - t_4} \times SR_{3※}/SD_{3※} \times RD_{3※} \quad (6.6)$$

6.3.3 Hypernetwork Modeling and Mechanism Evolution

The proposed hypernetwork model consists of three main stages: first, a three-layer QOUL hypernetwork is built from closed-loop feedback data from residents on urban disturbances [46–49]. The hypernetwork model integrates three subnetworks – Built Environment, Urban Performance/Public Perception, and Human Needs – interconnected via hyperedges (Figure 6.4). Unsupervised ML dynamically updates city-response data (QOUL-driven), while the hypernetwork's

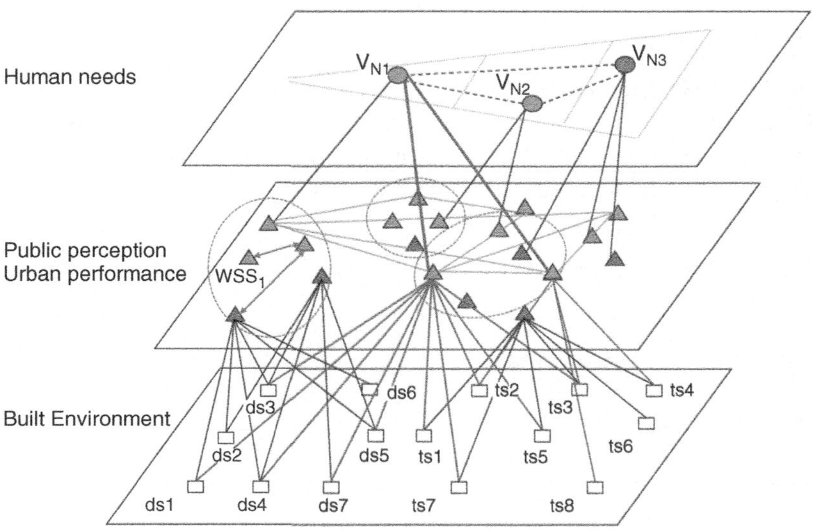

Figure 6.4 Diagram of QOUL hypernetwork.

top layer stratifies human needs into safety, social livelihood, and civic engagement layers, linking infrastructure to resident feedback.

6.3.3.1 Construction of Hypernetworks

1. Built Environment network ($B = V_B,\ E_{B-B}$): As shown in Equations 6.8 and 6.9, V_B denotes the set of nodes of the built environment such as pipelines, road sections, wells and gas stations, and E_{B-B} denotes whether the nodes are under the common jurisdiction of the municipal works management. This subnetwork tracks the built environment's real-time operations and maintenance status.
2. Urban response network ($UR = V_{UR},\ E_{UR-UR}$): As seen in Equations 6.10 and 6.11, V_{UR} represents the thematic set of urban functions perceived by residents under the cloudburst conditions, and E_{UR-UR} represents whether the functions in the urban lifeline system are interconnected (e.g. whether poor drainage system leads to flooding in the traffic system). This subnetwork translates objective built-environment data into perceived serviceability under disruptions, shaping human needs and triggering QOUL evaluations critical to urban resilience.
3. Human needs network ($N = V_N,\ E_{N-N}$): As shown in Equations 6.12 and 6.13, V_N represents the different levels of needs of urban residents, expressed as Safety and Health (basic livelihood), Social Livelihood (social livelihood), and Civic Engagement (high level livelihood). A direct urban response-to-human needs link was established to analyze evolving needs and hypernetwork dynamics, addressing complex, hard-to-quantify interactions between human need levels.

Hyperedges link heterogeneous nodes (e.g. Urban response, Human needs, Built Environment) into multinode groups, forming complex request networks. Figure 6.4 shows the structure of a QOUL hypernetwork, illustrating two examples of hyperedges, which can be presented as $SE_1 = \{V_{N1},\ DS_1,\ V_{B1},\ V_{B2},\ V_{B3},\ V_{B4}\}$ and $SE_2 = \{V_{N1},\ TS_1,\ V_{B1},\ V_{B2},\ V_{B6},\ V_{B7}\}$.

$$Model = \{B(V_B,\ E_{B-B}),\ UR(V_{UR},\ E_{UR-UR}),N(V_N,\ E_{N-N}),HE_{B\cup UR\cup N}\} \tag{6.7}$$

$$V_B = \{B_1,\ B_2,\ \cdots,\ B_n\} \tag{6.8}$$

$$E_{B-B} = \left\{\left(V_{B_i},\ V_{B_j}\right)\right\}\ \forall i,j = 1,\ 2,\ \cdots n \tag{6.9}$$

$$V_{UR} = \{UR_1,\ UR_2,\ \cdots,\ UR_n\} \tag{6.10}$$

$$E_{UR-UR} = \left\{\left(V_{UR_i},\ V_{UR_j}\right)\right\}\ \forall i,\ j = 1,\ 2,\ \cdots,\ n \tag{6.11}$$

$$V_N = \{N_1,\ N_2,\ \cdots,\ N_n\} \tag{6.12}$$

$$E_{N-N} = \left\{\left(V_{N_i}, V_{N_j}\right)\right\}\ \forall i,j = 1,\ 2,\ \cdots,\ n \tag{6.13}$$

A vital aspect of the hypernetwork modeling is the unique dynamic mechanism of the complex system. This study utilizes the Node component and Link component to identify the crucial structures within the hypernetwork and evaluate the topological dimension metrics. For further calculation, we utilize index matrix to represent the structure of hypergraph.

6.3.3.2 Evolutionary Models of Hypernetworks

The clustering coefficient quantifies node connectivity strength in hypernetworks, measured by the proportion of nodes forming hyper-triangles (via ≥ 2 hyperedges). Higher values indicate greater connectivity and hyper-triangle prevalence, key to small-world phenomena. This study calculates the average transitivity coefficient to analyze network topology.

$$C_H = \frac{6 \times \text{number of triangles}}{\text{number of paths of length 2}} \tag{6.14}$$

A hyper-triangle is defined as a sequence of three distinct vertices and three distinct hyperedges of the form V_A, E_1, V_B, E_2,V_c, E_3, V_A, where V_A, V_B, and V_c are mutually adjacent nodes and a 2-path is a path of length 2, comprising a sequence of the form v_i, E_p, v_j, E_q, v_k (it should be noted that in a path, all vertices and hyperedges must be distinct).

Hypernetwork evolution models reveal structural and dynamic development mechanisms. Unlike node-driven growth in complex networks, hypernetworks evolve through hyperedge expansion: (i) new hyper-triangles emerge, and old ones disappear in the hypernetwork at different time scales; (ii) information variation resulting from hyperedge changes in the hypernetwork can be used to explore the dynamics of the urban recovery process, which echoes with the rhythmic changes in nodes.

6.4 Results

6.4.1 Topics of QOUL People Concerned

6.4.1.1 Data Collection and Preprocessing

Urban precipitation trends have risen due to global warming and urban impacts, leading to more frequent rainfall events and significant socioeconomic consequences on metropolitan areas [50, 51]. This study analyzes QOUL during three cloudburst waves in B City using 220,567 citizen-generated service requests (June–August 2021), sourced from: (i) Smart urban governance platforms; (ii) District hotlines (16 districts); and (iii) Online request robots.

Each record includes dispatch/resolution times, content, location, departments, and feedback (Table 6.1). Unlike social media or IoT data, this dataset

Table 6.1 Samples of request data in B city.

No.	Dispatch time	Content	Location	Departments	Process result	Resolution time
7136	2021/7/9 5:57:00	A sewage well on the main road has collapsed, causing overflowed due to the rain...	Community M, District N	Municipal Drainage Group	Resolved, Satisfied with resolution	2021/7/13 7:38:00

reflects real-time citizen feedback on public service efficiency, links infrastructure supply–demand gaps to SDG-aligned needs, focuses on municipal issues (e.g. spikes in complaints during water/power outages). Data distribution aligns with population density (<5% deviation), ensuring representativeness.

6.4.1.2 LDA Topic Modeling

To gain more detailed insights into QOUL restoration, we further analyze lifeline systems and urban responses using LDA modeling, a generative probabilistic model for topic extraction [52]. The process includes data gathering, preprocessing, LDA model training, and topic visualization. Figure 6.5, visualized with pyLDAvis, displays circles representing topics, with circle size indicating significance. The panel in the right displays relevant terms and frequencies, and the optimal topic count is determined by comparing intertopic distance maps (additional thematic visualizations of lifeline systems in supplementary material).

Discrepancies between subjective perceptions and physical built environments are linked by desired urban responses. Topic modeling results are presented in Table 6.2, focusing on three layers of needs – basic safety and health, social livelihood, and civic engagement – within a framework detailed in Table 5.1.

The TS case study identifies four QOUL restoration priorities during cloudbursts: flooded roads disrupting transport and businesses, inadequate nighttime lighting heightening safety risks, inefficient bus operations (e.g. route planning), and poor road aesthetics reflecting civic expectations. The first two topics address safety and health needs, while the latter two reveal systemic inefficiencies and aspirational urban standards in Table 6.3, collectively guiding resilience in urban infrastructure [46].

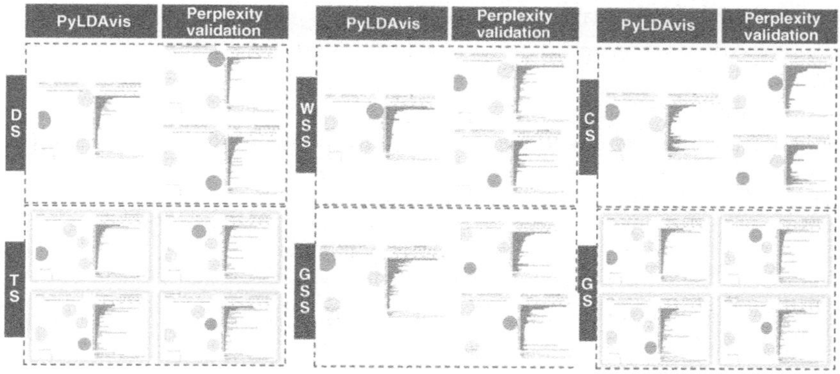

(a) Topics frequency for QOUL of resident for lifeline systems

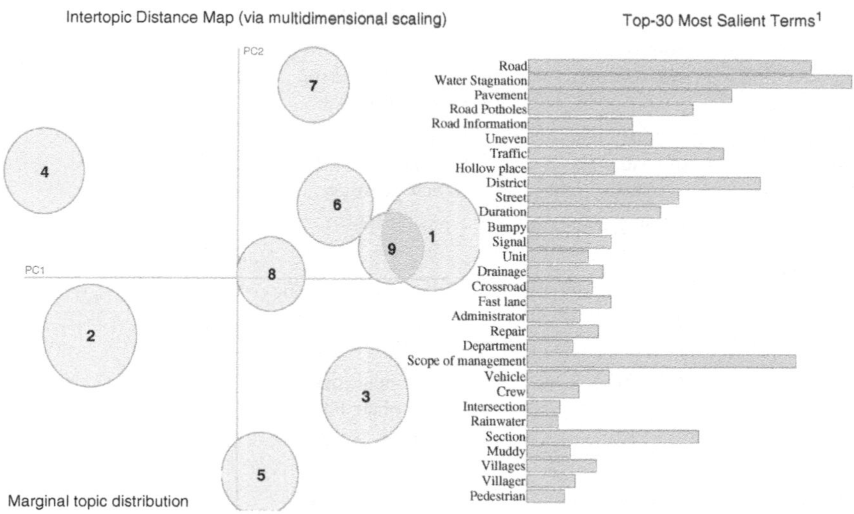

(b) Terms frequency for QOUL of resident for lifeline systems

Figure 6.5 Results of LDA.

6.4.2 Resilience of Lifeline Systems

Resilience quantification (Table 6.4) reveals QOUL dynamics under three cloudburst waves (Figure 6.6a). Baseline QOUL (predisruption) declined sharply during successive rainfall events: first wave (60 mm/h, July 12): QOUL plummeted to its lowest, recovering by July 16; second wave (50 mm/h, July 17): partial recovery interrupted by a third shock (35 mm/h), delaying baseline restoration until July 25.

Table 6.2 Main topics of quality of urban life (QOUL).

QOUL				
Lifeline system	**Subjective perception**	**Wanted urban response**	**Need layer**	**Reference**
Drainage system	Security hazards of drainage facilities	Well-operated drainage facilities without infiltration and attenuation	I	[1]
	Backwatering and house flooding		I	
	Flooded road	Desirable management of flood without surface gathered water	II	[1]
	Poor living environment	Living environment and place-making with cleanliness	III	
Transportation system	Security hazards of road	Full-equipped transport facilities for travelers' safety	I	[53]
	Blocked resident travel		I	
	Unreasonable bus operation	Well-operated traffic operation that meets daily travel needs	II	[54]
	Poor road appearance	Tidy appearance and unobstructed transport municipal road	III	[21]
Electricity system	Electricity and wiring safety hazards	Guaranteed water supply and safe water facilities	I I	[55]
	Unstable voltage	Continuous water supply, stable pressure, and flow	II	[56]
	Power outage		II	
	Unreasonable electricity expense	Price transparency, charge for electricity expense	III	[57]

QOUL				
Lifeline system	**Subjective perception**	**Wanted urban response**	**Need layer**	**Reference**
Water supply system	Burst water pipe	Well-operated electricity and safe facilities	I	[58]
	Water quality odor		I	
	Water outage	Continuous electricity supply, stable voltage	II	
	Unstable pressure and flow			
	Unreasonable water expense	Price transparency, charge for electricity expense	III	
Gas supply system	Security hazards of gas supply facilities	Well-operated gas and safe gas line facilities	I	
	Gas supply outage	Continuous gas supply and stable pressure	II	[59]
	Unreasonable gas expense	Price transparency, charge rationality for gas expense	III	
Communication system	Poor-quality network signal	High-quality network signal and base station	II II	[60]
	Unreasonable network expense	Charge rationality for calls and network	III	[61]

Table 6.3 Terms within topics extracted from transportation system.

	Topic 1	**Topic 2**	**Topic 3**	**Topic 4**
Topic	Security hazards of road	Blocked resident travel	Unreasonable bus operation	Poor road appearance
Corpus	*Stagnant water, flooded, road, water, pavement, rain, cloudburst,*	*Hazards, road, streetlight, danger, safety, off, street, community, square*	*Bus, road, bus stop, time, passenger, direction, line, license number, stop, sign*	*Potholes, mud, breaks, bumps, uneven, hole, pit, sidewalk, road, appearance*

Table 6.4 Quantification results of resilience under waves of cloudburst.

Urban resilience	Lifeline system	DS	TS	ES	WSS	GSS	CS
The first shock	0.52	0.31	0.47	0.71	0.67	0.98	0.99
The second shock	0.55	0.98	0.57	0.97	0.56	0.99	0.99
The third shock	0.99	1.00	0.98	1.00	1.00	1.00	1.00

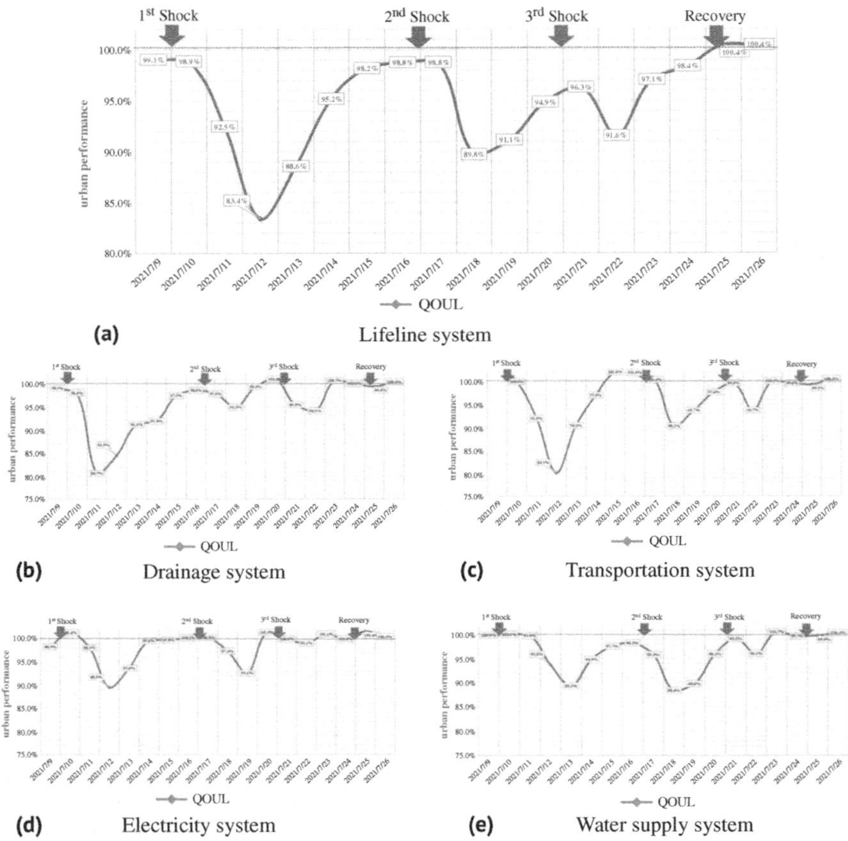

Figure 6.6 Resilience results.

Resilience scores rose progressively, indicating adaptive capacity despite assuming static resident perceptions during the flood season. The increasing resilience trend partly reflects citizens' sensitivity to early-stage rebound measures, which is corroborated by Chen et al. [62] on public opinion toward reopening measures during the three waves of COVID-19.

Drainage, transportation, and electricity systems mirrored overall urban resilience trends (rising scores), while gas/communication systems remained unaffected (Figure 6.6b–e). However, the water supply system (WSS) exhibited declining resilience (0.67–0.56) during successive cloudbursts, driven by outdated infrastructure: 42% of suburban/urban areas relied on self-provided wells due to incomplete municipal pipe upgrades, worsening water quality during floods.

6.4.3 Hypernetwork Modeling and Evolution Pattern

6.4.3.1 Hypernetwork Construction

The hypernetwork comprises three layers in Table 6.5: Human Needs (basic, social, high-level), Urban Response (22 LDA-derived nodes reflecting system serviceability), and Built Environment (18 nodes tracking physical infrastructure status).

In this study, we employed the hypernetwork model to identify hyperedges that influenced residents' QOUL during cloudburst events, as detailed in Table 6.6. This study highlights QOUL hyperedges that determine residents' needs and investigates how the underlying built environment entities impact urban functions.

6.4.3.2 Temporal Evolution Pattern

This study constructed time-sequence hypernetworks shown in Figure 6.7. We employed evolution coefficients to analyze the changing patterns of the

Table 6.5 Details of features extracted from hypernetwork.

Subnetwork	Vertex	Description
Human needs	V_{N1}	Safety and health (basic needs)
	V_{N2}	Social livelihood (social needs)
	V_{N3}	Civic engagement (high-level needs)
Urban response	V_{DS}	Urban response of drainage system
	V_{TS}	Urban response of transportation system
	V_{PSS}	Urban response of electricity system
	V_{WSS}	Urban response of water supply system
	V_{GSS}	Urban response of gas supply system
	V_{CS}	Urban response of communication system
Built environment	V_{B1-14}	The built environment attributes such as pipelines, road sections, wells, and gas stations

Table 6.6 The hyperedges of the QOUL hypernetwork.

Hyperedges	Human needs	Urban response	Built environment
HE_1	Safety and health	Security hazards of drainage facilities	$V_{B11} - V_{B13}, V_{B21} - V_{B22}$
HE_2	Safety and health	Back-watering and house flooding	$V_{B21} - V_{B22}$
HE_3	Social livelihood	Blocked resident travel	$V_{B11} - V_{B13}$
HE_4	Civic engagement	Poor living environment	$V_{B21} - V_{B22}, V_{B31} - V_{B33}$
HE_{5-20}	...	...	...
HE_{21}	Social livelihood	Poor-quality network signal	$V_{B17.1} - V_{B17.4}$
HE_{22}	Civic engagement	Unreasonable network expense	$V_{B18.1} - V_{B18.2}$

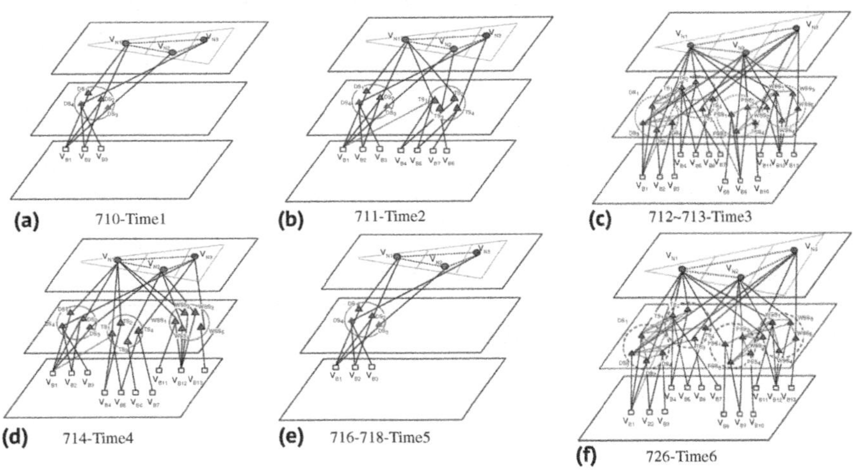

Figure 6.7 Time-sequence hypernetworks under waves of cloudburst.

hypernetwork. The clustering coefficient represents the closeness of connections between neighboring hyperedges. The coefficient reached 0.346 at T_2 (histogram), corresponding to the most rapid decline in QOUL resilience curve displayed in Figure 6.7. At T_3 of July 12–13, the clustering coefficient peaked at 0.375, with the majority of system recovery occurring during this period, resulting in a swift restoration of QOUL performance. The clustering coefficient reflects the gradient of the QOUL resilience curve.

Additionally, we need to evaluate specific changes in hyperedges to further explore temporal evolution patterns presented in Figure 6.8 and Table 6.7. As the

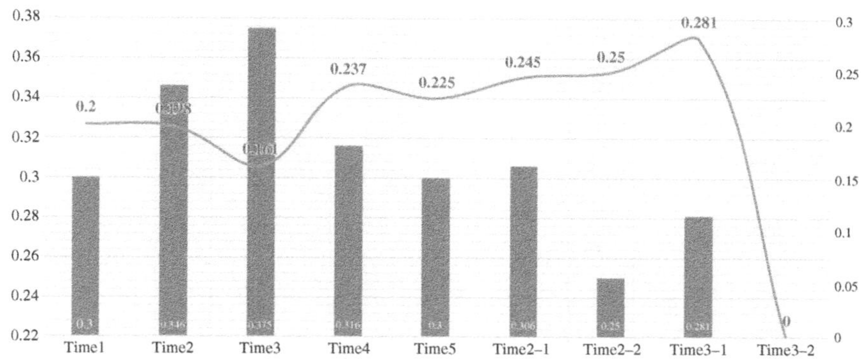

Figure 6.8 Results of clustering (histogram) and evolution coefficient (curve).

Table 6.7 Hypergraph index.

	T_1	T_2	T_3	T_4	T_5
C	0.300	0.346	0.375	0.316	0.300
E	0.200	0.198	0.161	0.237	0.225

time scale shifts from T_1 to T_5, new hyper-triangles emerge, and old ones vanish within the hypernetwork. The evolution coefficient reflects alterations in system transformations due to hyperedge variations, enabling us to investigate the driving hyperedge during the recovery process.

For instance, the coefficient reaches 0.198 by T_2, with the hyperedge $\{V_{N1}, DS_2, V_{B1}\}$ acting as the driving hyperedge in this transformation. We can further identify the critical function topic of security hazards of drainage facilities. During the recovery process, the coefficient increases to 0.237 by T_4, and driving hyperedges increase to three: $\{V_{N1} - TS_1 - V_{B1}\}$, $\{V_{N1} - TS_2 - V_{B9}\}$, and $\{V_{N1} - WSS_1 - V_{B12}\}$. This indicates that the water supply and electricity systems collaboratively become dominant.

6.4.3.3 The Underlying Built Environment

The cloudburst events in B on July 12 called for a critical evaluation of how QOUL can be impacted by the built environment. We examined specific hyperedges linked to each component within lifeline systems and their criticality at various times. The relationship between hyperedges and underlying built environment attributes can be observed from a time sequence perspective in Table 6.8 and Figure 6.9.

Table 6.8 The underlying built environment results of DS system.

Drainage system		T1	T2	T3		T4	T2-1	T2-2	Components	hyperedges
V_{B1}	Manhole cover	4.5%	4.2%	10.1%	7.7%	6.4%	5.6%	3.5%	**42.0%**	HE_1
	Rainwater grate	0.7%	1.5%	5.9%	1.9%	1.6%	0.6%	0.7%	13.0%	HE_2
	Drainage pipeline	1.0%	1.2%	4.7%	2.1%	0.7%	0.7%	0.6%	10.9%	HE_3
V_{B2}	Construction	0.7%	0.4%	1.7%	1.4%	0.5%	0.6%	0.3%	5.5%	HE_1
V_{B3}	Sewage well	1.7%	1.4%	6.9%	3.4%	2.0%	3.3%	1.5%	**20.1%**	HE_2
	Convenience facilities	0.4%	0.4%	1.2%	1.2%	1.1%	0.3%	0.6%	5.0%	HE_3
	Garbage and weeds	0.4%	0.5%	1.2%	0.3%	0.5%	0.4%	0.2%	3.4%	HE_1
Time		9.4%	9.5%	**31.7%**	18.0%	12.8%	11.4%	7.3%	100.0%	

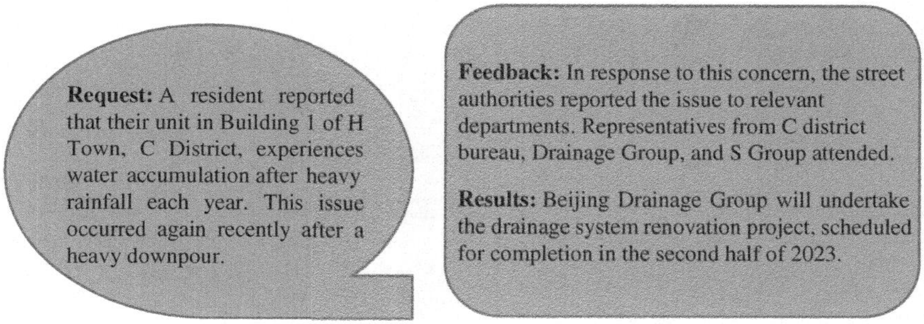

Figure 6.9 A closed-loop feedback data from residents on urban issues.

In the DS system, six hyperedges are coupled with 712 cloudburst waves, warranting attention to damaged rainwater grates (42.0%), which impact hyperedges $H_1 - H_3$, and sewage wells (20.1%), affecting hyperedge H_4. Furthermore, it is important to note that the drainage system components most exposed to challenges are concentrated at T_{3-1}. Notably, the QOUL curve reaches its lowest point at T_2, suggesting that residents' perception of the drainage system occurs before the actual damage onset. This finding highlights the necessity for engineering systems to incorporate the early perception of infrastructure issues by residents as an informative warning signal.

In the TS system, attention should be paid to road issues (50.0%), including collapsed roads (12.3%), potholed roads (31.7%), and damaged roads (20.1%), which affect hyperedges H_{4-7}, corresponding to the perception topic of security hazards of roads and flooded roads. The transportation system components facing the most significant challenges are concentrated at T_{3-1}. In line with this observation, the QOUL curve exhibits a minimum at T_{3-2}, suggesting that residents' perception of the transportation system closely aligns with its objective attributes. In the ES and WSS systems, attention should be focused on accessory issues, supply pipe issues, and private wells, affecting hyperedges H_{8-10} and H_{11-13}. The two system components present the most significant challenges, and the QOUL curve exhibits the minimum level simultaneously. This implies a strong link between residents' perception and the physical reality of objective attributes.

6.5 Discussion

6.5.1 Underlying Infrastructure of Human Needs

Urban lifeline infrastructure performance varies markedly between normal and disruptive conditions (Figure 6.10a). While systems operate smoothly under

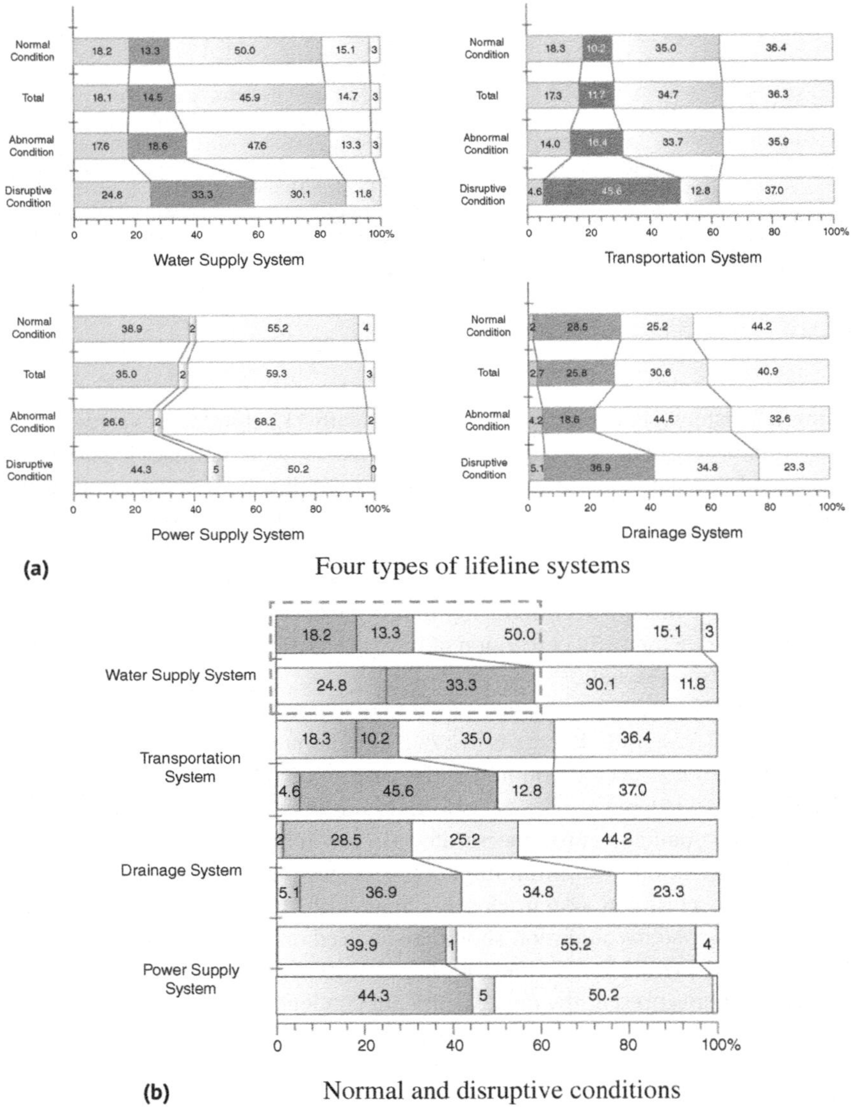

Figure 6.10 Buffering results of normal and disruptive conditions of lifeline systems.

normal circumstances, cloudbursts expose persistent vulnerabilities, particularly in drainage and transportation networks. Buffering analysis of service requests reveals distinct human need hierarchies during disruptions (Table 6.1): basic safety needs (gray, 76.1% prevalence in transportation systems), livelihood needs (black/white, declining during shocks), and civic needs.

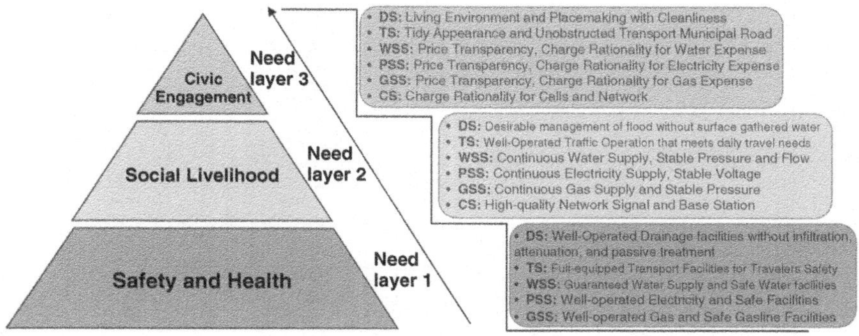

Figure 6.11 Hierarchy of needs.

During cloudbursts, basic safety demands surge (Figure 6.10b), with waterlogging incidents increasing from 10.2% to 45.6% due to substandard drainage. Concurrently, 44% of complaints highlighted transportation disruptions from road collapses, exacerbating QOUL declines [5, 63]. Resilience disparities across systems further illustrate this hierarchy (Figure 6.11): **Electricity systems** maintained higher resilience (26.7% of safety needs, resilience score 0.71) due to minimal storm damage. **Transportation systems** faltered (76.1% of safety needs, resilience score 0.47), reflecting infrastructural inadequacies.

Notably, suburban WSS exhibited anomalous resilience degradation (0.67 → 0.56) during sequential shocks, tied to reliance on vulnerable self-provided wells (42% of areas). These findings align with established hierarchies of disaster vulnerability, where basic needs dominate during crises, and underscore how resilient infrastructure sustains QOUL under chronic stressors [64].

6.5.2 Distribution of Spatial Features

To better understand the effects on downtown and suburban areas, we (i) quantify the extent of QOUL impact and (ii) evaluate the disparities in restoration efforts. Figure 6.12 depicts the difference in impact and recovery between downtown and suburban areas.

Spatial analysis reveals no significant QOUL difference between downtown and suburban areas during floods (t-test, Figure 6.12a), yet systemic disparities emerge:

QOUL Decline: drainage (−30%) and transportation (−25%) systems faced the steepest drops, versus water (−20%) and electricity systems.

Resilience Hierarchy: resilience scores ranked water supply (0.71) > electricity (0.67) > transportation (0.47) > drainage (0.31).

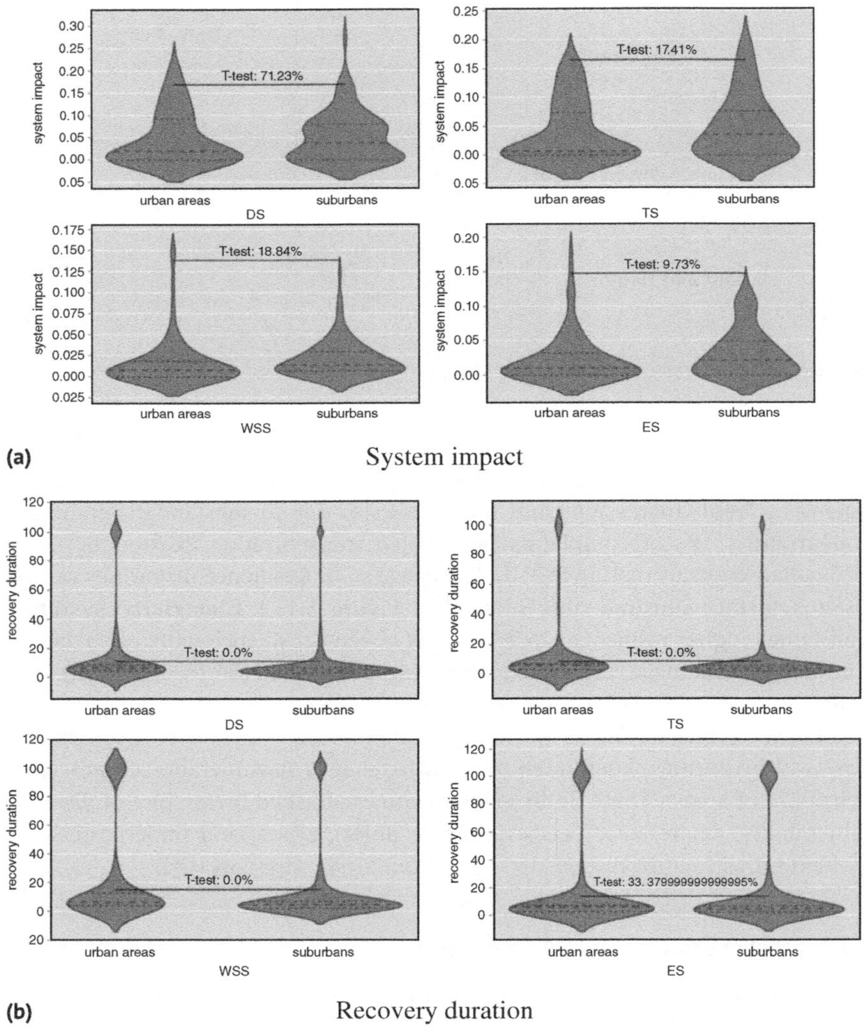

Figure 6.12 Distribution of assessment features for QOUL with different areas.

Suburban drainage exhibited a bimodal QOUL distribution (Figure 6.12a), driven by urban manhole issues (42.0%) and suburban sewage well failures (20.1%). Targeted interventions in hotspot areas (e.g. M Village's sewage pipe blockages) could mitigate these gaps.

Recovery durations diverged spatially: downtown areas lagged due to complex infrastructure interdependencies (Figure 6.12b). For example, B City's 2020 cloudburst caused 500 urban waterlogging incidents (130 severe), attributed to inadequate drainage. Conversely, the electricity system's uniform recovery (resilience: 0.71) highlighted comprehensive grid coverage, unaffected by spatial divides.

6.5.3 Human Agency Embedded in Hypernetwork

Resilient lifeline systems require collaboration across stakeholders (individuals, communities, institutions). Hypernetwork models map a network of nodes (agents) representing the built environment, urban performance, and human needs, interconnected via hyperedges. Metrics like clustering coefficients quantify connection strength, while hypernetwork evolution traces how these nodes (agents) interact during disasters – capturing human agency through feedback loops (e.g. resident perceptions shaping infrastructure priorities).

A symbiotic relationship emerges: human agency, embedded in the network of nodes (agents), identifies and resolves postdisaster needs, elevating QOUL. Enhanced QOUL, in turn, empowers agency by fostering environments where residents actively shape resilience strategies.

Livelihood resilience centers on this network of nodes (agents), prioritizing human-centric adaptation over technocratic frameworks. It leverages social capital (e.g. informal resource networks) to strengthen collective and individual capacities for risk response, ensuring accessible support during disruptions.

6.6 Conclusion

This study represents an important contribution to urban resilience in the Anthropocene, as it seeks to integrate human-centered urbanism into the development of a hypernetwork model that prioritizes the QOUL. By using text mining techniques to analyze request data, this study displays the mismatch between subjective human needs and objective built environment and utilizes QOUL to quantify changes in urban resilience under continuous stressor disturbances. A human-needs-driven hypernetwork model is further proposed to dynamically identify critical hyperedge alterations during resilience time-sequenced recovery, which unravel the functional and underlying built environment issues of urban systems, further aid precision urban governance.

Our findings reveal that QOUL serves as a vital link in connecting public perception with objective built environment. It is worth noting that the QOUL resilience curve in response to sustained shocks exhibits comparable trends of enhanced resilience for most lifeline systems, albeit with varying spatial effects and recovery durations. We reveal the infrastructure shortcomings of the municipal water supply network in B by analyzing its anomalous resilience. Furthermore, building upon these findings, urban rectification of waterlogging points can be more precise, as it highlights the value of updating vulnerable regions using citizen-perceived data.

The hypernetwork modeling results shed light on public opinion regarding recovery during waves of cloudburst and offer critical insights into the temporal evolution of urban resilience under a cascade of urban disasters. Through socially sensed request data from the urban governance platform, this study reveals the embodiment of QOUL in temporal evolution pattern of hypernetwork. Urban policymakers can gain insight to see the dynamic of urban resilience by accessing, utilizing, and further evaluating the lifeline infrastructures. The proposed hypernetwork model provides a valuable tool for assessing and improving resilience, laying a methodological foundation to inform efforts toward global resilience driven by human needs in the Anthropocene.

This study contributes to the development of evidence-based urban planning and management strategies that promote sustainable and resilient urban development in the Anthropocene. We harness QOUL to understand the nuanced relationship between built environment and subjective perception by engaging human needs with urban response to diverse disasters. Urban resilience in the Anthropocene is not a tightly defined concept but a broad heuristic tool to connect critical scholarship on human engagement and urban response through the critical idea of QOUL.

References

1 McClymont, K., Cunha, D.G.F., Maidment, C., et al. (2020). Towards urban resilience through sustainable drainage systems: a multi-objective optimisation problem. *Journal of Environmental Management* 275: 111173.

2 Shekhar, H., Schmidt, A.J., and Wehling, H.-W. (2019). Exploring wellbeing in human settlements—a spatial planning perspective. *Habitat International* 87: 66–74.

3 United Nations. (2016). Draft outcome document of the United Nations Conference on housing and sustainable urban development (Habitat III).

4 Huang, W., and Ling, M. (2018). System resilience assessment method of urban lifeline system for GIS. *Computers, Environment and Urban Systems* 71: 67–80.

5 Zhao, Z., Pan, S., Li, N., et al. (2022). A seismic emergency performance optimization model for infrastructure systems under demand differences: a case study in China. *Earthquake Engineering and Resilience* 1 (2): 196–210.

6 Stokols, D., Lejano, R.P., and Hipp, J. (2013). Enhancing the resilience of human–environment systems: a social ecological perspective. *Ecology and Society* 18 (1): 13–15.

7 Long, Y., Li, Z., Pan, S., et al. (2022). Assessing prehospital seismic resilience: an agent-based modeling approach. *IT Professional* 24 (3): 18–27.

8 Oliva, G., Panzieri, S., and Setola, R. (2010). Agent-based input–output interdependency model. *International Journal of Critical Infrastructure Protection* 3 (2): 76–82.

9 Crutzen, P.J. (2006). The "Anthropocene". In E. Ehlers and T. Krafft (Eds.), *Earth System Science in the Anthropocene* (pp. 13–18). Springer: Berlin Heidelberg.

10 Steffen, W., Richardson, K., Rockström, J., et al. (2015). Planetary boundaries: guiding human development on a changing planet. *Science* 347 (6223): 1259855.

11 Crutzen, P.J. (2021). Geology of Mankind (2002). In S. Benner, G. Lax, P.J. Crutzen, et al. (Eds.), *Paul J. Crutzen and the Anthropocene: A New Epoch in Earth's History* (pp. 23–25). Springer International Publishing.

12 Castree, N. (2014). The Anthropocene and geography I: the back story. *Geography Compass* 8 (7): 436–449.

13 Miller, C.A., Iles, A., and Jones, C.F. (2013). The social dimensions of energy transitions. *Science as Culture* 22 (2): 135–148.

14 Grove, K., and Adey, P. (2015). Security and the politics of resilience: an aesthetic response. *Politics* 35 (1): 78–84.

15 Elmqvist, T., Andersson, E., Frantzeskaki, N., et al. (2019). Sustainability and resilience for transformation in the urban century. *Nature Sustainability* 2 (4): 267–273.

16 Tanner, T., Lewis, D., Wrathall, D., et al. (2014). Livelihood resilience in the face of climate change. *Nature Climate Change* 5 (1): 23–26.

17 Wakefield, S. (2020). Urban resilience as critique: problematizing infrastructure in post-Sandy New York City. *Political Geography* 79: 102148.

18 Bai, X., van der Leeuw, S., O'Brien, K., et al. (2016). Plausible and desirable futures in the Anthropocene: a new research agenda. *Global Environmental Change* 39: 351–362.

19 Johannessen, Å., and Wamsler, C. (2017). What does resilience mean for urban water services? *Ecology and Society* 22 (1): 7–9.

20 Marans, R.W. (2015). Quality of urban life & environmental sustainability studies: future linkage opportunities. *Habitat International* 45: 47–52.

21 Wey, W.-M., and Huang, J.-Y. (2018). Urban sustainable transportation planning strategies for livable city's quality of life. *Habitat International* 82: 9–27.

22 Mouratidis, K. (2021). Urban planning and quality of life: a review of pathways linking the built environment to subjective well-being. *Cities* 115: 103229.

23 Cutter, S.L. (1985). *Rating Places: A Geographer's View on Quality of Life*. Association of American Geographers.

24 Kabisch, S., Koch, F., Gawel, E., et al. (2018). *Urban Transformations: Sustainable Urban Development Through Resource Efficiency, Quality of Life and Resilience* (Vol. 10). Springer.

25 Li, C., Huang, Y., Shen, Y., and Xu, L. (2022). Spatiotemporal patterns and mechanisms of street vending from the social sensing perspective: a comparison between law-enforcement reported and residents complain events. *Cities* 124: 103597.

26 Peng, X., Liang, Y., Xu, L., and Li, D. (2020). An approach for discovering Urban public management problem and optimizing urban governance based on citizen service hotline. *B Da Xue Xue Bao* 56 (4): 721–731.

27 Pan, S., Zhao, Z., Lim, H.W., et al. (2022). Restored quality of life-based approach (REQUALIFE) for urban seismic resilience assessment: case study. *International Journal of Disaster Risk Reduction* 79: 103170.

28 Diener, E., Oishi, S., and Tay, L. (2018). Advances in subjective well-being research. *Nature Human Behaviour* 2 (4): 253–260.

29 Feng, Y., You, H., Zhang, Z., et al. (2019). Hypergraph neural networks. *Proceedings of the AAAI Conference on Artificial Intelligence* 33 (01): 3558–3565.

30 Gao, Y., Feng, Y., Ji, S., and Ji, R. (2023). HGNN+: general hypergraph neural networks. *IEEE Transactions on Pattern Analysis and Machine Intelligence* 45 (3): 3181–3199.

31 Berge, C. (1973). *Graphs and Hypergraphs* (Trans. M. Edward). New York: American Elsevier.

32 Jayaram, N., and Srinivasan, K. (2008). Performance-based optimal design and rehabilitation of water distribution networks using life cycle costing. *Water Resources Research* 44 (1): 13–15.

33 Campbell, A., Converse, P.E., and Rodgers, W.L. (1976). *The Quality of American Life: Perceptions, Evaluations, and Satisfactions*. Russell Sage Foundation.

34 Yuan, F., Liu, R., Mao, L., and Li, M. (2021). Internet of people enabled framework for evaluating performance loss and resilience of urban critical infrastructures. *Safety Science* 134: 105079.

35 Hartmann, A.K. (2014). Large-deviation properties of resilience of transportation networks. *The European Physical Journal B* 87: 1–10.

36 Schlör, H., Venghaus, S., and Hake, J.-F. (2018). The FEW-Nexus city index–measuring urban resilience. *Applied Energy* 210: 382–392.

37 Maslow, A.H. (1943). A theory of human motivation. *Psychological Review* 50 (4): 370.

38 Yang, Z., Barroca, B., Weppe, A., et al. (2023). Indicator-based resilience assessment for critical infrastructures—a review. *Safety Science* 160: 106049.

39 Muñoz-Erickson, T.A., Meerow, S., Hobbins, R., et al. (2021). Beyond bouncing back? Comparing and contesting urban resilience frames in US and Latin American contexts. *Landscape and Urban Planning* 214: 104173.

40 Nan, C., and Sansavini, G. (2017). A quantitative method for assessing resilience of interdependent infrastructures. *Reliability Engineering & System Safety* 157: 35–53.

41 Liu, Y., Wang, X., Song, C., et al. (2023). Quantifying human mobility resilience to the COVID-19 pandemic: a case study of B, China. *Sustainable Cities and Society* 89: 104314.

42 Kontokosta, C.E., and Malik, A. (2018). The resilience to emergencies and disasters index: applying big data to benchmark and validate neighborhood resilience capacity. *Sustainable Cities and Society* 36: 272–285.

43 Podesta, C., Coleman, N., Esmalian, A., et al. (2021). Quantifying community resilience based on fluctuations in visits to points-of-interest derived from digital trace data. *Journal of the Royal Society Interface* 18 (177): 20210158.

44 Bruneau, M., Chang, S.E., Eguchi, R.T., et al. (2003). A framework to quantitatively assess and enhance the seismic resilience of communities. *Earthquake Spectra* 19 (4): 733–752.

45 Hong, B., Bonczak, B.J., Gupta, A., and Kontokosta, C.E. (2021). Measuring inequality in community resilience to natural disasters using large-scale mobility data. *Nature Communications* 12 (1): 1870.

46 Zhao, Z., Zhou, X., Zheng, Y., et al. (2024). Enhancing infrastructural dynamic responses to critical residents' needs for urban resilience through machine learning and hypernetwork analysis. *Sustainable Cities and Society* 106: 105366.

47 Zhao, Z., Li, Z., Tong, R., et al. (2025). Unveiling the spatial heterogeneity of factors influencing physical and perceived recovery disparities under extreme rainstorms: a geographically weighted machine learning approach. *Sustainable Cities and Society* 118: 106023.

48 Zhao, Z., Zhou, X., Lin, Z., et al. (2025). A resident-centric framework for postdisaster infrastructure recovery: characterizing hierarchical needs and fulfillment cycles to assess urban resilience. *Journal of Management in Engineering* 41 (3): 04025003.

49 Zhao, Z., Li, Z., Wang, T., et al. (2025). Post-disaster recovery planning for infrastructure systems based on residents' needs: a hypernetwork approach. *International Journal of Disaster Risk Reduction* 118: 105258.

50 Han, J.Y., Baik, J.J., and Lee, H. (2014). Urban impacts on precipitation. *Asia-Pacific Journal of Atmospheric Sciences* 50: 17–30.

51 Tabari, H. (2020). Climate change impact on flood and extreme precipitation increases with water availability. *Scientific Reports* 10 (1): 1–10.

52 De Oliveira Capela, F., and Ramirez-Marquez, J.E. (2019). Detecting urban identity perception via newspaper topic modeling. *Cities* 93: 72–83.

53 Pregnolato, M., Ford, A., Wilkinson, S.M., and Dawson, R.J. (2017). The impact of flooding on road transport: a depth-disruption function. *Transportation Research Part D: Transport and Environment* 55: 67–81.

54 An, K. (2020). Battery electric bus infrastructure planning under demand uncertainty. *Transportation Research Part C: Emerging Technologies* 111: 572–587.

55 Wu, J., Bai, Y., Fang, W., et al. (2021). An integrated quantitative risk assessment method for urban underground utility tunnels. *Reliability Engineering & System Safety* 213: 107792.

56 Yadav, M., Pal, N., and Saini, D.K. (2023). Low voltage ride through capability for resilient electrical distribution system integrated with renewable energy resources. *Energy Reports* 9: 833–858.

57 Arab, A., Khodaei, A., Han, Z., and Khator, S.K. (2015). Proactive recovery of electric power assets for resiliency enhancement. *IEEE Access* 3: 99–109.

58 Popescu, A.I. (2020). Long-term city innovation trajectories and quality of urban life. *Sustainability* 12 (24): 10587.

59 Cimellaro, G.P., Villa, O., and Bruneau, M. (2015). Resilience-based design of natural gas distribution networks. *Journal of Infrastructure Systems* 21 (1): 05014005.

60 World Health Organization. (2016). Report on quality universal health coverage and resilience technical workshop. *Convened by the WHO Service Delivery & Safety Department at the WHO Kobe Centre for Health Development Kobe, Japan 25–26 February 2016.*

61 Huang, Y., Li, J., Wu, G., and Fei, T. (2020). Quantifying the bias in place emotion extracted from photos on social networking sites: a case study on a university campus. *Cities* 102: 102719.

62 Chen, Y., Niu, H., and Silva, E.A. (2023). The road to recovery: sensing public opinion towards reopening measures with social media data in post-lockdown cities. *Cities* 132: 104054.

63 Pan, S., Zhao, Z., Lim, H.W., et al. (2022b). Restored quality of life-based approach (REQUALIFE) for urban seismic resilience assessment: quantitative method. *International Journal of Disaster Risk Reduction* 79: 103169.

64 Folke, C., Carpenter, S.R., Walker, B., et al. (2010). Resilience thinking: integrating resilience, adaptability and transformability. *Ecology and Society* 15 (4): 7–9.

7

Industrial IoT and Sustainable Development

Keshav Gupta[1], *Nishu Gupta*[2], *and Polat Goktas*[3,4,*]

[1] *Department of Computing Technologies, SRM Institute of Science & Technology, SRM Nagar, Kattankulathur, India*
[2] *VTT Technical Research Centre of Finland Ltd., Oulu, Finland*
[3] *UCD School of Computer Science, University College Dublin, Dublin, Ireland*
[4] *Faculty of Engineering and Natural Sciences, Sabanci University, Istanbul, Turkey*

*Corresponding author: polatgoktas@gmail.com

7.1 Foundations of Industrial Internet of Things in the Context of Sustainable Development

The Industrial Internet of Things (IIoT) represents a critical evolution in digital transformation, enabling interconnected systems to gather, analyze, and act on real-time data across industrial operations. In the context of sustainable development, IIoT contributes significantly to resource efficiency, circular economy practices, energy optimization, and transparency in production systems. Beier et al. [1] argue that IIoT is a catalyst for achieving industrial sustainability by enhancing productivity while reducing environmental impacts. Similarly, Manavalan and Jayakrishna [2] review how IoT-embedded supply chains support Industry 4.0 and align with sustainable development goals (SDGs). Hatzivasilis et al. [3] provide a forward-looking application of IIoT in wind energy using a novel communication protocol designed for circular economy practices.

IIoT technologies serve as enablers of data-driven environmental governance, empowering industries to align with international frameworks such as the UN SDGs and the European Green Deal [4, 5]. The integration of blockchain and edge computing with IIoT, as explored by Fraga-Lamas et al. [6] and Leng et al. [7],

AI and ML Techniques in IoT-based Communication: A Path to Sustainable Development Goals, First Edition. Edited by Sumita Mishra, Nishu Gupta and Polat Goktas.
 Companion Website: http://www.wiley.com/go/mishra_aiml

offers decentralized, traceable, and secure solutions, strengthening transparency and operational resilience. Thus, this chapter presents a comprehensive exploration of IIoT within the framework of sustainable development, beginning with smart manufacturing applications such as artificial intelligence (AI)-powered predictive maintenance and energy optimization to minimize downtime and enhance efficiency. It further investigates waste reduction and material efficiency through sensor-based monitoring and digital twins, supported by case studies from green factories embracing Industry 4.0 principles. On the environmental front, we detail how real-time analytics driven by big data and machine learning (ML) empower emissions tracking, footprint minimization, and sector-specific implementations in mining, textiles, and automotive. Cybersecurity, ethics, and governance are examined through resilient system architectures, responsible AI usage, and compliance with international standards. Finally, we spotlight future trends including digital twins, edge AI, and federated learning (FL), and highlight how circular economy integration and policy-driven roadmaps can enable resilient and scalable IIoT ecosystems aligned with global sustainability goals.

7.2 Smart Manufacturing and Resource Efficiency

As visualized in Figure 7.1, the IIoT sustainability cycle encapsulates the holistic role of IIoT in enabling sustainable transformation across industrial sectors. Starting with enhanced efficiency and waste reduction, IIoT systems streamline production through real-time monitoring and intelligent automation. By lowering energy consumption and minimizing environmental impacts, industries reduce their ecological footprint while boosting competitiveness. The cycle also reinforces smarter resource management, facilitating data-driven decisions in supply chains and operations. Ultimately, this technological framework promotes eco-friendly practices, encouraging industries to adopt long-term strategies aligned with sustainability goals such as the UN SDGs and the European Green Deal. The circular flow of benefits emphasizes the continuous, feedback-driven nature of IIoT implementations in building resilient and regenerative industrial systems.

7.2.1 AI-powered Predictive Maintenance and Energy Optimization

The integration of AI with IIoT systems has transformed the way industries manage energy and maintenance. Predictive maintenance allows for pre-emptive

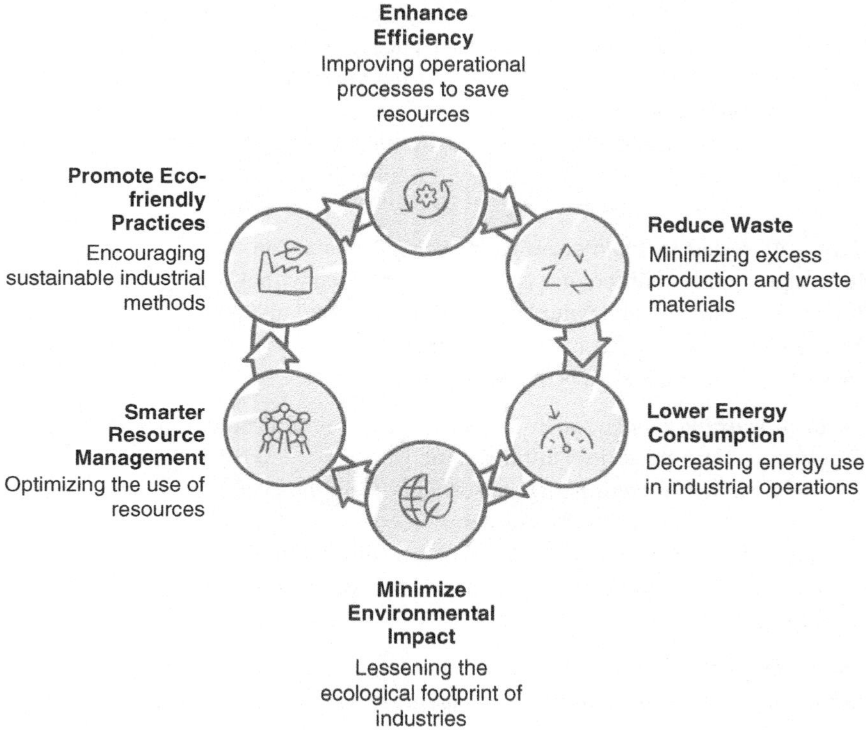

Figure 7.1 The industrial IoT (IIoT) sustainability cycle.

detection of machinery failure, lowering downtime and optimizing energy use – key aspects of sustainable industrial operations.

7.2.1.1 Fundamentals of Predictive Maintenance in IIoT Environments

Predictive maintenance (PdM) has become a foundational strategy within Industry 4.0 to minimize equipment failures and energy wastage. Unlike traditional reactive maintenance, which occurs postfailure, or preventive maintenance based on fixed schedules, predictive maintenance is data-driven and intelligent. It involves the collection and analysis of real-time data from IIoT-enabled sensors to detect anomalies, assess equipment health, and forecast failures before they occur.

Key to this evolution is the digital twin concept – virtual replicas of physical systems that can simulate behavior, predict anomalies, and optimize performance. Kerkeni et al. [8] developed a digital twin model specifically tailored for industrial machinery to enable continuous monitoring and early fault detection. Rana et al. [9]

emphasized the role of digital twins in facilitating sustainable industrial development by enabling scenario-based energy simulations and system diagnostics. Real-time analytics enabled by IIoT networks support continuous system evaluation. Bhargava et al. [10] showcased an AI-integrated IIoT platform for logistics systems, enhancing operational resilience and predictive maintenance capabilities.

7.2.1.2 AI Models for Maintenance and Energy Optimization

Modern predictive maintenance is deeply intertwined with AI and ML algorithms. These models process vast volumes of IIoT-generated data to detect patterns, classify anomalies, and predict machine failures.

Several AI techniques are being deployed:

- *Deep learning models*, such as convolutional neural networks (CNNs), are effective for time-series fault detection [11].
- *Attention mechanisms* help focus on significant data features, improving prediction accuracy in complex systems.
- *Edge AI*, where AI models are deployed directly on edge devices near the source of data generation, enhances real-time performance and reduces latency [12].

Agostinelli and Cumo [13] demonstrated the use of ML models for predictive maintenance in smart building management systems, integrating data from HVAC systems to reduce unnecessary energy use. Golec et al. [14] introduced the CAPTAIN testbed, enabling co-simulation for scalable AIoT environments, enhancing predictive performance in energy-critical systems. The integration of nonintrusive sensors with edge computing also allows for condition-based monitoring with lower energy overhead, as seen in biogas compressor systems [12].

7.2.1.3 Case Studies in Smart Manufacturing

Real-world case studies reinforce the practical value of AI-powered predictive maintenance. For example:

- **Smart Textiles**: Ranganathan et al. [15] developed an IoT-based moisture management system for fabric drying. The solution reduced energy consumption by adjusting drying cycles in real time, based on sensor feedback.
- **Cold Supply Chains**: Fatorachian and Pawar (2025) [16] analyzed Industry 4.0 tools in optimizing resource use in perishable goods transportation. Their AI-based framework minimized waste and energy use through dynamic route and storage adjustment.

- **Vehicular Logistics**: Bhargava et al. [10] implemented IIoT and AI in logistics, showcasing significant reductions in mechanical failures and emissions.
- **Energy Disaggregation**: Raiker et al. [17] demonstrated high-frequency energy sampling to identify individual energy-consuming components in industrial settings – providing data for targeted energy-saving interventions.
- **AI-driven Low Carbon Manufacturing**: Joshi and Sharma [18] provided a comprehensive review of algorithms used in low-carbon production environments, highlighting the synergy between predictive maintenance and decarbonization strategies.

7.2.1.4 Energy Efficiency and Circular Economy Synergies

Predictive maintenance contributes not only to cost savings but also to sustainability through enhanced energy efficiency and alignment with circular economy principles. Systems that can monitor their own degradation and schedule repairs or replacements only when needed extend equipment lifecycles and reduce material waste. Mazumdar et al. [19] emphasized real-world problems solved through AI in predictive systems, noting how predictive scheduling alone can reduce energy use by up to 20%. Similarly, Chen et al. [20] explored the broader impact of technological innovation on energy consumption and social equity, showing how AI can optimize resource use and promote sustainability in manufacturing. In addition, Golec et al. [21] showcased how simulation-based environments like CAPTAIN can train AI models to support sustainable energy management in serverless and scalable IIoT architectures.

7.2.2 Waste Reduction and Material Efficiency in Production Lines

The deployment of IIoT solutions has introduced novel paths to material efficiency and wastage reduction across manufacturing ecosystems. Driven by global sustainability demands and industrial competitiveness, manufacturing organizations are integrating smart systems, digital platforms, and real-time analytics to reduce waste streams, optimize material usage, and transition toward circular production models.

7.2.2.1 Digitalization and Circular Economy for Resource Optimization

Circular economy strategies aim to decouple industrial growth from resource consumption by designing out waste and keeping materials in use. IIoT platforms empower these strategies through real-time data collection, smart sensors, and process automation. Nudurupati et al. [22] highlighted how Indian small- and medium-sized enterprises (SMEs) adopted circular economy practices by leveraging digital platforms, which led to improved resource management, reduced waste, and enhanced sustainability performance. These systems enabled traceability, inventory control, and performance feedback loops. Mboli et al. [23] presented an IoT-enabled decision support system that facilitated real-time decision-making in circular economy business models. By embedding sensors into key operational nodes, companies could proactively identify inefficiencies and adjust operations to reduce material losses. Singh and Powder [24] further emphasized the importance of supply chain partnerships, arguing that shared resource flows and collaborative material management are essential for reducing cumulative waste across networks.

7.2.2.2 Material Flow Cost Accounting and Lifecycle-based Efficiency

Material Flow Cost Accounting (MFCA) enables as a powerful tool for identifying hidden inefficiencies. Schmidt [25] extended MFCA by combining it with environmental material flow analysis, enabling organizations to trace the economic and ecological costs of wasted materials and energy during production. This approach aligns with lifecycle thinking and supports sustainability assessments of products and processes. For instance, Broadbent [26] emphasized steel's high recyclability, promoting its role in a closed-loop system with measurable environmental and financial benefits. Similarly, Tecchio et al. [27] recommended the integration of circularity metrics into product policy and eco-design standards, providing systemic support for material efficiency.

7.2.2.3 IoT-supported Process Innovation and Additive Manufacturing

Emerging digital technologies, especially 3D printing (additive manufacturing), offer significant potential for waste reduction. Additive manufacturing builds objects layer by layer, minimizing excess material compared to subtractive methods. Olawumi et al. [28] showcased how recycled materials such as plastics and composites could be reused in 3D printing processes, converting waste into functional products. Balogun and Oladapo [28] evaluated the energy demands of 3D printing, suggesting optimized operation strategies for reducing power consumption during fabrication. In machining and fabrication processes, Gupta and Laubscher [29] proposed sustainable cutting strategies and lubrication methods to reduce material scrap and hazardous waste, particularly in titanium machining. These digital innovations are enhanced by IoT monitoring, which enables real-time feedback on production parameters and tool wear.

7.2.2.4 Sustainable Materials and Waste Repurposing in Construction and Manufacturing

Construction and manufacturing industries generate vast amounts of waste. Integrating recycled components into industrial inputs offers both economic and ecological benefits. Rojo-López et al. [30] experimented with quaternary blends of Portland cement, metakaolin, biomass ash, and granite powder to develop sustainable self-compacting concrete. Their study demonstrated how smart material combinations can achieve desired performance while minimizing virgin raw material consumption. Jones et al. [31] focused on UK construction firms and suggested strategic frameworks to enhance waste minimization and energy conservation, highlighting a mix of behavior change, training, and smart technology adoption.

7.2.2.5 Operational Scheduling and Supply Chain Efficiency

IIoT-based production scheduling and analytics can significantly improve material use by reducing delays, inventory surplus, and overproduction. Le Hesran et al. [32] conducted a comprehensive review on waste-aware scheduling algorithms, advocating for flexible systems that adjust in real time to production changes. Srai et al. [33] developed a capability maturity model for assessing sustainable supply networks. Their research emphasized aligning operations strategy with sustainability goals by evaluating supplier capabilities, coordination efficiency, and systemic resource optimization.

7.2.2.6 Overcoming Barriers to Smart, Circular Transitions

Despite promising technologies, many firms struggle to implement efficient waste management due to systemic, technical, or cultural challenges. Sharma et al. [34] identified key barriers to digital circular economy adoption in the food sector, including lack of awareness, insufficient investment, and workforce resistance. Ayçin and Kaya [35] applied fuzzy DEMATEL to categorize these barriers and visualize their interdependencies in Turkey's zero waste initiative. Their model enables policymakers to prioritize interventions for maximum systemic impact. Glavič [36] provided a macro perspective on the evolution of sustainable consumption and production, identifying coordination, regulatory alignment, and capacity-building as ongoing challenges. Hegab et al. [37] further proposed a framework of strategies and performance indicators to advance sustainable production systems across sectors.

7.2.3 Case Studies: Green Factories and Industry 4.0 Best Practices

Industry 4.0 technologies, such as big data, IoT, blockchain, cyber-physical systems, and digital twins, are key enablers in establishing green factories. These technologies support real-time monitoring, predictive analytics, intelligent automation, and sustainable decision-making, fostering operational excellence and ecological responsibility. This section presents real-world case studies and best practices demonstrating how industrial entities are successfully adopting smart, sustainable practices.

7.2.3.1 Distributed and Modular Smart Factories

Green factories thrive on adaptability and decentralization. Rauch et al. [38] proposed a distributed manufacturing network model where smart and agile mini-factories use local resources and energy-efficient configurations to reduce emissions from logistics and transport. These modular production setups are designed to be quickly reconfigurable and interoperable via IIoT platforms, supporting low-carbon production closer to demand hubs. Waibel et al. [39] further emphasized that smart production systems, through sensor-based feedback and autonomous scheduling, enhance sustainability by reducing idle machine time, energy wastage, and redundant operations. Their study highlights how smart production not only boosts throughput but also minimizes environmental footprint.

7.2.3.2 Human-centric and Ergonomic Manufacturing Practices

Industry 4.0 technologies are increasingly applied with a human-centered lens to improve worker safety, engagement, and overall social sustainability. Siew et al. [40] explored the integration of human-oriented disassembly and maintenance systems. Their work demonstrates that placing humans at the core of automated systems, through ergonomic feedback, co-bot (collaborative robot) configurations, and adaptive interfaces, leads to sustainable outcomes in terms of safety and job satisfaction. Eriksson et al. [41] moved beyond traditional Industry 4.0, proposing an Industry 5.0 framework that combines human-centric values with advanced technologies. Peruzzini and Pellicciari [42] developed a human factor assessment model that integrates social KPIs (Key Performance Indicators) into sustainability metrics for smart factory systems.

7.2.3.3 Digital Twins, AI, and Smart Process Optimization

Digital twin systems offer a digital replica of physical assets, enabling continuous monitoring, predictive maintenance, and energy optimization. Park et al. [43] detailed procedures for operating digital twins at the work-center level, highlighting their role in identifying inefficiencies and enabling real-time corrective actions that lead to energy and resource savings. Ma et al. [44] integrated big data analytics with digital twin architectures in energy-intensive industries to reduce emissions and enhance resource utilization. Meanwhile, Qi et al. [45] addressed implementation challenges of IIoT-based analytics in smart factories, calling for robust data governance frameworks to ensure consistency, accuracy, and sustainable decision-making.

7.2.3.4 Blockchain Applications in Sustainable Manufacturing

Blockchain technology provides a secure and transparent framework for supply chain traceability and material accountability. Su et al. [46] identified technical challenges and proposed a fuzzy decision-making support system to integrate blockchain into Industry 4.0 for sustainability. Their approach ensured traceability of material flow, enhancing compliance with environmental regulations and improving consumer trust. Jamwal et al. [47] combined blockchain with circular economy principles and Industry 4.0 tools, presenting a robust framework for handling data transparency and lifecycle tracking in sustainable manufacturing.

7.2.3.5 Circular Economy Practices and Lifecycle Sustainability

Green factories are increasingly evaluated based on their ability to close material loops and implement lifecycle-based optimization strategies. Selicati and Cardinale [48] showed that coupling exergy analysis with lifecycle assessment provided a comprehensive view of energy flows and environmental burdens, supporting green manufacturing decisions. Jamwal et al. [49] developed a sustainability assessment framework to evaluate Industry 4.0 readiness for circular production. Their multidimensional model considers economic, environmental, and social factors – an essential tool for industries aiming to embed sustainability at their core. In addition, Kerdlap et al. [50] provided an in-depth analysis of zero-waste manufacturing in Singapore, identifying barriers and proposing strategies for circular economy transitions, including digital infrastructure investment and policy alignment.

7.2.3.6 Sector-specific Case Studies

In the *textile industry*, González et al. [51] proposed big data-driven sustainability solutions, developing predictive models for optimizing water and energy use during dyeing and finishing processes. Their case study illustrates the power of Industry 4.0 in traditionally high-impact sectors. In *metalworking*, Baptista et al. (2018) [52] implemented the MAESTRI efficiency framework to monitor performance through the Total Efficiency Index, evaluating material, energy, and time use. It demonstrated tangible resource savings through digital monitoring and control loops. In addition, in *discrete manufacturing*, Zhang et al. [53] proposed an augmented Lagrangian coordination method for optimizing energy allocation in smart manufacturing services. Their model aligns production objectives with energy conservation targets, especially in multiagent systems. Jamwal et al. [49] also presented a comprehensive case for Indian SMEs, showing how integrated digitalization and circular practices can enhance competitiveness and sustainability. Their case, included in Table 7.1, illustrates how predictive models can guide sustainable production decisions and reduce environmental loads without sacrificing efficiency.

Table 7.1 Selected Industry 4.0 and green factory case studies.

Topic	Sector	Contribution	Key reference
Distributed Smart Factories	Manufacturing	Agile modular mini-factories	[38]
Human-centric Design	Cross-sector	Human-centered Industry 5.0 vision	[41]
Digital Twins & Energy Optimization	Precision Engineering	Real-time work center control	[43]
Blockchain for Traceability	Smart Manufacturing	Fuzzy blockchain decision support	[46]
Lifecycle Sustainability	Automotive	Coupling LCA & exergy for evaluation	[48]
Textile Big Data Sustainability	Textile	Big data in energy and water use	[51]

Note: The content in this table reflects a synthesized overview of core findings and practical insights drawn from the cited empirical studies.

7.3 Data-driven Decision-making and Environmental Monitoring

7.3.1 Integration of Big Data and ML for Real-time Analytics

In the era of Industry 4.0, the fusion of big data and ML within the IIoT ecosystems is transforming the manufacturing landscape. These technologies empower organizations to make real-time decisions, enhance operational efficiency, reduce waste, and predict system failures before they occur. This subsection presents the conceptual evolution, architectural frameworks, applied strategies, and use cases across sectors. Table 7.2 presents a curated overview of representative studies that demonstrate the diverse applications of big data and ML in real-time industrial analytics. The table highlights domain-specific innovations, ranging from predictive maintenance and sensor drift detection to autonomous farming and smart irrigation. Each case illustrates how data-driven technologies are operationalized within IIoT ecosystems to enhance sustainability, responsiveness, and decision-making efficiency.

7.3.1.1 Conceptual Foundations: Convergence of IIoT, Big Data, and ML

At the heart of real-time analytics lies the seamless convergence between data acquisition systems, computational platforms, and adaptive algorithms. As Ren et al. [54] highlight, industrial IoT sensors deployed in edge environments

Table 7.2 Representative case studies on big data and ML integration for real-time IIoT analytics.

Application domain	Key technologies	Contribution	Key references
Edge Computing & Smart Sensors	Edge AI, Big Data, IIoT	Developed a scalable sensor monitoring platform using edge computing and AI	[54, 73]
Precision Livestock Farming	Autonomic Computing, ML, IoT	Introduced a self-regulating big data framework for livestock process optimization	[55]
IoT Sensor Drift Detection	Drift Analytics, Streaming Data	Proposed methods for identifying temporal sensor drift patterns in IIoT	[59]
Digital Twin for PdM	Digital Twin, Predictive Analytics, RUL Estimation	Designed optimal RUL estimation using twin-based statistical models	[63]
Dairy Industry Optimization	AI, Real-time Monitoring, Dairy 4.0	Demonstrated the integration of Industry 4.0 in dairy production systems	[66]
Smart Irrigation Systems	IoT, ML, Data-driven Irrigation	Systematic review highlighting IoT-based precision irrigation in sub-Saharan Africa	[67]

Note: The entries in this table provide a structured overview of real-world implementations, highlighting critical insights and innovations as documented in the selected empirical studies.

generate massive amounts of heterogeneous data requiring fast preprocessing and intelligent interpretation. They developed a platform integrating edge computing and AI for scalable monitoring and decision support in industrial environments. García et al. [55] took this further with an autonomic computing approach in the beef production industry. Their system self-regulates data pipelines and applies ML models for precision livestock farming, illustrating how autonomous control loops can be driven by big data analytics.

To ensure operational robustness, Oks et al. [56] provided a structured taxonomy of cyber-physical systems in Industry 4.0, categorizing their role in integrating data, computation, and actuation at scale. This aligns with Hassan et al. [57], who surveyed ML applications for IIoT systems and emphasized the need for hybrid architectures that can process both structured and unstructured industrial data streams in real time. As shown in Figure 7.2, the AI/ML-driven decision loop represents a foundational model in IIoT systems for achieving real-time

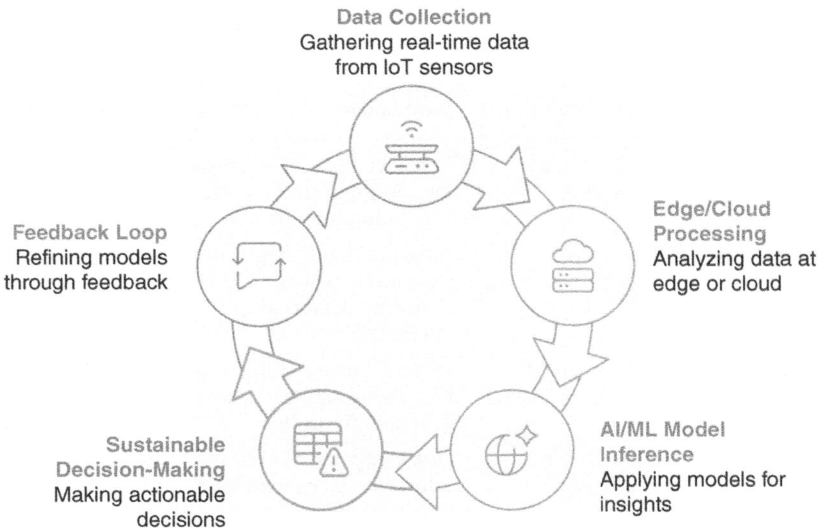

Figure 7.2 AI/ML-driven decision loop in industrial IoT (IIoT) systems.

responsiveness and sustainability. The process begins with data collection from IoT-enabled sensors, followed by edge or cloud-based processing for initial analysis. Subsequently, ML models are applied to generate predictive or prescriptive insights. These insights inform sustainable decision-making, such as optimizing energy use or predicting equipment failure. A feedback loop then refines these models based on operational outcomes, making the entire system more intelligent and adaptive over time. This closed-loop architecture is critical for aligning IIoT operations with circular economy principles and the broader goals of sustainable industrial transformation.

7.3.1.2 Real-time Monitoring and Smart Sensor Analytics

Sensor-driven monitoring is a foundational pillar of Industry 4.0. As industrial systems become more complex and connected, sensing technologies and real-time analytics must evolve accordingly. Liskiewicz et al. [58] described breakthroughs in tribological sensing that allow continuous assessment of wear, lubrication, and surface interaction, all in real time, greatly aiding predictive maintenance efforts. Addressing temporal variance in sensor outputs, Munirathinam [59] proposed drift detection analytics that account for evolving data patterns and prevent false positives. Similarly, Chiu et al. [60] proposed an integrative ML model embedded in a cyber-physical system, significantly improving fault detection accuracy and system responsiveness. Lin and Chen [61] designed a computational intelligence-based decision support system specifically for injection molding, demonstrating how live machine feedback and ML algorithms can work together to achieve zero-defect manufacturing.

7.3.1.3 Predictive Maintenance and Anomaly Detection

ML and big data form the backbone of next-generation PdM systems, where real-time data is used to forecast equipment failures and optimize resource planning. Rousopoulou et al. [62] implemented a cognitive analytics framework to monitor injection molding machines. By continuously evaluating sensor data, their model detected anomalies and proactively recommended actions. Anis et al. [63] offered a case where digital twin systems combined with optimal remaining useful life (RUL) estimation models provided proactive maintenance strategies in complex manufacturing settings. Complementarily, Chouliaras and Sotiriadis [64] developed a resource usage-based real-time anomaly detection system for NoSQL databases using ML techniques, demonstrating the applicability of real-time analytics in backend infrastructure as well. Innovatively, Saidy et al. [65] applied statistical process control (SPC) charts within PdM systems. Their unconventional use of SPC in predictive analytics highlights how traditional methods, when infused with ML capabilities, can support intelligent fault management.

7.3.1.4 Industry-specific Applications and Innovations

The fusion of big data and ML is not limited to manufacturing – it has transformed sectors such as food production, agriculture, law, and even tourism:

- **Dairy Production**: Hassoun et al. [66] introduced "Dairy 4.0," illustrating how sensors and ML algorithms optimize milk quality and yield, creating value across the dairy supply chain.
- **Smart Agriculture**: Wanyama et al. [67] conducted a systematic review revealing how real-time sensor data supports smart irrigation systems in sub-Saharan Africa – enhancing water efficiency and sustainability.
- **Legal Informatics**: Agrawal et al. [68] developed a data-centric framework that uses big data analytics to streamline regulatory processes, monitor legal compliance, and digitize workflows.
- **Smart Tourism**: Tsaih and Hsu [69] conceptualized an AI-based tourism framework, where predictive models suggest experiences and manage flow, showcasing the wide reach of real-time analytics beyond factory floors.

7.3.1.5 Challenges, Limitations, and Future Research Pathways

Despite their benefits, real-time big data and ML systems face challenges, including:

- **Data heterogeneity and noise**, requiring robust preprocessing.
- **Latency constraints**, especially in mission-critical applications.
- **Interpretability and trust**, as most ML models still function as black boxes.

Rathore and Sarin [70] explored the design principles of next-generation analytics platforms for biopharmaceuticals, advocating for systems that are both computationally lean and explainable. Yan and Li [71], through bibliometric analysis, identified key research trends, pointing to the increasing need for federated learning, privacy-preserving analytics, and cross-platform data fusion. Agrawal and Nargund [72] emphasized deep learning as a transformative element in manufacturing, while simultaneously calling for its integration with human-in-the-loop architectures and sustainability metrics to ensure transparent and ethical AI adoption.

7.3.2 Applications in Sectors such as Mining, Textiles, and Automotive

Industrial IoT and data-driven solutions are rapidly being adopted across various domains to enhance sustainability, operational intelligence, and efficiency. Sectors such as mining, textiles, and automotive have distinct production demands, yet they converge in leveraging IIoT, AI, and big data technologies to address issues such as resource optimization, emissions tracking, predictive maintenance, and personalized production. This subsection details how these technologies are being applied and evaluated across these three sectors.

7.3.2.1 Mining and Energy-intensive Industries

Mining and process manufacturing represent some of the most energy- and resource-intensive industrial sectors. The integration of IIoT and big data analytics enables optimization of material flows, equipment utilization, and predictive maintenance, which are crucial for sustainability in mining operations. Ma et al. [44] proposed a digital twin and big data-driven framework specifically designed for energy-intensive manufacturing environments such as mining and metallurgy. Their system utilizes real-time energy consumption data to adjust operations dynamically, leading to reduced energy waste and lower carbon emissions. Reis and Kenett [74] evaluated the value of information (VoI) in process industries by measuring how data-centric strategies affect decision quality in the chemical sector. Their work laid the foundation for data-informed decision frameworks in continuous-flow industries like mining. Wen et al. [75] further introduced a deep transfer learning model for fault diagnosis in mining equipment, improving predictive maintenance accuracy through advanced sparse autoencoders.

7.3.2.2 Textile and Fashion Industry

The textile industry faces major sustainability challenges related to overproduction, water usage, and fast-changing consumer preferences. By adopting IIoT and

AI-driven systems, the sector is shifting toward more personalized, efficient, and responsive production. Sharma et al. [76] developed an intelligent recommendation system that utilizes consumer data to generate personalized fashion designs, reducing material waste by matching production with actual demand. Their sensor-based system represents a shift toward data-informed garment manufacturing. In addition, Syafrudin et al. [77] built a real-time open-source data processing architecture tailored for sustainable manufacturing, which has direct applicability to textile factories focused on just-in-time (JIT) models and waste minimization. Fareri et al. [78] further used text mining methods to study how Industry 4.0 technologies impact job profiles in the textile and apparel sector, revealing a clear trend toward upskilled roles and automation-supported creativity.

7.3.2.3 Automotive and Smart Manufacturing

The automotive industry has been a frontrunner in adopting smart technologies for predictive maintenance, quality control, and adaptive production. With the rise of electric and autonomous vehicles, sustainability and data-driven operations have become even more critical. Manimuthu et al. [79] presented a federated AI-based automobile assembly model, incorporating smart contracts to securely share data across stakeholders. This approach enhances transparency, real-time fault detection, and flexible supply chain orchestration. Additionally, Bazi et al. [80] developed a hybrid CNN-BiLSTM model to monitor tool wear in automotive manufacturing, improving machining accuracy and reducing energy usage. Lorenz et al. [81] used process mining in make-to-stock automotive production to identify process inefficiencies and redesign workflows – demonstrating how IIoT analytics improves both productivity and sustainability. Cerquitelli et al. [82] also offered a comprehensive review of methodologies and tools for data-driven manufacturing, many of which have been applied to automotive production lines for optimization of predictive scheduling and real-time analytics.

7.3.2.4 Cross-sectoral Innovation and Interoperability

Although sector-specific applications are evolving, many technologies are converging toward modular, interoperable, and scalable platforms. Ma et al. [83] proposed a predictive production planning system based on big data analytics, particularly targeting energy-intensive automotive and mining sectors. Chiang et al. [84] highlighted the broader implications of big data in chemical engineering, many of which are transferable to automotive materials processing, from paint to adhesives. Fisher et al. [85] further provided practical considerations in deploying data-driven systems for process manufacturing, stressing the importance of robust data pipelines, domain-specific model tuning, and operational safety, all of which are applicable across the mining, textile, and automotive industries.

7.4 Cybersecurity, Ethics, and Governance in Sustainable IIoT Systems

7.4.1 Risks and Resilience: Addressing Cybersecurity in Interconnected Industrial Systems

As industrial environments evolve into hyperconnected ecosystems, the cybersecurity landscape faces growing complexity and urgency. The IIoT brings not only benefits in automation and efficiency but also expanded attack surfaces and systemic vulnerabilities [86, 87]. This section explores current strategies, architectural innovations, and frameworks for managing cybersecurity risks and building resilience across IIoT systems.

7.4.1.1 Evolving Threat Landscape in IIoT

The convergence of cyber and physical domains introduces multidimensional risks that traditional IT systems were not designed to handle. Ward and Janczewski [88] highlighted the importance of multidomain risk factor identification through knowledge synthesis, revealing threats specific to interconnected assets. Adaros-Boye et al. [89] proposed a continuous assessment method tailored to industrial contexts, noting that dynamic asset environments require ongoing risk evaluation rather than periodic audits. Petrenko et al. [90] supplemented this with methodological recommendations for practical cyber risk management workflows in IIoT-heavy sectors.

7.4.1.2 Architectural Solutions for Resilient IIoT Systems

Architectural innovation is crucial to enhancing IIoT system security. El-Hajj and Gebremariam [91] introduced the ASCON-based digital twin security model, which strengthens resilience by isolating and monitoring interactions in real time. Similarly, Vijayakumaran et al. [92] designed a next-generation cybersecurity architecture with layered defenses, adaptive control systems, and AI-based anomaly detection. Sen and Song [93] outlined a networked industrial control system architecture designed for secure IIoT integration, while Nakkeeran and Mathi [94] proposed a generalized security framework that addresses common attack vectors such as spoofing, denial-of-service, and man-in-the-middle exploits.

7.4.1.3 AI and Blockchain-driven Defense Mechanisms

AI and blockchain technologies are rapidly becoming integral to modern industrial security infrastructures. Li et al. [95] discussed how DL models can detect and respond to anomalies faster than rule-based systems. Rahman and Hossain [96] envisioned software-defined security (SDS) architectures for 6G-enabled IIoT, where security configurations adapt dynamically based on real-time data analysis.

Umer et al. [97] proposed a blockchain-based provenance framework integrated with AI to ensure transparency and traceability in cloud manufacturing. This fusion enables trust in distributed IIoT environments, where physical asset data must be verified across multiple platforms. In addition, Zanasi et al. [98] introduced a flexible zero trust architecture, emphasizing continuous authentication, micro-segmentation, and role-based access as cornerstones of IIoT cybersecurity.

7.4.1.4 Standards, Policies, and Continuous Risk Management

Cybersecurity in IIoT requires alignment with global standards and the development of continuous monitoring policies. Buja et al. [99] conducted a systematic review of existing cybersecurity standards for IIoT, identifying gaps and inconsistencies across protocols. Adaros-Boye et al. [100] elaborated a methodological view of continuous risk management, integrating operational metrics into decision-making tools. Hansch et al. [101] proposed a framework for deriving impact-driven security requirements, linking business-critical processes to cyber protection needs. Additionally, Ahmad et al. [102] provided a comprehensive survey on communication security in "Industry X," stressing the need for interoperable security protocols across cloud, edge, and on-premise systems. Kalinin et al. [103] offered a unique angle by connecting cybersecurity assessments to smart city infrastructure risks – illustrating the societal scale of industrial cyber threats.

7.4.1.5 Toward Proactive, Resilient, and Adaptive Cybersecurity

Looking ahead, the emphasis is shifting from reactive defence to proactive resilience and adaptive systems. Tariq [104] suggested a zero-tolerance cybersecurity paradigm for enterprise-specific IIoT deployments, while Gopal et al. [105] pointed out the emerging fin-tech cybersecurity challenges as industrial and financial networks increasingly overlap. The growing integration of IIoT into national infrastructure, logistics, energy systems, and healthcare mandates a cybersecurity mindset grounded in resilience, redundancy, and autonomy, a holistic view where systems self-monitor, self-heal, and co-evolve with threats.

7.4.2 Policies and Global Standards Supporting Sustainable IIoT Adoption

The rapid integration of IIoT technologies into manufacturing, logistics, and energy systems has prompted a surge in global policy development and standardization. For IIoT systems to truly enable sustainable development, they must comply with comprehensive frameworks that address energy efficiency, cybersecurity, data governance, and interoperability. This subsection explores current global policy efforts, emerging governance models, and regulatory mechanisms that underpin sustainable IIoT growth.

7.4.2.1 Frameworks for Governance, ESG Alignment, and Sustainable Production

A growing number of countries and international agencies are adopting Environment, Social, and Governance (ESG) metrics as foundational principles for IIoT regulation. Chen et al. [106] introduced a big data-driven methodology to monitor machine-level energy consumption in IIoT contexts, aligned with ESG reporting goals. Liu et al. [107] analyzed the dual nature of IIoT adoption, noting that while it promotes green transformation, it can also exacerbate digital and energy resource inequalities, highlighting the need for balanced, policy-aligned deployments. Arfelis et al. [108] further provided a case of IIoT integration in circular economy-based plastics management, showing how industry-specific policy design and compliance frameworks facilitate sustainable resource use.

7.4.2.2 Standardization Initiatives and Interoperability Protocols

The success of IIoT systems depends on interoperable communication standards and regulatory consistency across regions. Barbieri et al. [109] analyzed federated learning implementations and their energy footprints, advocating for standardized data exchange formats and regulatory thresholds to ensure both privacy and sustainability. Martins et al. [110] emphasized the integration of network slicing, ML, and sustainability criteria into IIoT communication architectures. Their work recommends policy guidance for energy-aware AI implementation in real-time industrial networks. Additionally, Yang and Xiong [111] proposed a dual access control framework for IIoT cloud environments, aiming to meet future regulatory standards on decentralization and efficiency.

7.4.2.3 Sector-specific Policy Developments

Policies are increasingly tailored to sectoral needs. Almadani and Mostafa [112] introduced a communication model for agro-industries based on IIoT protocols to enhance data flow and traceability – key requirements under smart agriculture regulations. Rafiq et al. [113] explored emergency data transmission and offloading strategies in energy networks, aligned with energy policy frameworks in 6TiSCH-based IoT systems. Philipp et al. [114] also studied green seaport transformation policies, revealing how automation, renewable energy, and AI control systems can co-evolve with logistics governance models.

7.4.2.4 Policy-driven Digital Identity, Privacy, and Ethical AI

The convergence of IIoT and AI raises concerns about data sovereignty, ethical decision-making, and identity management. Magd et al. [115] advocated for explainable AI (XAI) frameworks that align with ethical governance policies, particularly in society 5.0 contexts. Onamade and Okpaegbe [116] examined the role of self-sovereign identities (SSI) in blockchain-secured IIoT systems, suggesting that privacy-by-design policies can future-proof identity frameworks. Alabi et al. [117] also investigated the influence of digital strategies on SME sustainability efforts, showing how regulatory incentives (e.g. digital transition grants) influence IIoT adoption among smaller firms.

7.4.2.5 Global and Regional Policy Alignment with Agenda 2030

International policy frameworks such as the UN Agenda 2030 and SDGs increasingly shape national strategies for IIoT innovation. Lopez et al. [118] proposed the use of small satellite technologies to support industrial policies in Latin America aligned with Agenda 2030. Andersen et al. [119] evaluated the intersection of big data and IIoT in governance, highlighting how national strategies for smart industry and digital sovereignty influence infrastructure investments. Furthermore, Abir et al. [120] envisioned Industry 4.0 as a pathway toward secure, green, and circular economies, calling for global data governance protocols to ensure fairness, transparency, and shared innovation benefits.

7.5 Future Outlook: Scalable and Resilient IIoT Ecosystems for 2030 and Beyond

The evolution of IIoT is tightly coupled with the rapid advancement of digital twins, edge AI, and FL. These technologies are enabling next-generation industrial systems to be more autonomous, scalable, secure, and sustainable. This section outlines the role and applications of these emerging technologies in the context of sustainable industrial operations.

7.5.1 Digital Twins for Predictive and Sustainable Operations

Digital twins, virtual representations of physical assets, are foundational in achieving operational efficiency, predictive maintenance, and real-time sustainability feedback loops. Augustine [121] categorized real-world industrial use cases of digital twins, such as production line simulations and remote asset monitoring. These models enable energy-efficient configurations and proactive fault mitigation. Tang et al. [122] applied digital twin-assisted deep reinforcement learning for dynamic resource allocation in network slicing scenarios within Industry 4.0, improving responsiveness while reducing computational waste. Similarly, Bandara and Buics [123] presented a comprehensive review of digital twins in sustainable supply chains, identifying enabling technologies and challenges for adoption. Samanta et al. [124, 125] further showcased an integrated model combining digital twins and blockchain to secure smart city infrastructure, while Shah et al. [126] documented cross-domain use cases for digital twins in predictive analytics and remote control systems.

7.5.2 Edge AI for Distributed and Low-latency Intelligence

Edge AI is transforming IIoT by bringing intelligence closer to data sources, significantly reducing energy consumption and latency. Bellavista and Mora [127] described how edge cloud platforms enable distributed AI in IIoT applications, referencing the IoTwins project for smart manufacturing deployments. Yacelga et al. [128] conducted an in-depth review on fog computing and its synergy with edge AI for industrial use, highlighting challenges such as interoperability and power efficiency. Sharma et al. [129] also emphasized ML and DL models deployed at the network edge to secure 5G-enabled IIoT environments, making real-time threat detection feasible without cloud reliance. Furthermore, Bhadra et al. [130] proposed a novel convergence of cognitive IoT and robotic process automation that leverages edge intelligence for dynamic industrial workflows, reinforcing the concept of Industry 4.0+.

7.5.3 Federated Learning and Privacy-preserving AI in IIoT

As data privacy and governance become central to sustainable digital ecosystems, FL offers a distributed model training framework without sharing raw data across sites. While direct implementation examples are still nascent, Eswaran et al. [131] proposed integration strategies for FL within Industry 6.0 to promote AI-driven sustainability and data autonomy. Gilles et al. [132] addressed the protection of geo-distributed computing systems, a core enabler of federated learning, by suggesting security protocols that can be embedded at edge nodes to manage privacy-preserving collaboration. Vanickis et al. [133] provided early architectural considerations for zero-trust networking, which underpins secure federated architectures, especially in multitenant industrial environments.

7.6 Conclusion

The integration of IIoT technologies within the sustainable development paradigm marks a critical turning point for modern industry. Through this chapter, we have illustrated how IIoT acts as both a technological backbone and a strategic enabler of greener, more resilient industrial operations. From energy optimization and predictive maintenance in smart manufacturing to waste reduction through circular design, the deployment of IIoT has the potential to transform how we produce, consume, and regenerate industrial resources. By examining data-driven environmental monitoring systems, we highlighted how IIoT, big data, and ML collectively enable organizations to monitor emissions, reduce carbon footprints, and respond rapidly to environmental feedback. Our discussion on sectoral applications in mining, textiles, and automotive further reinforces the adaptability of IIoT across diverse industrial contexts. In parallel, we addressed the increasing need for robust cybersecurity frameworks, ethical AI use, and harmonized policy guidelines to ensure that IIoT deployments are secure, equitable, and aligned with ESG objectives. Finally, our exploration of emerging innovations, such as digital twins, edge AI, and FL, indicates that the future of IIoT lies in decentralized, privacy-preserving, and intelligent systems that are deeply integrated with global sustainability goals. Ultimately, this chapter contributes to the growing discourse on aligning advanced technologies with the principles of sustainable development. It offers practical insights, theoretical underpinnings, and strategic guidance for policymakers, researchers, and industrial stakeholders working toward a climate-conscious and tech-enabled future.

References

1 Beier, G., Niehoff, S., and Xue, B. (2018). More sustainability in industry through industrial internet of things? *Applied Sciences* 8 (2): 219.

2 Manavalan, E., and Jayakrishna, K. (2019). A review of Internet of Things (IoT) embedded sustainable supply chain for industry 4.0 requirements. *Computers & Industrial Engineering* 127: 925–953.

3 Hatzivasilis, G., Fysarakis, K., Soultatos, O., et al. (2018). The industrial Internet of Things as an enabler for a circular economy Hy-LP: a novel IIoT protocol, evaluated on a wind park's SDN/NFV-enabled 5G industrial network. *Computer Communications* 119: 127–137.

4 Narula, S., Puppala, H., Kumar, A., et al. (2021). Applicability of Industry 4.0 technologies in the adoption of global reporting initiative standards for achieving sustainability. *Journal of Cleaner Production* 305: 127141.

5 Tiwari, K., and Khan, M.S. (2020). Sustainability accounting and reporting in the Industry 4.0. *Journal of Cleaner Production* 258: 120783.

6 Fraga-Lamas, P., Lopes, S.I., and Fernández-Caramés, T.M. (2021). Green IoT and edge AI as key technological enablers for a sustainable digital transition towards a smart circular economy: an Industry 5.0 use case. *Sensors* 21 (17): 5745.

7 Leng, J., Chen, Z., Huang, Z., et al. (2022). Secure blockchain middleware for decentralized IIoT towards industry 5.0: a review of architecture, enablers, challenges, and directions. *Machines* 10 (10): 858.

8 Kerkeni, R., Khlif, S., Mhalla, A., and Bouzrara, K. (2024). Digital twin applied to predictive maintenance for Industry 4.0. *Journal of Nondestructive Evaluation, Diagnostics and Prognostics of Engineering Systems* 7 (4): 041008.

9 Rana, R., and Bhambri, P. (2024). Digital twin for sustainable industrial development. *AI-Driven Digital Twin and Industry 4.0* 241–253.

10 Bhargava, A., Bhargava, D., Kumar, P.N., et al. (2022). Industrial IoT and AI implementation in vehicular logistics and supply chain management for vehicle mediated transportation systems. *International Journal of System Assurance Engineering and Management* 13 (Suppl 1): 673–680.

11 De Luca, R., Ferraro, A., Galli, A., et al. (2023). A deep attention based approach for predictive maintenance applications in IoT scenarios. *Journal of Manufacturing Technology Management* 34 (4): 535–556.

12 Li, Z. (2023). Condition-based monitoring of biogas compressor using nonintrusive sensors and machine learning. In: *Residuals and Biosolids 2023.* Water Environment Federation.

13 Agostinelli, S., and Cumo, F. (2022). Machine learning approach for predictive maintenance IN an advanced building management system. *Energy Production and Management in the 21st Century V: The Quest for Sustainable Energy* 255: 131.

14 Golec, M., Wu, H., Ozturac, R., et al. (2024). Captain: a testbed for co-simulation of scalable serverless computing environments for AIoT enabled predictive maintenance in industry 4.0. *IEEE Internet of Things Journal* 12 (16): 32283–32295.

15 Ranganathan, C.S., Riazulhameed, A.A.M.A., Saravanan, S., et al. (2024). IoT based real-time moisture management in fabric drying processes. In: *2024 First International Conference on Innovations in Communications, Electrical and Computer Engineering (ICICEC)* (pp. 1–6). IEEE.

16 Fatorachian, H., and Pawar, K. (2025). Enhancing waste management in cold supply chains via digital transformation. *Procedia Computer Science* 253: 167–176.

17 Raiker, G.A., Loganathan, U., Agrawal, S., et al. (2020). Energy disaggregation using energy demand model and IoT-based control. *IEEE Transactions on Industry Applications* 57 (2): 1746–1754.

18 Joshi, S., and Sharma, M. (2024). Intelligent algorithms and methodologies for low-carbon smart manufacturing: review on past research, recent developments and future research directions. *IET Collaborative Intelligent Manufacturing* 6 (1): e12094.

19 Mazumdar, B.D., Mishra, M., Ghatak, A., and Verma, V.K. (2024). AI for industry 4.0 with real-world problems. In: *Data Analytics and Artificial Intelligence for Predictive Maintenance in Smart Manufacturing* (pp. 326–352). CRC Press.

20 Chen, M., Sinha, A., Hu, K., and Shah, M.I. (2021). Impact of technological innovation on energy efficiency in Industry 4.0 era: moderation of shadow economy in sustainable development. *Technological Forecasting and Social Change* 164: 120521.

21 Nudurupati, S.S., Budhwar, P., Pappu, R.P., et al. (2022). Transforming sustainability of Indian small and medium-sized enterprises through circular economy adoption. *Journal of Business Research* 149: 250–269.

22 Mboli, J.S., Thakker, D., and Mishra, J.L. (2022). An internet of things-enabled decision support system for circular economy business model. *Software: Practice and Experience* 52 (3): 772–787.

23 Singh, P.J., and Power, D. (2009). The nature and effectiveness of collaboration between firms, their customers and suppliers: a supply chain perspective. *Supply Chain Management: An International Journal* 14 (3): 189–200.

24 Schmidt, M. (2015). The interpretation and extension of Material Flow Cost Accounting (MFCA) in the context of environmental material flow analysis. *Journal of Cleaner Production* 108: 1310–1319.

25 Broadbent, C. (2016). Steel's recyclability: demonstrating the benefits of recycling steel to achieve a circular economy. *The International Journal of Life Cycle Assessment* 21: 1658–1665.

26 Tecchio, P., McAlister, C., Mathieux, F., and Ardente, F. (2017). In search of standards to support circularity in product policies: a systematic approach. *Journal of Cleaner Production* 168: 1533–1546.

27 Olawumi, M.A., Oladapo, B.I., Ikumapayi, O.M., and Akinyoola, J.O. (2023). Waste to wonder to explore possibilities with recycled materials in 3D printing. *Science of the Total Environment* 905: 167109.

28 Balogun, V.A., and Oladapo, B.I. (2016). Electrical energy demand modeling of 3D printing technology for sustainable manufacture. *International Journal of Engineering* 29 (7): 1–8.

29 Gupta, K., and Laubscher, R.F. (2017). Sustainable machining of titanium alloys: a critical review. *Proceedings of the Institution of Mechanical Engineers, Part B: Journal of Engineering Manufacture* 231 (14): 2543–2560.

30 Rojo-López, G., Nunes, S., González-Fonteboa, B., and Martínez-Abella, F. (2020). Quaternary blends of Portland cement, metakaolin, biomass ash and granite powder for production of self-compacting concrete. *Journal of Cleaner Production* 266: 121666.

31 Jones, J., Jackson, J., Tudor, T., and Bates, M. (2012). Strategies to enhance waste minimization and energy conservation within organizations: a case study from the UK construction sector. *Waste Management & Research* 30 (9): 981–990.

32 Le Hesran, C., Ladier, A.L., Botta-Genoulaz, V., and Laforest, V. (2019). Operations scheduling for waste minimization: a review. *Journal of Cleaner Production* 206: 211–226.

33 Srai, J.S., Alinaghian, L.S., and Kirkwood, D.A. (2013). Understanding sustainable supply network capabilities of multinationals: a capability maturity model approach. *Proceedings of the Institution of Mechanical Engineers, Part B: Journal of Engineering Manufacture* 227 (4): 595–615.

34 Sharma, M., Joshi, S., and Govindan, K. (2023). Overcoming barriers to implement digital technologies to achieve sustainable production and consumption in the food sector: a circular economy perspective. *Sustainable Production and Consumption* 39: 203–215.

35 Ayçin, E., and Kayapinar Kaya, S. (2021). Towards the circular economy: analysis of barriers to implementation of Turkey's zero waste management using the fuzzy DEMATEL method. *Waste Management & Research* 39 (8): 1078–1089.

36 Glavič, P. (2021). Evolution and current challenges of sustainable consumption and production. *Sustainability* 13 (16): 9379.

37 Hegab, H., Shaban, I., Jamil, M., and Khanna, N. (2023). Toward sustainable future: strategies, indicators, and challenges for implementing sustainable production systems. *Sustainable Materials and Technologies* 36: e00617.

38 Rauch, E., Dallasega, P., and Matt, D.T. (2017). Distributed manufacturing network models of smart and agile mini-factories. *International Journal of Agile Systems and Management* 10 (3–4): 185–205.

39 Waibel, M.W., Steenkamp, L.P., Moloko, N., and Oosthuizen, G.A. (2017). Investigating the effects of smart production systems on sustainability elements. *Procedia Manufacturing* 8: 731–737.

40 Siew, C.Y., Chang, M.M.L., Ong, S.K., and Nee, A.Y. (2020). Human-oriented maintenance and disassembly in sustainable manufacturing. *Computers & Industrial Engineering* 150: 106903.

41 Eriksson, K.M., Olsson, A.K., and Carlsson, L. (2024). Beyond lean production practices and Industry 4.0 technologies toward the human-centric Industry 5.0. *Technological Sustainability* 3 (3): 286–308.

42 Peruzzini, M., and Pellicciari, M. (2017). A human factors assessment model for sustainable manufacturing. *International Journal of Agile Systems and Management* 10 (3–4): 206–230.

43 Park, K.T., Lee, D., and Noh, S.D. (2020). Operation procedures of a work-center-level digital twin for sustainable and smart manufacturing. *International Journal of Precision Engineering and Manufacturing-Green Technology* 7 (3): 791–814.

44 Ma, S., Ding, W., Liu, Y., et al. (2022). Digital twin and big data-driven sustainable smart manufacturing based on information management systems for energy-intensive industries. *Applied Energy* 326: 119986.

45 Qi, Q., Xu, Z., and Rani, P. (2023). Big data analytics challenges to implementing the intelligent Industrial Internet of Things (IIoT) systems in sustainable manufacturing operations. *Technological Forecasting and Social Change* 190: 122401.

46 Su, D., Zhang, L., Peng, H., et al. (2023). Technical challenges of blockchain technology for sustainable manufacturing paradigm in Industry 4.0 era using a fuzzy decision support system. *Technological Forecasting and Social Change* 188: 122275.

47 Jamwal, A., Agrawal, R., and Sharma, M. (2022). A framework to overcome blockchain enabled sustainable manufacturing issues through circular economy and industry 4.0 measures. *International Journal of Mathematical, Engineering and Management Sciences* 7 (6): 764.

48 Selicati, V., and Cardinale, N. (2021). The benefits in coupling exergy analysis and life cycle assessment in the context of sustainable manufacturing for Industry 4.0: a real industrial case. *International Journal of Heat & Technology* 39 (1): 12–22.

49 Jamwal, A., Agrawal, R., Sharma, M., et al. (2021). Developing a sustainability framework for Industry 4.0. *Procedia Cirp* 98: 430–435.

50 Kerdlap, P., Low, J.S.C., and Ramakrishna, S. (2019). Zero waste manufacturing: a framework and review of technology, research, and implementation barriers for enabling a circular economy transition in Singapore. *Resources, Conservation and Recycling* 151: 104438.

51 González, G.A., Hernández, A.S., Vásquez, M.G., et al. (2022). Sustainable manufacturing in the fourth industrial revolution: a big data application proposal in the textile industry. *Journal of Industrial Engineering and Management* 15 (4): 614–636.

52 Baptista, A.J., Lourenço, E.J., Silva, E.J., et al. (2018). MAESTRI efficiency framework: the concept supporting the total efficiency index. Application case study in the metalworking sector. *Procedia CIRP* 69: 318–323.

53 Zhang, G., Wang, G., Chen, C.H., et al. (2021). Augmented Lagrangian coordination for energy-optimal allocation of smart manufacturing services. *Robotics and Computer-Integrated Manufacturing* 71: 102161.

54 Ren, S., Kim, J.S., Cho, W.S., et al. (2021). Big data platform for intelligence industrial IoT sensor monitoring system based on edge computing and AI. In: *2021 International Conference on Artificial Intelligence in Information and Communication (ICAIIC)* (pp. 480–482). IEEE.

55 García, R., Aguilar, J., Toro, M., et al. (2023). Autonomic computing in a beef-production process for precision livestock farming. *Journal of Industrial Information Integration* 31: 100425.

56 Oks, S.J., Jalowski, M., Lechner, M., et al. (2022). Cyber-physical systems in the context of Industry 4.0: a review, categorization and outlook. *Information Systems Frontiers* 26: 1731–1772.

57 Hassan, M.B., Ahmed, E.S.A., and Saeed, R.A. (2021). Machine learning for industrial IoT systems. In: *Handbook of Research on Innovations and Applications of AI, IoT, and Cognitive Technologies* (pp. 336–358). IGI Global.

58 Liskiewicz, T., Sherrington, I., Khan, T., and Liu, Y. (2023). Advances in sensing for real-time monitoring of tribological parameters. *Tribology International* 189: 108965.

59 Munirathinam, S. (2021). Drift detection analytics for IoT sensors. *Procedia Computer Science* 180: 903–912.

60 Chiu, M.C., Tsai, C.D., and Li, T.L. (2020). An integrative machine learning method to improve fault detection and productivity performance in a cyber-physical system. *Journal of Computing and Information Science in Engineering* 20 (2): 021009.

61 Lin, J.S., and Chen, K.H. (2024). A novel decision support system based on computational intelligence and machine learning: towards zero-defect manufacturing in injection molding. *Journal of Industrial Information Integration* 40: 100621.

62 Rousopoulou, V., Nizamis, A., Vafeiadis, T., et al. (2020). Predictive maintenance for injection molding machines enabled by cognitive analytics for Industry 4.0. *Frontiers in Artificial Intelligence* 3: 578152.

63 Anis, M.D., Taghipour, S., and Lee, C.G. (2020). Optimal RUL estimation: a state-of-art digital twin application. In: *2020 Annual Reliability and Maintainability Symposium (RAMS)* (pp. 1–7). IEEE.

64 Chouliaras, S., and Sotiriadis, S. (2019). Real-time anomaly detection of NoSQL systems based on resource usage monitoring. *IEEE Transactions on Industrial Informatics* 16 (9): 6042–6049.

65 Saidy, C., Xia, K., Kircaliali, A., et al. (2020). The application of statistical quality control methods in predictive maintenance 4.0: an unconventional use of statistical process control (SPC) charts in health monitoring and predictive analytics. In: *Advances in Asset Management and Condition Monitoring: COMADEM 2019* (pp. 1051–1061). Cham: Springer International Publishing.

66 Hassoun, A., Garcia-Garcia, G., Trollman, H., et al. (2023). Birth of Dairy 4.0: opportunities and challenges in adoption of fourth industrial revolution technologies in the production of milk and its derivatives. *Current Research in Food Science* 7: 100535.

67 Wanyama, J., Bwambale, E., Kiraga, S., et al. (2024). A systematic review of fourth industrial revolution technologies in smart irrigation: constraints, opportunities, and future prospects for sub-Saharan Africa. *Smart Agricultural Technology* 7: 100412.

68 Agrawal, S., Sahu, A., and Kumar, G. (2022). A conceptual framework for the implementation of Industry 4.0 in legal informatics. *Sustainable Computing: Informatics and Systems* 33: 100650.

69 Tsaih, R.H., and Hsu, C.C. (2018). Artificial intelligence in smart tourism: a conceptual framework. *Proceedings of the 18th International Conference on Electronic Business* (pp. 124133). ICEB, Guilin, China, December 2–6.

70 Rathore, A.S., and Sarin, D. (2024). What should next-generation analytical platforms for biopharmaceutical production look like? *Trends in Biotechnology* 42 (3): 282–292.

71 Yan, J.H., and Li, B.L. (2020). Research hotspots and tendency of intelligent manufacturing. *Chinese Science Bulletin* 8: 684–694.

72 Agrawal, K., and Nargund, N. (2024). Deep learning in Industry 4.0: transforming manufacturing through data-driven innovation. In: *International Conference on Distributed Computing and Intelligent Technology* (pp. 222–236). Cham: Springer Nature Switzerland.

73 Agrawal, K., Goktas, P., Sahoo, B., et al. (2024). IoT-based service allocation in edge computing using game theory. In: *International Conference on Distributed Computing and Intelligent Technology* (pp. 45–60). Cham: Springer Nature Switzerland.

74 Reis, M.S., and Kenett, R. (2018). Assessing the value of information of data-centric activities in the chemical processing Industry 4.0. *AIChE Journal* 64 (11): 3868–3881.

75 Wen, L., Gao, L., and Li, X. (2017). A new deep transfer learning based on sparse auto-encoder for fault diagnosis. *IEEE Transactions on Systems, Man, and Cybernetics: Systems* 49 (1): 136–144.

76 Sharma, S., Koehl, L., Bruniaux, P., et al. (2021). Development of an intelligent data-driven system to recommend personalized fashion design solutions. *Sensors* 21 (12): 4239.

77 Syafrudin, M., Fitriyani, N.L., Li, D., et al. (2017). An open source-based real-time data processing architecture framework for manufacturing sustainability. *Sustainability* 9 (11): 2139.

78 Fareri, S., Fantoni, G., Chiarello, F., et al. (2020). Estimating Industry 4.0 impact on job profiles and skills using text mining. *Computers in Industry* 118: 103222.

79 Manimuthu, A., Venkatesh, V.G., Shi, Y., et al. (2022). Design and development of automobile assembly model using federated artificial intelligence with smart contract. *International Journal of Production Research* 60 (1): 111–135.

80 Bazi, R., Benkedjouh, T., Habbouche, H., et al. (2022). A hybrid CNN-BiLSTM approach-based variational mode decomposition for tool wear monitoring. *The International Journal of Advanced Manufacturing Technology* 119 (5): 3803–3817.

81 Lorenz, R., Senoner, J., Sihn, W., and Netland, T. (2021). Using process mining to improve productivity in make-to-stock manufacturing. *International Journal of Production Research* 59 (16): 4869–4880.

82 Cerquitelli, T., Pagliari, D.J., Calimera, A., et al. (2021). Manufacturing as a data-driven practice: methodologies, technologies, and tools. *Proceedings of the IEEE* 109 (4): 399–422.

83 Ma, S., Zhang, Y., Lv, J., et al. (2020). Big data driven predictive production planning for energy-intensive manufacturing industries. *Energy* 211: 118320.

84 Chiang, L., Lu, B., and Castillo, I. (2017). Big data analytics in chemical engineering. *Annual Review of Chemical and Biomolecular Engineering* 8 (1): 63–85.

85 Fisher, O.J., Watson, N.J., Escrig, J.E., et al. (2020). Considerations, challenges and opportunities when developing data-driven models for process manufacturing systems. *Computers & Chemical Engineering* 140: 106881.

86 Yagiz, M.A., MohajerAnsari, P., Pesé, M.D., and Goktas, P. (2024). Transforming In-vehicle network intrusion detection: VAE-based knowledge distillation meets explainable AI. In: *Proceedings of the Sixth Workshop on CPS & IoT Security and Privacy* (pp. 93–103). Association for Computing Machinery.

87 Yagiz, M.A., and Goktas, P. (2025). LENS-XAI: redefining lightweight and explainable network security through knowledge distillation and variational autoencoders for scalable intrusion detection in cybersecurity. *arXiv preprint arXiv:2501.00790.*

88 Ward, G., and Janczewski, L. (2021). Using knowledge synthesis to identify multi-dimensional risk factors in IoT assets. In: *International Conference on Advances in Cyber Security* (pp. 176–197). Singapore: Springer Singapore.

89 Adaros Boye, C., Kearney, P., and Josephs, M. (2018). Cyber-risks in the industrial internet of things (IIoT): towards a method for continuous assessment. In: *Information Security: 21st International Conference, ISC 2018,* Guildford, UK, September 9–12, 2018, *Proceedings 21* (pp. 502–519). Springer International Publishing.

90 Petrenko, A.A., Petrenko, S.A., Makoveichuk, K.A., and Olifirov, A.V. (2021). Methodological recommendations for the cyber risks management. In: *CEUR Workshop Proceedings* (Vol. 2914, pp. 234–247).

91 El-Hajj, M., and Gebremariam, T.H. (2024). Enhancing resilience in digital twins: ASCON-based security solutions for Industry 4.0. *Network* 4 (3): 260–294.

92 Vijayakumaran, C., Muthusenthil, B., and Manickavasagam, B. (2020). A reliable next generation cyber security architecture for industrial internet of things environment. *International Journal of Electrical and Computer Engineering* 10 (1): 387.

93 Sen, S., and Song, L. (2021). An IIoT-based networked industrial control system architecture to secure industrial applications. In: *2021 IEEE Industrial Electronics and Applications Conference (IEACon)* (pp. 280–285). IEEE.

94 Nakkeeran, M., and Mathi, S. (2021). A generalized comprehensive security architecture framework for IoT applications against cyber-attacks. In: *Artificial Intelligence and Technologies: Select Proceedings of ICRTAC-AIT 2020* (pp. 455–471). Singapore: Springer Singapore.

95 Li, Y., Zuo, Y., Song, H., and Lv, Z. (2021). Deep learning in security of internet of things. *IEEE Internet of Things Journal* 9 (22): 22133–22146.

96 Rahman, M.A., and Hossain, M.S. (2022). A deep learning assisted software defined security architecture for 6G wireless networks: IIoT perspective. *IEEE Wireless Communications* 29 (2): 52–59.

97 Umer, M.A., Belay, E.G., and Gouveia, L.B. (2024). Leveraging artificial intelligence and provenance blockchain framework to mitigate risks in cloud manufacturing in industry 4.0. *Electronics* 13 (3): 660.

98 Zanasi, C., Russo, S., and Colajanni, M. (2024). Flexible zero trust architecture for the cybersecurity of industrial IoT infrastructures. *Ad Hoc Networks* 156: 103414.

99 Buja, A., Apostolova, M., Luma, A., and Januzaj, Y. (2022). Cyber security standards for the industrial internet of things (IIoT)—a systematic review. In: *2022 International Congress on Human-Computer Interaction, Optimization and Robotic Applications (HORA)* (pp. 1–6). IEEE.

100 Adaros-Boye, C., Kearney, P., and Josephs, M. (2019). Continuous risk management for industrial IoT: a methodological view. In: *International Conference on Risks and Security of Internet and Systems* (pp. 34–49). Cham: Springer International Publishing.

101 Hansch, G., Schneider, P., and Brost, G.S. (2019). Deriving impact-driven security requirements and monitoring measures for industrial IoT. In: *Proceedings of the 5th on Cyber-Physical System Security Workshop* (pp. 37–45). Association for Computing Machinery.

102 Ahmad, I., Rodriguez, F., Kumar, T., et al. (2024). Communications security in Industry X: a survey. *IEEE Open Journal of the Communications Society* 5: 982–1025.

103 Kalinin, M., Krundyshev, V., and Zegzhda, P. (2021). Cybersecurity risk assessment in smart city infrastructures. *Machines* 9 (4): 78.

104 Tariq, U. (2022). Zero-tolerance security paradigm for enterprise-specific industrial internet of things. *Electronics* 11 (23): 3953.

105 Gopal, S., Gupta, P., and Minocha, A. (2023). Advancements in fin-tech and security challenges of banking industry. In: *2023 4th International Conference on Intelligent Engineering and Management (ICIEM)* (pp. 1–6). IEEE.

106 Chen, L., Zheng, J., Zhu, L., and Liu, Y. (2022). A big data methodology for analyzing machine energy consumption based on IIoT technology under ESG concept. In: *2022 IEEE 2nd International Conference on Mobile Networks and Wireless Communications (ICMNWC)* (pp. 1–8). IEEE.

107 Liu, L., Liu, L., Liu, K., and Jiménez-Zarco, A.I. (2024). Double-edged sword of industrial internet of things: empirical evidence from green transformation of enterprises. *Heliyon* 10 (19): e38595.

108 Arfelis, S., Rainer, J., and González-Pérez, D.M. (2022). Architecture development to incorporate Industry 4.0 solutions to plastics management: circular Economy. In: *Advances in Automation and Robotics Research: Proceedings of the 3rd Latin American Congress on Automation and Robotics, Monterrey, Mexico 2021* (pp. 121–127). Springer International Publishing.

109 Barbieri, L., Kianoush, S., Nicoli, M., et al. (2025). A close look at the communication efficiency and the energy footprints of robust federated learning in industrial IoT. *IEEE Internet of Things Journal* 12 (11): 15130–15150.

110 Martins, J.S., Carvalho, T.C., Moreira, R., et al. (2023). Enhancing network slicing architectures with machine learning, security, sustainability and experimental networks integration. *IEEE Access* 11: 69144–69163.

111 Yang, M., and Xiong, H. (2025). Efficient and decentralized dual access control for cloud-based industrial internet of things. *IEEE Internet of Things Journal* 12 (12): 21082–21090.

112 Almadani, B., and Mostafa, S.M. (2021). IIoT based multimodal communication model for agriculture and agro-industries. *IEEE Access* 9: 10070–10088.

113 Rafiq, A., Muthanna, M.S.A., Muthanna, A., et al. (2022). Intelligent edge computing enabled reliable emergency data transmission and energy efficient offloading in 6TiSCH-based IIoT networks. *Sustainable Energy Technologies and Assessments* 53: 102492.

114 Philipp, R., Prause, G., Olaniyi, E.O., and Lemke, F. (2021). Towards green and smart seaports: renewable energy and automation technologies for bulk cargo loading operations. *Rigas Tehniskas Universitates Zinatniskie Raksti* 25 (1): 650–665.

115 Magd, H., Ansari, M.S.A., and Khan, S.A. (2024). Need for explainable artificial intelligence ethnic decision-making in Society 5.0. In: *XAI Based Intelligent Systems for Society 5.0* (pp. 103–127). Elsevier.

116 Onamade, A., and Okpaegbe, E. (2024). The use of self-sovereign identity in blockchain systems. In: *Leveraging Blockchain Technology* (pp. 117–129). CRC Press.

117 Alabi, A.T., Ubandawaki, A.T., and Issa, S.O. (2025). The role of digital technology in the sustainability strategies of SMEs. In: *The Future of Small Business in Industry 5.0* (pp. 347–378). IGI Global Scientific Publishing.

118 Lopez, Y.I.P., Jayachandran, A.V.T., Diaz, F.C., and Flores, N.C.B. (2019). Implementing small satellite technologies to formulate industrial policy in Latin America region based in the UN-Agenda 2030. In: *Proceedings of the International Astronautical Congress, IAC* (pp. IAC-19_B4_1_11_x54343).

119 Andersen, D.L., Ashbrook, C.S.A., and Karlborg, N.B. (2020). Significance of big data analytics and the internet of things (IoT) aspects in industrial development, governance and sustainability. *International Journal of Intelligent Networks* 1: 107–111.

120 Abir, A.Z.M.N. (2024). Industry 4.0 technologies for a data-driven, secure, green and circular manufacturing economy. In: *2024 IEEE Green Technologies Conference (GreenTech)* (pp. 26–31). IEEE.

121 Augustine, P. (2020). The industry use cases for the digital twin idea. In: *Advances in Computers* (Vol. 117, No. 1, pp. 79–105). Elsevier.

122 Tang, L., Du, Y., Liu, Q., et al. (2023). Digital-twin-assisted resource allocation for network slicing in Industry 4.0 and beyond using distributed deep reinforcement learning. *IEEE Internet of Things Journal* 10 (19): 16989–17006.

123 Bandara, L.V., and Buics, L. (2024). Digital twins in sustainable supply chains: a comprehensive review of current applications and enablers for successful adoption. *Engineering Proceedings* 79 (1): 64.

124 Samanta, S., Sarkar, A., and Bulo, Y. (2022). Secure smart city infrastructure using digital twin and blockchain. In: *Advances in Information Communication Technology and Computing: Proceedings of AICTC 2021* (pp. 367–375). Singapore: Springer Nature Singapore.

125 Samanta, S., Sarkar, A., and Bulo, Y. (2022). Secure 6G communication in smart city using blockchain. In: *Emerging Technologies in Data Mining and Information Security: Proceedings of IEMIS 2022, Volume 1* (pp. 487–496). Singapore: Springer Nature Singapore.

126 Shah, I.A., Sial, Q., Jhanjhi, N.Z., and Gaur, L. (2023). Use cases for digital twin. In: *Digital Twins and Healthcare: Trends, Techniques, and Challenges* (pp. 102–118). IGI Global.

127 Bellavista, P., and Mora, A. (2019). Edge cloud as an enabler for distributed AI in industrial IoT applications: the experience of the IoTwins project. In: *AI&IoT@ AI* IA* (pp. 1–15).

128 Yacelga, A.L., Arevalo, N.B., and Zambrano, L.A. (2023). Fog computing in the industrial internet of things: challenges, trends, and strategies. *Fusion: Practice & Applications* 13 (2): 91–105.

129 Sharma, P., Jain, S., Gupta, S., and Chamola, V. (2021). Role of machine learning and deep learning in securing 5G-driven industrial IoT applications. *Ad Hoc Networks* 123: 102685.

130 Bhadra, P., Chakraborty, S., and Saha, S. (2023). Cognitive IoT meets robotic process automation: the unique convergence revolutionizing digital transformation in the Industry 4.0 era. In: *Confluence of Artificial Intelligence and Robotic Process Automation* (pp. 355–388). Singapore: Springer Nature Singapore.

131 Eswaran, U., Eswaran, V., Eswaran, V., and Murali, K. (2024). Empowering the factory of the future: integrating artificial intelligence, machine learning, and IoT innovations in industry 6.0. In: *Evolution and Advances in Computing Technologies for Industry 6.0* (pp. 1–21). CRC Press.

132 Gilles, O., Faura, D., and Pérez, D.G. (2023). Securing communication in the field: protecting geo-distributed computing in an untrusted environment. In: *C&ESAR* (pp. 21–36).

133 Vanickis, R., Jacob, P., Dehghanzadeh, S., and Lee, B. (2018). Access control policy enforcement for zero-trust-networking. In: *2018 29th Irish Signals and Systems Conference (ISSC)* (pp. 1–6). IEEE.

8

Energy Management and Smart Grid Communication

Polat Goktas[1,2] and Shaibu Ibrahim[3,]*

[1] *UCD School of Computer Science, University College Dublin, Dublin, Ireland*
[2] *Faculty of Engineering and Natural Sciences, Sabanci University, Istanbul, Turkey*
[3] *Acciona Energia USA Global LLC, United States*

*Corresponding author: ishaibu136@gmail.com

8.1 Introduction to Smart Grids and Sustainable Energy Management

The international energy situation is changing with a paradigm shift led by climate change, urbanization, energy security, and the push for the accomplishment of the Sustainable Development Goals (SDGs), in particular SDG 7 (Affordable and Clean Energy), SDG 9 (Industry, Innovation and Infrastructure), and SDG 13 (Climate Action). At the center of this shift lies the convergence of smart grid technologies and sustainable energy management systems, which together promise an intelligent, reliable, low-carbon energy future [1, 2]. Smart grid communication technologies are increasingly recognized as enablers of several SDGs, particularly SDGs 7, 9, 11, and 13, as visualized in Figure 8.1.

Smart grid technologies support the achievement of multiple SDGs, including SDG 7 (Affordable and Clean Energy), SDG 9 (Industry, Innovation, and Infrastructure), SDG 11 (Sustainable Cities and Communities), and SDG 13 (Climate Action). These connections highlight how intelligent energy systems foster resilience, innovation, inclusivity, and environmental sustainability.

Smart grids represent the digital evolution of traditional power networks, integrating information and communication technologies to enable real-time

AI and ML Techniques in IoT-based Communication: A Path to Sustainable Development Goals, First Edition. Edited by Sumita Mishra, Nishu Gupta and Polat Goktas.

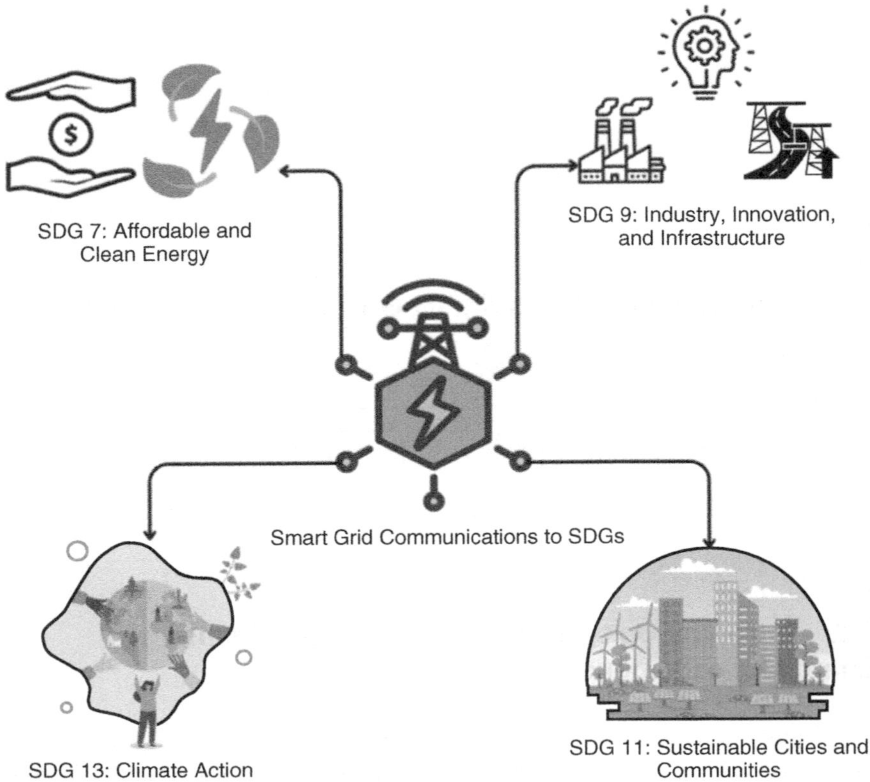

Figure 8.1 Mapping smart grid communication to key Sustainable Development Goals (SDGs).

monitoring, bidirectional power flows, and automated control. These systems provide a platform for integrating distributed energy resources (DERs) such as solar, wind, and energy storage systems [3, 4]. By supporting demand response, predictive analytics, and consumer participation, smart grids are the backbone of the emerging smart energy ecosystem [5, 6]. Beyond technological enhancements, smart grids play a critical role in decarbonizing energy supply, enhancing resilience to climate impacts, and optimizing the use of renewables [7, 8]. Their ability to unify electricity, heating, transport, and industrial systems through sector coupling opens up pathways to achieve 100% renewable energy targets [9, 10].

Despite their potential, smart grid deployment faces challenges including data privacy, cyber-physical security, interoperability, regulatory fragmentation, and technical integration of intermittent renewable sources [11–13]. These challenges necessitate a more intelligent and adaptive grid architecture, supported by advancements in Internet of Things (IoT), machine learning (ML), and artificial intelligence (AI). The role of big data in smart grid analytics is also critical, enabling insights for grid optimization, asset management, and forecasting [6]. A sustainable energy transition requires harmonized technical innovation, policy alignment, and stakeholder engagement [14].

This chapter aims to provide a comprehensive and integrative perspective on the intersection of smart grid communication and sustainable energy management. It addresses cutting-edge technologies, real-world applications, and emerging frameworks that define next-generation energy systems. Key contributions include:

- An exploration of how *IoT technologies* facilitate real-time communication and decision-making in grid environments.
- Insight into the role of *ML* for predictive analytics in energy forecasting and anomaly detection.
- A detailed review of *AI-based optimization methods* for load balancing and distributed energy management.
- Discussion of *cybersecurity, data governance, and regulatory frameworks*, highlighting opportunities and challenges.
- A forward-looking synthesis of *digital twins, blockchain, and federated learning (FL)* as pillars of future smart grid resilience.

In the following sections, this chapter explores key enablers of modern smart grids: the role of IoT in enabling real-time communication and grid responsiveness; ML approaches for predictive energy management; AI-powered optimization strategies for grid operations, stability, and DER integration; and the overarching challenges and future directions, including cybersecurity, regulatory alignment, and the integration of advanced technologies such as FL, blockchain, and digital twins.

8.2 Role of IoT in Smart Grid Communication

8.2.1 IoT Architecture for Real-time Grid Monitoring

The growing sophistication of the power system supplied by DERs, prosumer loads, and variable renewable resources requires real-time monitoring as a key element of contemporary grid management. IoT-based systems can provide fine-grained visibility, data gathering, and control, adding both resilience and sustainability.

8.2.1.1 Architectural Layers and Smart Grid Integration

IoT architectures designed for smart grid systems are typically organized into a three-layer model: the perception layer, the network layer, and the application layer. Each layer plays a pivotal role in ensuring seamless data flow and system intelligence [15].

- The *perception layer* encompasses sensors, smart meters, phasor measurement units (PMUs), and actuator devices that collect data on grid parameters such as current, voltage, frequency, and power consumption [16].
- The *network layer* transmits this data using various wired and wireless communication protocols while ensuring efficiency and scalability [17, 18].
- The *application layer* provides visualization, data analytics, and grid control through dashboards, AI models, and automated systems [19].

This layered architecture enables real-time monitoring, supports demand-response mechanisms, and enhances the resilience of both centralized and distributed energy systems [20].

8.2.1.2 Real-time Sensing and Data Acquisition

Real-time sensing is the backbone of IoT-enabled grid monitoring. With the deployment of smart sensors at transmission and distribution nodes, it becomes possible to detect voltage sags, energy losses, grid faults, and load imbalances in milliseconds [21]. Smart meters record electricity usage and transmit time-series data at frequent intervals, enabling fine-grained visibility into consumption behavior. Modern IoT deployments support plug-and-play integration of sensors and meters with minimal manual intervention, thus accelerating deployment. Data acquisition systems now incorporate self-diagnosing and self-healing mechanisms, improving fault tolerance and reducing human involvement in grid maintenance [22]. In microgrids, these systems are further optimized for localized energy balancing, load forecasting, and integration with renewable sources such as solar and wind [23].

8.2.1.3 Network and Communication Protocols

The network layer is responsible for transmitting collected data to cloud or edge-based processing systems. Multiple communication technologies are used to support diverse requirements across latency, range, and energy efficiency.

- *Short-range protocols* like ZigBee and BLE are suited for in-home energy management.
- *Low-power wide-area networks (LPWANs)* such as LoRaWAN and NB-IoT offer long-range, low-energy transmission, ideal for rural smart grid nodes.
- *5G technology*, with its ultra-low latency, is emerging as a key enabler for real-time control in mission-critical grid environments [24].

For lightweight and reliable communication, protocols such as MQTT (publish-subscribe model), CoAP (constrained environments), and HTTP/2 are increasingly adopted [22]. These ensure energy-efficient, asynchronous communication even under constrained bandwidth conditions. However, the use of diverse protocols introduces interoperability challenges, requiring standardized translation and integration mechanisms [25].

8.2.1.4 Edge Computing and AI-driven Control

Traditional cloud-centric architectures are insufficient for latency-sensitive energy systems. Therefore, the shift toward edge and fog computing has gained momentum. Edge devices process data locally, reducing transmission delays and improving response times for applications like fault detection, voltage stabilization, and predictive load management [26]. AI techniques embedded at the edge allow on-device analytics for anomaly detection, load forecasting, and operational optimization. These approaches support resilient demand-side management and make the grid more adaptable to uncertainties in energy supply and demand [27].

Moreover, digital twins – virtual replicas of physical assets – are being developed for real-time simulations of grid behavior. These models offer predictive insights and simulate failures, allowing proactive management and better energy planning [15]. Blockchain technologies are also being integrated for secure and traceable peer-to-peer energy trading, particularly in DER-rich environments [28].

8.2.1.5 Security, Privacy, and Interoperability

The highly distributed and dynamic nature of IoT-enabled grids introduces numerous cybersecurity and privacy risks. These include:

- *Eavesdropping and spoofing* of sensor data
- *Denial-of-service (DoS)* attacks targeting communication nodes
- *Data tampering* leading to misinformed control decisions [28, 29]

Security solutions must be lightweight and scalable. Lightweight cryptographic protocols, like those based on Elliptic curve cryptography, are preferred due to their minimal computational burden [30]. Intrusion detection systems (IDS) based on ML and deep learning (DL) are also gaining traction for identifying abnormal behavior in smart grid communication [31]. Privacy concerns related to fine-grained consumption data can be mitigated using data aggregation techniques, anonymization, and FL, which allows collaborative AI training without exposing sensitive data [22]. Finally, interoperability remains a critical issue due to the heterogeneity of vendors, hardware, and communication standards. Efforts are underway to harmonize protocols and develop hierarchical authentication frameworks to secure mobile and cross-domain energy infrastructures [32, 33].

8.2.2 Data Acquisition, Edge Computing, and Device Integration

As smarter grids turn into extremely dynamic and distributed cyber-physical systems, the functions of data acquisition, edge computing, and device integration become more and more critical to facilitating real-time responsiveness, operational efficiency, and secure energy transactions.

8.2.2.1 Foundations of IoT-based Data Acquisition in Smart Grids

Data acquisition forms the foundational layer of intelligent energy systems by enabling real-time access to physical parameters from diverse components of the power grid. Through IoT-enabled smart meters, PMUs, voltage and current sensors, and environmental monitoring units, granular datasets are continuously captured and transmitted for processing. The primary objective extends beyond ensuring operational visibility; it also encompasses improving situational awareness, which is essential for accurate load forecasting, timely fault localization, and effective demand-side energy optimization [34]. As traditional Supervisory Control and Data Acquisition (SCADA) systems give way to distributed and autonomous data collection infrastructures, IoT architectures support decentralized intelligence using lightweight sensors and embedded systems [35]. However, such pervasive monitoring introduces new requirements for real-time transmission, synchronization, and energy-efficient data capture mechanisms, especially in areas with limited connectivity [36].

8.2.2.2 Edge Computing for Low-latency Processing and Analytics

The exponential growth of data from grid-connected IoT devices necessitates local, near-source data processing. This is where edge computing, the paradigm of offloading computation from centralized cloud systems to localized edge nodes, becomes essential [37]. By performing analytics close to where data is generated, edge computing significantly reduces latency and improves system responsiveness. This is particularly critical for applications such as *Real-time grid fault detection, Dynamic voltage regulation, Predictive equipment maintenance*, and *Intrusion or anomaly detection.*

These capabilities are enhanced by embedding ML models directly into edge devices, creating intelligent energy nodes capable of context-aware decision-making [38, 39]. Furthermore, edge-enabled architectures benefit from integration with 5G, Software Defined Networking (SDN), and Network Function Virtualization (NFV), allowing for dynamic orchestration of computing resources [40]. Still, challenges such as power limitations, hardware constraints, and secure data isolation continue to be key research directions.

8.2.2.3 Device Integration and Heterogeneity in Smart Energy Ecosystems

The effective deployment of IoT systems in the smart grid relies on seamless integration of diverse physical devices, including sensors, actuators, gateways, and smart appliances. However, these devices often vary in hardware capability, communication standards, and security support, making heterogeneous device integration a significant barrier [41]. Middleware platforms and IoT gateways are used to harmonize communication, providing translation services between protocols such as ZigBee, LoRaWAN, NB-IoT, and Wi-Fi. Additionally, interoperability frameworks like oneM2M and edge abstraction layers help abstract hardware differences, enabling plug-and-play functionality and easier lifecycle management [42].

Security becomes a top concern when integration is not securely managed. Poorly integrated or legacy devices can become entry points for attackers, making device-level identity verification, key management, and remote firmware updates critical features [43]. Increasingly, blockchain-based identity and attestation mechanisms are being proposed to ensure trusted device enrolment [34].

8.2.2.4 Privacy-preserving Data Handling at the Edge

The real-time nature of data generated within smart grid infrastructures, particularly from residential smart meters and consumer-end devices, often entails the collection of granular, personally identifiable information. Such data can inadvertently reveal sensitive behavioral patterns, including occupancy, lifestyle habits, and energy usage routines. As a result, privacy protection has emerged as a critical concern, especially under the purview of stringent data protection regulations such as the General Data Protection Regulation (GDPR). To mitigate these risks, edge computing plays a pivotal role by serving as a privacy-preserving computational layer. Instead of transmitting raw data to centralized servers or cloud environments, edge devices perform localized data processing and transmit only aggregated, anonymized, or insight-level information. This approach significantly reduces the system's attack surface, minimizes exposure to unauthorized access, and enhances compliance with data minimization principles [44].

Advancements in privacy-enhancing technologies are further reinforcing this edge-centric model:

- **Federated Learning (FL)**: Enables decentralized model training across multiple edge nodes without the need to share raw data, thereby preserving data locality and privacy.
- **Differential Privacy (DP)**: Introduces mathematically calibrated noise into datasets or analytical outputs, ensuring that individual user contributions remain indistinguishable within aggregate statistics.

- **Encrypted ML (EML)**: Facilitates the training and inference of ML models on encrypted inputs, enabling secure computation without data exposure.

These methods are increasingly being integrated with edge computing frameworks to support privacy-respecting analytics in key domains such as demand forecasting, dynamic pricing, personalized energy services, and billing systems [45]. Collectively, these innovations enable the development of trustworthy and ethically responsible smart grid systems, where user privacy is preserved without compromising system intelligence or operational efficiency.

8.2.3 Use Cases: Load Forecasting, Outage Management, Demand-side Response

Implementation of IoT in smart grid infrastructure has resulted in massive improvement in operation intelligence and real-time response. Load forecasting, outage management, and demand-side response (DSR) are the three highest applications with greatest impact. These use cases demonstrate the critical role of secure and privacy-conscious communication in modern grid ecosystems, as summarized in Table 8.1.

8.2.3.1 Load Forecasting

Load forecasting is indispensable for grid stability, resource allocation, and the integration of renewables. As demonstrated by Ibrahim et al. [46], short-term load forecasting models using ML significantly outperform traditional methods by incorporating weather data, usage history, and user behavior. The deployment of low-cost smart power meters [47] has made granular load monitoring accessible,

Table 8.1 Summary of use cases, technologies, and security considerations.

Use case	Key technologies	Security/privacy considerations	Key references
Load Forecasting	Smart meters, ML models, LPWAN, IoET, big data platforms	Encrypted transmission, edge analytics, federated ML	[46–49]
Outage Management	IoT sensors, blockchain, EV data streams, PMUs	Immutable logging, real-time coordination	[51, 52]
Demand-side Response	Smart meters, adaptive scheduling, AI-based dashboards	Real-time automation, user privacy protocols	[53, 54]

Note: The content presented in this table is derived from a synthesis of key insights and empirical findings reported across the cited literature.

even in resource-constrained regions. The Internet of Energy Things (IoET) paradigm further advances these capabilities by integrating LPWAN technologies, facilitating efficient data transfer across wide geographical areas [48]. Big data ecosystems support hybrid forecasting frameworks that scale across utilities and regions [49], while intelligent agents are deployed at substations for autonomous, real-time predictions [50].

8.2.3.2 Outage Management

Outage detection and restoration form the backbone of smart grid reliability. The use of IoT sensors and decentralized detection algorithms enables near-instantaneous fault identification, as noted by Yaghmaee and Hejazi [51]. These are complemented by blockchain-supported alert systems that provide immutable fault logging and faster coordination among grid nodes. Savari et al. [52] demonstrate how real-time electric vehicle load data can be integrated into predictive maintenance and outage detection systems. Cloud-embedded ML models can localize outages and recommend optimal restoration strategies. Meanwhile, enhanced grid visualization supported by micro-PMUs and mobile communication networks offers dynamic fault tracking and decision support [53].

8.2.3.3 Demand-side Response

DSR mechanisms empower consumers to actively engage with the grid by shifting or reducing their load in response to pricing or peak signals. Ahammed and Khan [54] illustrate an IoT-based smart metering system in a developing country that uses real-time data to manage both power quality and DSR participation. Edge-based automation allows appliances to adjust consumption dynamically, while preserving data privacy. DSR platforms are increasingly AI-driven, incorporating adaptive energy scheduling, personalized user insights, and grid-aware optimization [46]. Frameworks for sustainable energy management now embed DSR algorithms that adjust in real time based on both grid-side and consumer-side metrics [53].

8.3 ML for Predictive Energy Management

8.3.1 ML Models for Consumption Forecasting, Anomaly Detection, and Fault Prediction

With evolved smart grid technologies becoming maturely smart and decentralized networks, ML turned out to be an essential tool in the extraction of actionable intelligence from energy consumption patterns. With ML methods, grid operators are able to forecast energy demand, identify anomaly in real time, and foresee

Table 8.2 ML applications in smart grid operations: techniques, benefits, and representative studies.

Application area	ML techniques used	Key benefits	Selected references
Consumption Forecasting	ANN, LASSO, SVR, Gray Models, Ensemble Learning	Non-linear modeling, high accuracy, adaptability	[55–58]
Anomaly Detection	Autoencoders, Clustering, PCA, Decision Trees	Theft detection, demand profiling, cyber resilience	[59, 60]
Fault Prediction	SVM, RNN, Decision Trees, Encoder-Decoders, XAI tools	Early fault alerts, reduced downtime, interpretability	[61–63]

Note: The content presented in this table is derived from a synthesis of key insights and empirical findings reported across the cited literature.

equipment failure prior to when they manifest themselves as interruptions. This functionality implies proactive energy management, enhanced grid resilience, and cost reduction. Table 8.2 provides an overview of widely used ML models across three major application areas – consumption forecasting, anomaly detection, and fault prediction – along with their respective benefits and representative studies.

8.3.1.1 Consumption Forecasting

ML models for electricity consumption forecasting have become vital for grid planning, renewable integration, and DSR. These models outperform traditional time-series methods by capturing nonlinear and context-sensitive consumption behaviors. Jain et al. [55] applied LASSO regression to building-level data, demonstrating the potential of feature selection in high-dimensional datasets. More recent work by Ding et al. [56] introduced adaptive gray models with seasonality-aware parameters, yielding excellent results in residential sectors. Hybrid and DL models are also gaining prominence. For instance, Sekhar and Dahiya [57] presented a robust ANN framework for real-time energy prediction in buildings, while Shen et al. [58] deployed ensemble models combining support vector regression (SVR) and random forests to manage noise and improve generalization. These techniques ensure more reliable day-ahead scheduling, load shifting, and integration with demand response programs.

8.3.1.2 Anomaly Detection

ML-driven anomaly detection allows operators to proactively identify irregular consumption patterns or cybersecurity threats without relying on manual thresholds. Shaikh et al. [59] utilized unsupervised learning methods, such as autoencoders and clustering, to detect abnormal consumption spikes in smart meter data. Data clustering and classification approaches like k-means and decision trees have also been successfully applied to detect outliers in grid telemetry. Culaba et al. [60] used clustering to group consumption profiles and detect variations based on appliance usage behavior. Anomaly detection in high-dimensional data often leverages dimensionality reduction techniques, such as PCA or t-SNE, to visualize and isolate abnormal trends in building and household consumption. These models are essential in identifying threats such as meter tampering, electricity theft, or cyberattacks that mimic normal load profiles but disrupt system integrity.

8.3.1.3 Fault Prediction for Proactive Maintenance

Predictive maintenance has emerged as one of the most impactful applications of ML in grid infrastructure. By anticipating faults in transformers, lines, or circuit breakers, utilities can reduce downtime and maintenance costs. Taha and Mansour [61] introduced an ML-based diagnostic model using decision trees for transformer failure classification. Similarly, Saravanan et al. [62] applied support vector machines (SVMs) to sensor-based fault datasets, achieving high accuracy in real-time applications. DL models such as encoder–decoder architectures [63] and recurrent neural networks (RNNs) are also being applied for long-term failure prediction and operational risk assessment. Explainable AI (XAI) techniques, like SHAP values and decision trees, are helping interpret these models, ensuring greater trust and adoption in utility operations.

8.3.2 Supervised Versus Unsupervised Approaches in Grid Intelligence

The greater access to high-resolution energy data from IoT sensors has opened up new avenues to employ ML for energy management and smart grid communication. Applications fall largely into supervised learning on labeled datasets and unsupervised learning that discovers latent patterns with no target outputs. Both methods have unique benefits for the application, which spans from appliance recognition and load forecasting to load profiling and anomaly detection.

8.3.2.1 Supervised Learning in Grid Operations

Supervised learning models are trained on historical data where the outcomes (labels) are known. These models excel in tasks like power consumption forecasting, equipment fault classification, and appliance usage identification. Popular techniques include ANNs, SVMs, Random Forests, and k-Nearest Neighbors (k-NNs) [64, 65]. For example, ANNs have been used to classify appliance-level consumption data based on low-frequency readings, enabling energy disaggregation and smarter load control [65]. Similarly, k-NN has been applied in residential demand estimation models to enhance energy-saving strategies [64]. These models can be deployed in demand response platforms, fault diagnosis systems, and predictive maintenance routines with high reliability and interpretability when sufficient labeled data is available. However, their performance is limited by the availability and quality of labeled training datasets, which are often expensive and time-consuming to generate in energy systems.

8.3.2.2 Unsupervised Learning in Smart Grid Analytics

In contrast, unsupervised learning algorithms identify patterns, clusters, or anomalies in data without predefined output labels. These methods are particularly well-suited for load profiling, customer segmentation, and anomaly detection, areas where labeled datasets are scarce or non-existent. Techniques such as k-means clustering, hierarchical clustering, and principal component analysis (PCA) have been applied to detect energy usage patterns across consumer groups [66]. For instance, load profile clustering can uncover hidden behavioral patterns in residential or commercial consumption, supporting dynamic tariff design and microgrid optimization. Moreover, unsupervised learning plays a crucial role in cybersecurity, where real-time anomaly detection systems are used to monitor grid communication networks for suspicious activities without requiring prior attack signatures. Despite its flexibility, unsupervised learning lacks direct interpretability and requires domain expertise for meaningful post-analysis of clusters or anomalies. Hybrid models and semi-supervised techniques are increasingly being explored to combine the strengths of both approaches.

8.3.2.3 Comparative Outlook and Use Case Suitability

Supervised and unsupervised learning serve complementary roles in grid intelligence. Supervised methods are ideal for applications where historical outcomes are available and accuracy is paramount, such as load forecasting and fault detection. Unsupervised models, meanwhile, are indispensable for exploratory data analysis, data-driven decision-making, and privacy-preserving analytics in decentralized systems. As shown in Table 8.3, each approach offers trade-offs in terms of data requirements, model interpretability, and deployment complexity. The selection of learning strategy depends on both the problem context and data availability.

8.3.3 Case Studies and Real-world Deployments

The transition from theoretical modelling to practical applications of energy management and smart grid communication technologies is gaining momentum across the globe. A growing number of real-world deployments demonstrate how

Table 8.3 Comparison of supervised and unsupervised ML approaches in smart grid applications.

Learning approach	Common ML techniques	Key use cases	Advantages	Selected references
Supervised Learning	ANN, SVM, Random Forest, k-NN	Load forecasting, fault detection, appliance recognition	High accuracy; well-suited for classification/regression	[64, 65]
Unsupervised Learning	K-means, Hierarchical Clustering, PCA	Load profiling, anomaly detection, customer segmentation	No labeled data required; useful for pattern discovery	[66]

Note: This table is based on a synthesis of findings from selected smart grid ML applications in the cited literature.

AI, ML, and IoT systems are supporting efficiency, sustainability, and predictive control in various energy contexts, from industrial facilities and wind farms to smart buildings and decentralized microgrids.

8.3.3.1 Predictive Maintenance and Operational Optimization

A core application area for AI in real-world deployments is predictive maintenance, which aims to reduce operational costs and system downtime. In solar energy systems, Qureshi et al. [67] applied explainable DL to detect panel defects in hot climates using aerial thermographic imaging. This approach improved inspection accuracy and interpretability, crucial for field teams working under extreme environmental conditions. Similarly, in central heating systems, Behzad et al. [68] implemented condition-based monitoring using sensor data and ML algorithms. Their work achieved notable improvements in sustainability performance, demonstrating how AI can drive energy efficiency in legacy systems. In wind energy, Zhang et al. [69] and Sobral et al. [70] developed maintenance frameworks that considered seasonal uncertainty and operational patterns, optimizing maintenance schedules for improved availability and reduced costs. Shin et al. [71] and Sayal et al. [72] further demonstrated how predictive analytics can be embedded into renewable energy infrastructures such as smart substations and hybrid systems, resulting in higher system reliability and minimal manual intervention.

8.3.3.2 Energy Management Systems for Sustainability and Net-zero Goals

The deployment of smart energy management systems (EMS) has become central to achieving decarbonization and sustainability targets. Chen et al. [73] introduced a lifelong learning-based EMS for building-level control, adapting dynamically to usage patterns to support net-zero emissions goals. This model reflects a shift from static optimization to continuous, intelligent adaptation. In a broader policy and planning context, Tabaku et al. [74] examined AI-based EMS for carbon emission reduction and SDG alignment. Their deployment, centered on Albania's energy infrastructure, demonstrated how real-time load optimization and renewable prioritization can be harmonized through AI to achieve national climate targets. Jalli et al. [75] also emphasized the role of big data analytics in large-scale renewable energy projects. Their work showcased how meteorological, market, and grid data streams can be fused to support energy forecasting and operational decisions across regions.

8.3.3.3 Business and Microgrid Applications of Smart Energy Analytics

In commercial contexts, Shatat et al. [76] applied energy consumption analytics to optimize resource use in business environments. By analyzing load patterns, the system offered actionable insights for reducing peak loads and aligning energy use with operational schedules – demonstrating the scalability of AI in enterprise settings. At the microgrid level, Meydani et al. [77] explored intelligent energy scheduling and optimization techniques. Their real-world deployment considered cost, emissions, and system constraints, balancing local generation with load demands. Sharma et al. [78] supported this direction by highlighting DL architectures applicable across domains, particularly for system forecasting and dynamic resource allocation. These projects collectively demonstrate the versatility and adaptability of AI/ML approaches across grid scales and economic sectors, offering robust pathways to operational efficiency and sustainable communication infrastructures.

8.4 AI-enabled Optimization in Grid Operations and Load Balancing

8.4.1 AI Techniques for Grid Stability and Load Distribution

Grid modernization is critically dependent on the stability of the grid and optimal dispatching of loads in real time. As grids increasingly utilize increasingly large amounts of DERs, electric vehicles, and bidirectional power flow, conventional control methods are no longer adequate. AI provides scalable, adaptive, and distributed control methodologies for dynamic supply–demand balancing, voltage control, and fault tolerance.

8.4.1.1 AI for Planning and Restructuring Power Distribution Networks

At the planning stage, AI facilitates multiobjective optimization for network design that considers system reliability, cost, carbon impact, and fault tolerance.

Ahmed and Salama [79] introduced a model combining optimization algorithms with reliability constraints, enabling planners to design robust networks capable of accommodating renewable integration. Filipiak [80] developed an evolutionary algorithm and classifier system for post-fault network reconfiguration, highlighting how AI can rapidly find optimal switch settings to restore service. Piao et al. [81] further proposed the concept of virtual microgrids, applying AI for dynamic topology management to strengthen grid resilience and autonomy. Their approach supports fault isolation and improved DER coordination, ensuring continuity during disruptions.

8.4.1.2 Real-time Load Forecasting and Distribution

Real-time load forecasting is critical for grid stability. Tent–Chebyshev chaotic maps combined with improved Particle Swarm Optimization (PSO) models offer high accuracy in reactive power prediction [82]. These models allow dynamic voltage adjustment based on predicted load behaviors. In a complementary study, Wan et al. [83] optimized battery energy storage allocation to manage reactive power variations and support voltage stability in distribution systems. Moreover, neural networks, decision trees, and SVMs are widely used to forecast loads and guide dynamic reallocation of power flows. These models are often deployed at substations or edge nodes, enabling decentralized, autonomous control of feeder line loading [84].

8.4.1.3 Islanding Detection and Adaptive Load Curtailment

AI also plays a crucial role in ensuring grid survivability during blackouts or component failures. Gbadega and Balogun [84] presented an islanding detection strategy using DL, which identifies the onset of grid separation and triggers intelligent load curtailment. Such adaptive systems prevent cascading failures and maintain operational zones even during emergencies. Further, situation awareness algorithms improve response time and fault recovery in large-scale systems. Ghaemi et al. [85] developed a real-time monitoring approach that leverages ML to prioritize corrective actions across the network based on fault criticality and grid impact.

8.4.2 Optimization in DERs and Microgrids

The proliferation of DERs, such as rooftop solar, wind turbines, fuel cells, and battery storage, has fundamentally changed how modern power systems are managed. In parallel, microgrids have emerged as localized energy systems capable of operating autonomously or in coordination with the main grid. The growing complexity of these decentralized systems demands advanced optimization strategies to ensure cost-efficiency, grid reliability, and environmental sustainability.

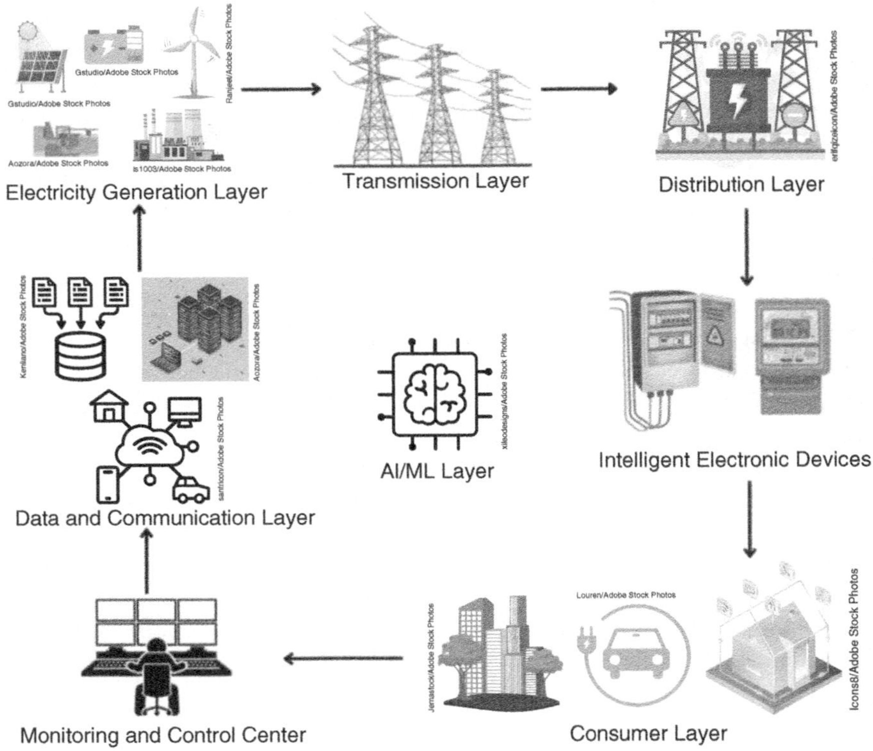

Figure 8.2 Layered architecture of an intelligent energy management system in smart grids.

AI-driven techniques, multiobjective optimization models, and semantic interoperability frameworks are being integrated into DER and microgrid operations for generation scheduling, energy storage control, load balancing, and fault-tolerant dispatch. To support the complexity of decentralized energy systems, smart grids are increasingly structured into interoperable layers that span from generation to end-user consumption, enabling real-time responsiveness and optimization (see Figure 8.2).

The framework integrates various layers from electricity generation to the consumer end, incorporating intelligent electronic devices, AI/ML analytics, data communication, and centralized monitoring and control systems. The bidirectional flow of information supports real-time optimization, automation, and decision-making across generation, transmission, distribution, and consumption.

8.4.2.1 Multiobjective and Predictive Optimization Strategies

Modern DER optimization involves balancing multiple, often conflicting, objectives such as minimizing operational cost, emissions, and power losses while maximizing energy reliability and grid stability. Soares et al. [86] introduced a multidimensional signaling method to guide metaheuristic algorithms in solving complex scheduling and dispatch problems. Babayomi et al. [87] extended this by proposing predictive control models for DERs that incorporate real-time forecasts of load, weather, and market signals. Hybrid approaches that fuse evolutionary computation with neural networks have shown superior performance in adaptive optimization, as seen in the work of Kefayat et al. [88], who combined ant colony optimization with artificial neural networks for load balancing and generation planning.

8.4.2.2 Energy Storage and Demand Response Optimization

Energy storage systems enhance microgrid flexibility by enabling peak shaving, load shifting, and emergency backup. Keerthisinghe et al. [89] proposed a decision-policy-based energy management system that simultaneously manages PV generation and battery storage for residential users, ensuring cost-effectiveness and user comfort. Demand response (FR) programs can be optimized by using intelligent algorithms to dynamically schedule consumer loads based on grid conditions. Bilal et al. [90] reviewed computational intelligence approaches that enable fast, scalable optimization of DR strategies in microgrids. Fose et al. [91] also highlighted reinforcement learning frameworks capable of learning optimal DR schedules while adapting to grid and market signals.

8.4.2.3 Semantic Interoperability and Holonic Architectures

As DER ecosystems grow more complex, semantic interoperability becomes crucial for ensuring seamless communication among heterogeneous devices. Howell et al. [92] introduced a holonic optimization model that uses semantic web technologies to unify data exchange across DERs, enhancing system-wide optimization and coordination. Nastic et al. [42] reinforced the importance of IoT middleware and edge abstractions to enable optimization at device and system levels.

8.4.2.4 Resilience-oriented Control and Islanding Operations

Microgrids must often operate in islanded mode during grid faults or natural disasters. Ali et al. [93] developed hybrid optimization models for islanded microgrid scheduling, which include dynamic reconnection strategies and voltage stabilization. Meena et al. [94] optimized islanding detection and DER scheduling using heuristic methods, ensuring continuity of supply during grid separation. Sanseverino et al. [95] further demonstrated multiagent reinforcement learning approaches to optimize autonomous decision-making in isolated grid environments.

8.5 Challenges, Opportunities, and Future Outlook

8.5.1 Cybersecurity, Data Privacy, and Interoperability Concerns

As smart grids transform into highly interconnected systems facilitated by DERs, cloud computing, and AI, data privacy, interoperability, and cybersecurity have become essential challenges. It is crucial that energy communication networks are cyber-secure, ensure user confidentiality, and operate efficiently over heterogeneous systems as the building block to achieve a secure and sustainable digital energy future.

8.5.1.1 Cybersecurity Challenges in Smart Grids

Smart grids are increasingly vulnerable to cyberattacks due to their reliance on open communication protocols, cloud-based SCADA systems, and distributed architectures. Cyberattacks can range from malware injection and phishing to coordinated DoS attacks and advanced persistent threats [96]. ML-based IDS are now being used to identify abnormal behavior patterns in grid communications [97]. Additionally, blockchain and homomorphic encryption schemes are being explored to secure peer-to-peer energy transactions and sensitive load data [98]. However, legacy infrastructure and fragmented protocols continue to pose barriers to universal security standardization.

8.5.1.2 Data Privacy Risks and Governance

The large-scale collection of real-time consumption data from smart meters and IoT sensors raises serious privacy concerns. Detailed usage data can inadvertently reveal user behavior, occupancy patterns, and socioeconomic status [99]. Privacy-preserving frameworks such as FL, differential privacy, and edge analytics are being developed to protect consumer data while enabling intelligent forecasting and control [100]. Additionally, compliance with international privacy regulations like GDPR requires utilities to implement data minimization, informed consent, and transparent usage policies [101].

8.5.1.3 Interoperability and Standards in Smart Grid Communication

Interoperability is key for integrating heterogeneous devices and platforms across the smart grid ecosystem. The adoption of standardized data models, such as the IEC 61970/61968 Common Information Model (CIM), facilitates consistent information exchange and semantic understanding between systems [102]. Despite progress, gaps remain in translating standards into interoperable implementations. Mutule et al. [103] reported on European Union Horizon 2020 projects contributing to standardization but noted uneven adoption across member states. Tools like OpenADR and IEC 61850 have advanced automation and control integration but require continuous validation and harmonization [104]. Recent trends emphasize the need for interoperability testing, semantic validation engines, and platform-agnostic interfaces to reduce vendor lock-in and enhance modularity in system design [105].

8.5.1.4 Toward Secure and Interoperable Energy Systems

To ensure a secure and privacy-preserving future, smart grid communication must integrate cyber resilience, data ethics, and interoperable architectures as design principles. Future work should focus on:

- Zero-trust security models and AI-based threat prediction
- Cross-layer encryption standards for distributed control
- Federated data ecosystems with built-in privacy mechanisms
- Interoperable middleware and semantic data models

By combining technological innovation with regulatory alignment, smart grids can achieve secure, user-centric, and scalable energy communication infrastructure.

8.5.2 Regulatory Frameworks and Standardization Issues

As smart grids become more interconnected and reliant on data-driven technologies, the establishment of comprehensive regulatory frameworks and standardization protocols becomes essential. These frameworks enable the interoperability, security, efficiency, and scalability of power systems across countries and operators. However, the diversity in national standards, varying implementation maturity, and evolving technologies present several challenges.

8.5.2.1 Evolution of Global Smart Grid Standards

International organizations such as IEC, IEEE, and NIST have led initiatives to develop reference models and common languages for smart grid operations. The CIM developed by IEC aims to standardize the semantic representation of network elements and processes in energy management systems [106]. Similarly, the IEEE 802.15.4g standard focuses on enhancing smart metering communications, especially for low-rate wireless personal area networks [107]. Comparative studies, such as those by Miao et al. [108], have analyzed standardization roadmaps from major institutions, including SGCC (China), NIST (USA), and IEC, identifying commonalities and gaps in communication, cybersecurity, and data interoperability.

8.5.2.2 Challenges in Framework Alignment and Adoption

One major challenge in regulatory harmonization is the asynchronous adoption of standards across regions. While some countries have rapidly deployed standardized platforms, others still rely on legacy systems, leading to integration bottlenecks [109]. Regulatory fragmentation also limits cross-border energy trade, particularly in contexts such as global energy interconnection initiatives [110]. Technical and political considerations often delay or prevent the full implementation of frameworks. Rana et al. [111] emphasized the difficulty of integrating renewables and storage under existing utility regulations that were originally designed for centralized systems. Additionally, lack of clarity in data ownership and cybersecurity responsibilities creates uncertainty. National frameworks may differ in assigning liability for grid failures caused by cyber intrusions or communication errors [112–114].

8.5.2.3 Standardization and Grid Modernization

The modernization of the power grid relies on frameworks that are flexible and scalable. Newer standards such as those discussed by Sharma and Saini [115] in power line communication, and Samad and Bienert [116] in the context of smart manufacturing, demonstrate the need for convergence between energy systems and industrial automation. Moreover, initiatives like Advanced Metering Infrastructure (AMI) standardization, discussed by Bennett and Highfill [117], have catalyzed efforts toward improving data granularity and network stability.

8.5.2.4 Cybersecurity and Compliance Integration

Cybersecurity and standardization are two hand-in-hand principles. As Chhaya et al. [118] and Bui et al. [119] point out, regulatory frameworks must incorporate dynamic security provisions that adapt to evolving threats. Smart grid cyber policies need to address not only device-level encryption but also grid-wide threat detection, response mechanisms, and recovery protocols. Kaainoa and Jenkins [120] stress the importance of turning grid data into actionable information, which requires harmonized data models and metadata standards. Without these, even secure communication may fail to provide meaningful insights for control decisions.

8.5.3 Future Directions

With advancements in the digitalization of power systems, the future of energy management and smart grid communication could be defined by a convergence of emerging technologies, data governance, and solid architectures. Some of the new technologies being developed and deployed include digital twins, FL, blockchain integration, and self-healing systems, which are already starting to redefine

Table 8.4 Emerging technologies and research trends in smart grid energy management.

Technology	Key applications	Expected contribution	References
Digital Twin	Grid simulations, asset diagnostics	Real-time fault detection, proactive maintenance	[124, 125]
Federated Learning & XAI	Privacy-aware learning, trust enhancement	Transparent analytics, regulatory compliance	[128, 129]
Blockchain	P2P energy trading, billing automation	Tamper-proof transactions, decentralized market participation	[126]
Smart Manufacturing	Industrial energy optimization	Load scheduling, predictive control, demand flexibility	[121, 122]
Sustainable Production	Green factories, circular economy	Emission reduction, integration with clean energy systems	[122, 127]

the way energy networks are monitored, optimized, and secured. Table 8.4 summarizes the emerging technologies most likely to shape the future trajectory of smart grids, based on current studies.

8.5.3.1 Smart Manufacturing and Energy Optimization

Smart manufacturing offers unprecedented opportunities for energy efficiency and real-time optimization. By incorporating IoT sensors, digital platforms, and big data analytics, manufacturing systems can better align with grid requirements. For instance, Meng et al. [121] emphasize the role of intelligent systems in achieving resource-efficient production through smart scheduling, predictive maintenance, and energy feedback control. Likewise, Cui et al. [122] discuss how manufacturing big data ecosystems enable dynamic load balancing by aligning energy demand with production cycles. These advances enable factories to participate in DSR, reduce peak consumption, and enhance operational flexibility. In countries like India, sustainable manufacturing initiatives are gaining traction through government-backed energy optimization programs [123]. This evolution toward cleaner production aligns industries with smart grid goals, allowing greater integration of DERs and enhanced coordination with utility services.

8.5.3.2 Digital Twin Platforms and Virtual Grid Simulation

Digital twins, virtual representations of physical systems, are transforming how energy infrastructure is monitored, tested, and controlled. Choi et al. [124] propose a model for diagnosing and evaluating smart factories using digital twin environments, where data is continuously synchronized between physical and virtual layers. These simulations allow grid operators to predict and respond to faults, assess system behavior under variable conditions, and test optimization strategies without affecting real-world operations. In addition, Fathy et al. [125] apply digital twin platforms for maintenance planning, showing their efficacy in simulating degradation patterns, component failures, and energy dispatch logic. These capabilities enhance resilience and reduce operational costs, especially when coupled with AI for intelligent decision-making and self-healing capabilities.

8.5.3.3 Blockchain and Secure Energy Transactions

Blockchain is gaining prominence as a backbone technology for secure, decentralized energy exchange. Wu et al. [126] present a blockchain-based P2P trading framework where prosumers can transact energy directly, with full traceability and without intermediaries. These systems enable real-time auditing, automated billing via smart contracts, and elimination of double-spending issues. In distributed microgrids and community energy systems, blockchain ensures data immutability and operational transparency. When integrated with AI-based trust models, such frameworks can dynamically assess transaction credibility and execute incentive-based DSR programs securely.

8.5.3.4 Sustainable Production and Renewable Integration

The convergence of smart grids and sustainable manufacturing plays a vital role in decarbonization strategies. Sarkar et al. [127] describe how multitype biofuel manufacturing systems optimize energy utilization and production scheduling through AI-based decision support. These systems minimize energy waste and support flexible production aligned with grid constraints. In addition, Gupta et al. [123] emphasize green practices such as lean production, material reuse, and carbon auditing, which are now being embedded into energy analytics tools to guide policy and corporate sustainability reporting. These trends align industries with net-zero pathways and smart grid participation.

8.5.3.5 Federated and Explainable AI for Privacy and Transparency

With increasing concerns over data privacy and model accountability, FL and XAI are emerging as enablers of trustworthy grid intelligence. Sarker et al. [128] explore explainable FL models that train locally on edge devices, enabling collaborative learning without centralizing sensitive data. Khalil et al. [129] also apply FL for privacy-preserving load forecasting in residential environments. Their approach ensures GDPR-compliant data handling while maintaining model performance. XAI methods, such as SHAP or LIME, offer transparency into how AI models make decisions, improving operator trust and facilitating regulatory approval for AI deployments in energy systems.

8.6 Conclusion

With the world focusing on achieving both energy security and sustainability, smart grids have come to be a cornerstone pillar in the power system modernization process. This chapter has presented an integrated overview of how smart grid communication, facilitated by advanced technologies such as IoT, AI, and ML, is transforming traditional energy systems into intelligent, adaptive, and resilient infrastructures. From real-time monitoring to predictive load balancing, these technologies enable decentralized and responsive energy management, especially in the context of increasing DER integration and electrification. However, with growing digital interconnectivity comes a host of challenges. Cybersecurity threats, data privacy concerns, and interoperability limitations are not merely technical issues but also policy and governance challenges. This chapter highlighted the importance of embedding security and compliance at every layer of the grid architecture to ensure system trustworthiness and long-term operability. In the future, the convergence of FL, blockchain-based energy trading, and digital twin modelling holds promising directions for creating future-proofed energy systems. These innovations emphasize privacy preservation, automation, and proactive control, features critical to scaling up smart grids in both urban and rural contexts. Our chapter contributes by synthesizing these developments, offering both technical depth and strategic insight into designing the next generation of sustainable energy infrastructures. The way forward lies in collaborative innovation, across disciplines, sectors, and nations, to realize smart grids that are not only technically advanced but also equitable, secure, and aligned with global climate targets.

References

1 Solomon, B.D., and Krishna, K. (2011). The coming sustainable energy transition: history, strategies, and outlook. *Energy Policy* 39 (11): 7422.

2 Al-Shetwi, A.Q. (2022). Sustainable development of renewable energy integrated power sector: trends, environmental impacts, and recent challenges. *Science of the Total Environment* 822: 153645.

3 Zakeri, B., and Syri, S. (2015). Electrical energy storage systems: a comparative life cycle cost analysis. *Renewable and Sustainable Energy Reviews* 42: 569.

4 Tan, K.M., Babu, T.S., Ramachandaramurthy, V.K., et al. (2021). Empowering smart grid: a comprehensive review of energy storage technology. *Journal of Energy Storage* 39: 102591.

5 Rahbar, K., Chai, C.C., and Zhang, R. (2018). Energy cooperation optimization in microgrids with renewable energy integration. *IEEE Transactions on Smart Grid* 9 (2): 1482.

6 Peng, X., Deng, D., Cheng, S., et al. (2015). Key technologies of electric power big data. *Proceedings of the Chinese Society of Electrical Engineering* 35 (3): 503.

7 Bogdanov, D., Gulagi, A., Fasihi, M., and Breyer, C. (2021). Full energy sector transition towards 100% renewable energy supply: integrating power, heat, transport and industry sectors including desalination. *Applied Energy* 283: 116273.

8 Lund, H., Østergaard, P.A., Chang, M., et al. (2018). The status of 4th generation district heating: research and results. *Energy* 164: 147–159.

9 Breyer, C., Khalili, S., Bogdanov, D., et al. (2022). On the history and future of 100% renewable energy systems research. *IEEE Access* 10: 78176–78218.

10 Mathiesen, B.V., Lund, H., Connolly, D., et al. (2015). Smart energy systems for coherent 100% renewable energy and transport solutions. *Applied Energy* 145: 139–154.

11 Yang, X., Song, Y., Wang, G., and Wang, W. (2010). A comprehensive review on the development of sustainable energy strategy and implementation in China. *IEEE Transactions on Sustainable Energy* 1 (2): 57.

12 Hepbasli, A. (2008). A key review on exergetic analysis and assessment of renewable energy resources for a sustainable future. *Renewable and Sustainable Energy Reviews* 12 (3): 593–661.

13 Bale, C.S., Varga, L., and Foxon, T.J. (2015). Energy and complexity: new ways forward. *Applied Energy* 138: 150–159.

14 Lund, H., Andersen, A.N., Østergaard, P.A., et al. (2012). From electricity smart grids to smart energy systems—a market operation based approach and understanding. *Energy* 42 (1): 96–102.

15 Ghasempour, A. (2019). Internet of things in smart grid: architecture, applications, services, key technologies, and challenges. *Inventions* 4 (1): 22.

16 Franco, P., Martinez, J.M., Kim, Y.C., and Ahmed, M.A. (2021). IoT based approach for load monitoring and activity recognition in smart homes. *IEEE Access* 9: 45325–45339.

17 Wang, W., Chen, Q., Yin, Z., et al. (2022). Blockchain and PUF-based lightweight authentication protocol for wireless medical sensor networks. *IEEE Internet of Things Journal* 9 (11): 8883.

18 Du, H., Wang, J., Niyato, D., et al. (2023). Rethinking wireless communication security in semantic internet of things. *IEEE Wireless Communications* 30 (3): 36.

19 Aldossari, S.M., and Chen, K.-C. (2019). Machine learning for wireless communication channel modeling: an overview. *Wireless Personal Communications* 106 (1 Special Issue): 41.

20 Albarakati, A.J., Boujoudar, Y., Azeroual, M., et al. (2022). Microgrid energy management and monitoring systems: a comprehensive review. *Frontiers in Energy Research* 10: 1097858.

21 Aljafari, B., Vasantharaj, S., Indragandhi, V., and Vaibhav, R. (2022). Optimization of DC, AC, and hybrid AC/DC microgrid-based IoT systems: a review. *Energies* 15 (18): 6813.

22 Chen, C.M., Chen, L., Huang, Y., et al. (2021). Lightweight authentication protocol in edge-based smart grid environment. *EURASIP Journal on Wireless Communications and Networking* 2021: 1–18.

23 Prieto González, L., Fensel, A., Gómez Berbís, J.M., et al. (2021). A survey on energy efficiency in smart homes and smart grids. *Energies* 14 (21): 7273.

24 Borgaonkar, R., Anne Tøndel, I., Zenebe Degefa, M., and Gilje Jaatun, M. (2021). Improving smart grid security through 5G enabled IoT and edge computing. *Concurrency and Computation: Practice and Experience* 33 (18): e6466.

25 Lee, E., Seo, Y.D., Oh, S.R., and Kim, Y.G. (2021). A survey on standards for interoperability and security in the internet of things. *IEEE Communications Surveys & Tutorials* 23 (2): 1020–1047.

26 Elsisi, M., Su, C.L., and Ali, M.N. (2023). Design of reliable IoT systems with deep learning to support resilient demand side management in smart grids against adversarial attacks. *IEEE Transactions on Industry Applications* 60 (2): 2095–2106.

27 Parvin, K., Hannan, M.A., Mun, L.H., et al. (2022). The future energy internet for utility energy service and demand-side management in smart grid: current practices, challenges and future directions. *Sustainable Energy Technologies and Assessments* 53: 102648.

28 Kilthau, M., Asman, M., Karmann, A., et al. (2023). Integrating peer-to-peer energy trading and flexibility market with self-sovereign identity for decentralized energy dispatch and congestion management. *IEEE Access* 11: 145395–145420.

29 Page, A., Hijazi, S., Askan, D., et al. (2016). Research directions in cloud-based decision support systems for health monitoring using Internet-of-Things driven data acquisition. *International Journal of Serving Computing* 4 (4): 18–34.

30 Naru, E.R., Saini, H., and Sharma, M. (2017). A recent review on lightweight cryptography in IoT. In: *2017 International Conference on I-SMAC (IoT in Social, Mobile, Analytics and Cloud) (I-SMAC)* (pp. 887–890). IEEE.

31 Thakkar, A., and Lohiya, R. (2021). A review on machine learning and deep learning perspectives of IDS for IoT: recent updates, security issues, and challenges. *Archives of Computational Methods in Engineering* 28 (4): 3211–3243.

32 Saadeh, M., Sleit, A., Sabri, K.E., and Almobaideen, W. (2018). Hierarchical architecture and protocol for mobile object authentication in the context of IoT smart cities. *Journal of Network and Computer Applications* 121: 1–19.

33 Minoli, D., Sohraby, K., and Occhiogrosso, B. (2017). IoT considerations, requirements, and architectures for smart buildings—energy optimization and

next-generation building management systems. *IEEE Internet of Things Journal* 4 (1): 269–283.
34 Ferrag, M.A., Derdour, M., Mukherjee, M., et al. (2018). Blockchain technologies for the internet of things: research issues and challenges. *IEEE Internet of Things Journal* 6 (2): 2188–2204.
35 Premsankar, G., Di Francesco, M., and Taleb, T. (2018). Edge computing for the internet of things: a case study. *IEEE Internet of Things Journal* 5 (2): 1275–1284.
36 He, D., Kumar, N., Zeadally, S., et al. (2017). Efficient and privacy-preserving data aggregation scheme for smart grid against internal adversaries. *IEEE Transactions on Smart Grid* 9 (4): 3313–3321.
37 Barakabitze, A.A., Ahmad, A., Mijumbi, R., and Hines, A. (2020). 5G network slicing using SDN and NFV: a survey of taxonomy, architectures and future challenges. *Computer Networks* 167: 106984.
38 Shi, W., Cao, J., Zhang, Q., et al. (2016). Edge computing: vision and challenges. *IEEE Internet of Things Journal* 3 (5): 637–646.
39 Agrawal, K., Goktas, P., Sahoo, B., et al. (2024). IoT-based service allocation in edge computing using game theory. In: *International Conference on Distributed Computing and Intelligent Technology* (pp. 45–60). Cham: Springer Nature Switzerland.
40 Taleb, T., Samdanis, K., Mada, B., et al. (2017). On multi-access edge computing: a survey of the emerging 5G network edge cloud architecture and orchestration. *IEEE Communications Surveys & Tutorials* 19 (3): 1657–1681.
41 Alrawais, A., Alhothaily, A., Hu, C., and Cheng, X. (2017). Fog computing for the internet of things: security and privacy issues. *IEEE Internet Computing* 21 (2): 34–42.
42 Nastic, S., Sehic, S., and Dustdar, S. (2017). Provisioning software-defined IoT cloud systems. *IEEE International Conference on Cloud Engineering (IC2E)* (pp. 268–273). IEEE.
43 Roman, R., Lopez, J., and Mambo, M. (2018). Mobile edge computing, fog et al.: a survey and analysis of security threats and challenges. *Future Generation Computer Systems* 78: 680–698.
44 Lu, Y., Huang, X., Dai, Y., et al. (2020). Blockchain and federated learning for privacy-preserved data sharing in industrial IoT. *IEEE Transactions on Industrial Informatics* 16 (6): 4177–4186.
45 Zhou, Z., Chen, X., Li, E., et al. (2019). Edge intelligence: paving the last mile of artificial intelligence with edge computing. *Proceedings of the IEEE* 107 (8): 1738–1762.
46 Ibrahim, B., Rabelo, L., Gutierrez-Franco, E., and Clavijo-Buritica, N. (2022). Machine learning for short-term load forecasting in smart grids. *Energies* 15 (21): 8079.

47 Abate, F., Carratù, M., Liguori, C., and Paciello, V. (2019). A low cost smart power meter for IoT. *Measurement* 136: 59–66.

48 Song, Y., Lin, J., Tang, M., and Dong, S. (2017). An internet of energy things based on wireless LPWAN. *Engineering* 3 (4): 460–466.

49 Hossain, E., Khan, I., Un-Noor, F., et al. (2019). Application of big data and machine learning in smart grid, and associated security concerns: a review. *IEEE Access* 7: 13960–13988.

50 Spanò, E., Niccolini, L., Di Pascoli, S., and Iannacconeluca, G. (2014). Last-meter smart grid embedded in an Internet-of-Things platform. *IEEE Transactions on Smart Grid* 6 (1): 468–476.

51 Yaghmaee, M.H., and Hejazi, H. (2018). Design and implementation of an Internet of Things based smart energy metering. In: *2018 IEEE International Conference on Smart Energy Grid Engineering (SEGE)* (pp. 191–194). IEEE.

52 Savari, G.F., Krishnasamy, V., Sathik, J., et al. (2020). Internet of Things based real-time electric vehicle load forecasting and charging station recommendation. *ISA Transactions* 97: 431–447.

53 Ullah, M., Narayanan, A., Wolff, A., and Nardelli, P.H. (2022). Industrial energy management system: design of a conceptual framework using IoT and big data. *IEEE Access* 10: 110557–110567.

54 Ahammed, M.T., and Khan, I. (2022). Ensuring power quality and demand-side management through IoT-based smart meters in a developing country. *Energy* 250: 123747.

55 Jain, R.K., Damoulas, T., and Kontokosta, C.E. (2014). Towards data-driven energy consumption forecasting of multi-family residential buildings: feature selection via the lasso. In: *International Conference on Computing in Civil and Building Engineering (2014)* (pp. 1675–1682) American Society of Civil Engineers, Orlando.

56 Ding, S., Tao, Z., Li, R., and Qin, X. (2022). A novel seasonal adaptive grey model with the data-restacking technique for monthly renewable energy consumption forecasting. *Expert Systems with Applications* 208: 118115.

57 Sekhar, C., and Dahiya, R. (2023). Robust framework based on hybrid deep learning approach for short term load forecasting of building electricity demand. *Energy* 268: 126660.

58 Shen, Y., Ma, Y., Deng, S., et al. (2021). An ensemble model based on deep learning and data preprocessing for short-term electrical load forecasting. *Sustainability* 13 (4): 1694.

59 Shaikh, AK., Nazir, A., Khan, I., and Shah, AS. (2022). Short term energy consumption forecasting using neural basis expansion analysis for interpretable time series. *Scientific Reports* 12 (1): 22562.

60 Culaba, A.B., Del Rosario, A.J.R., Ubando, A.T., and Chang, J.S. (2020). Machine learning-based energy consumption clustering and forecasting for mixed-use buildings. *International Journal of Energy Research* 44 (12): 9659–9673.

61 Taha, I.B., and Mansour, D.A. (2021). Novel power transformer fault diagnosis using optimized machine learning methods. *Intelligent Automation & Soft Computing* 28 (3): 739–752.

62 Saravanan, D., Hasan, A., Singh, A., et al. (2020). Fault prediction of transformer using machine learning and DGA. In: *2020 IEEE International Conference on Computing, Power and Communication Technologies (GUCON)* (pp. 1–5). IEEE.

63 Dorado Rueda, F., Durán Suárez, J., and del Real Torres, A. (2021). Short-term load forecasting using encoder-decoder wavenet: application to the French grid. *Energies* 14 (9): 2524.

64 Radha, K.S., Priya, R., and Jeevitha, K. (2023). Energy management based on k-nearest neighbour approach in residential application. In: *Proceedings of the 3rd International Conference on Artificial Intelligence and Smart Energy, ICAIS 2023* (pp. 966–971). IEEE.

65 Paradiso, F., Paganelli, F., Luchetta, A., et al. (2013). ANN-based appliance recognition from low-frequency energy monitoring data. In: *2013 IEEE 14th International Symposium on a World of Wireless, Mobile and Multimedia Networks, WoWMoM*. IEEE.

66 Nunna, H.S.V.S.K., Surash, A., Zhanatbekov, A., et al. (2020). Prediction of power demand in residential areas using the load profile clustering technique. In: *9th IEEE International Conference on Power Electronics, Drives and Energy Systems, PEDES 2020*.

67 Qureshi, U.R., Rashid, A., Altini, N., et al. (2025). Explainable intelligent inspection of solar photovoltaic systems with deep transfer learning: considering warmer weather effects using aerial radiometric infrared thermography. *Electronics* 14 (4): 755.

68 Behzad, M., Kim, H., Behzad, M., and Behambari, H.A. (2019). Improving sustainability performance of heating facilities in a central boiler room by condition-based maintenance. *Journal of Cleaner Production* 206: 713–723.

69 Sobral Zhang, H., Tian, Z., and Zuo, M. (2024). Component-level wind farm maintenance considering seasonal uncertainties. In: *2024 IEEE International Conference on Prognostics and Health Management, ICPHM* (pp. 146–151). IEEE.

70 Sobral, J., Datia, N., Pato, M., and Soares, C.G. (2025). Machine learning algorithms to use in the development of maintenance strategies for wind turbines. In: *Proceedings of the 6th International Conference on Renewable Energies Offshore, RENEW 2024*, 491. CRC Press.

71 Shin, W., Han, J., and Rhee, W. (2021). AI-assistance for predictive maintenance of renewable energy systems. *Energy* 221: 119775.

72 Sayal, A., Chaithra, N., Jha, J., and Allagari, N. (2024). AI-based predictive maintenance strategies for improving the reliability of green power systems.

In: *Digital Technologies to Implement the UN Sustainable Development Goals* (W. Leal Filho, S. Kautish, T. Wall, et al. (eds). World Sustainability Series. Springer, Cham. https://doi.org/10.1007/978-3-031-68427-2_2.

73 Chen, S., Ge, W., Liang, X., et al. (2024). Lifelong learning with deep conditional generative replay for dynamic and adaptive modeling towards net zero emissions target in building energy system. *Applied Energy* 353: 122189.

74 Tabaku, E., Vyshka, E., Kapçiu, R., et al. (2025). Utilizing artificial intelligence in energy management systems to improve carbon emission reduction and sustainability. *Jurnal Ilmiah Ilmu Terapan Universitas Jambi* 9 (1.0): 393.

75 Jalli, R., Kumar, P.P., Nuvvula, R.S., et al. (2023). Data science applications in renewable energy: leveraging big data for sustainable solutions. In: *2023 12th International Conference on Renewable Energy Research and Applications (ICRERA)* (pp. 465–471*)*. IEEE.

76 Shatat, A., Shatat, A., Mobin, M., and Theeb, Y. (2024). Enhancing energy consumption in business through data analysis. In: *2024 International Conference on Decision Aid Sciences and Applications, DASA*.

77 Meydani, A., Shahinzadeh, H., Nafisi, H., and Gharehpetian, G.B. (2024). Synchrophasor technology applications and optimal placement of micro-phasor measurement unit (μPMU): part II. In: *2024 28th International Electrical Power Distribution Conference (EPDC)* (pp. 1–18). IEEE.

78 Sharma, V.S., Mahajan, S., Nayyar, A., and Pandit, A. (2024). Deep learning in engineering, energy and finance—principles and applications. In: *Deep Learning in Engineering, Energy and Finance*. CRC Press.

79 Ahmed, H.M., and Salama, M.M.A. (2019). A multi-objective optimization model for the planning of power distribution networks considering system reliability. In: *2019 IEEE Canadian Conference of Electrical and Computer Engineering (CCECE)* (pp. 1–5). IEEE.

80 Filipiak, S. (2010). Application of evolutionary algorithm and classifier system in optimisation of electric power distribution networks post-fault configuration. *International Review of Electrical Engineering* 5(3): 1151–1158.

81 Piao, L., Xue, F., Wu, Q., et al. (2021). Active planning for power distribution networks based on virtual microgrids. In: *2021 IEEE Sustainable Power and Energy Conference (iSPEC)* (pp. 1743–1748). IEEE.

82 Zhang, J., and Ye, J. (2024). Reactive power optimization based on Tent-Chebyshev switching mapping particle swarm optimization. In: *2024 4th International Conference on Energy Engineering and Power Systems (EEPS)* (pp. 420–424). IEEE.

83 Wan, T., Tao, Y., Qiu, J., and Lai, S. (2021). Data-driven hierarchical optimal allocation of battery energy storage system. *IEEE Transactions on Sustainable Energy* 12 (4): 2097–2109.

84 Gbadega, P.A., and Balogun, O.A. (2024). An Islanding detection and load curtailment strategy for radial distribution networks using Squid Game optimizer algorithm. In: *2024 International Conference on Advanced Mechatronic Systems (ICAMechS)* (pp. 268–273). IEEE.

85 Ghaemi, A., Safari, A., Afsharirad, H., and Shayeghi, H. (2022). Situational awareness and deficiency warning system in a smart distribution network based on stacking ensemble learning. *Applied Soft Computing* 128: 109427.

86 Soares, J., Ghazvini, M.A.F., Silva, M., and Vale, Z. (2016). Multi-dimensional signaling method for population-based metaheuristics: solving the large-scale scheduling problem in smart grids. *Swarm and Evolutionary Computation* 29: 13–32.

87 Babayomi, O., Zhang, Z., Dragicevic, T., et al. (2023). Smart grid evolution: predictive control of distributed energy resources—a review. *International Journal of Electrical Power & Energy Systems* 147: 108812.

88 Kefayat, M., Lashkar Ara, A., and Nabavi Niaki, S.A. (2015). A hybrid of ant colony optimization and artificial neural network for short-term load forecasting. *Energy Conversion and Management* 92: 134–146.

89 Keerthisinghe, C., Chapman, A.C., and Verbič, G. (2018). Energy management of PV-storage systems: policy-based reinforcement learning. *IEEE Transactions on Industrial Informatics* 15 (6): 3301–3312.

90 Bilal, M., Algethami, A.A., Imdadullah, and Hameed, S. (2024). Review of computational intelligence approaches for demand response optimization in microgrids. *IEEE Access* 12: 367–379.

91 Fose, N., Singh, A.R., Krishnamurthy, S., and Ratshitanga, M. (2024). Empowering distribution system operators: a review of reinforcement learning applications in demand response. *Heliyon* 10 (2): e22223.

92 Howell, S., Rezgui, Y., Hippolyte, J.L., and Mourshed, M. (2016). Semantic interoperability for holonic energy optimization of connected smart homes and distributed energy resources. In: *eWork and eBusiness in Architecture, Engineering and Construction: ECPPM 2016* (pp. 259–267). CRC Press.

93 Ali, M., Abdulgalil, M.A., Habiballah, I., and Khalid, M. (2023). Optimal scheduling of isolated microgrids with hybrid renewables and energy storage systems considering demand response. *IEEE Access* 11: 80266–80273.

94 Meena, N.K., Yang, J., and Zacharis, E. (2019). Optimal planning and operational management of Islanded microgrids: a review. *Energy Procedia* 158: 2480–2485.

95 Sanseverino, E.R., Di Silvestre, M.L., Mineo, L., et al. (2016). A multi-agent system reinforcement learning based optimal power flow for islanded microgrids. In: *2016 IEEE 16th International Conference on Environment and Electrical Engineering (EEEIC)* (pp. 1–6). IEEE.

96 El Mrabet, Z., Kaabouch, N., El Ghazi, H., and El Ghazi, H. (2018). Cyber-security in smart grid: survey and challenges. *Computers & Electrical Engineering* 67: 469–482.

97 Silva, B.N., Khan, M., and Han, K. (2020). Towards sustainable smart cities: a review of trends, architectures, components, and open challenges in smart grid. *Sustainable Cities and Society* 60: 102252.

98 Mitrea, D., Toderean, L., Cioara, T., et al. (2023). Smart contracts and homomorphic encryption for private P2P energy trading and demand response on blockchain. *Heliyon* 9 (11): e22357.

99 Binjubeir, M., Ahmed, A.A., Ismail, M.A.B., et al. (2019). Comprehensive survey on big data privacy protection. *IEEE Access* 8: 20067–20079.

100 Rahman, R., Kumar, N., and Nguyen, D.C. (2024). Electrical load forecasting in smart grid: a personalized federated learning approach. *arXiv preprint arXiv:2411.10619*.

101 Martinez, J., Ruiz, A., Puelles, J., et al. (2019). Smart grid challenges through the lens of the European General Data Protection Regulation. In: *International Conference on Information Systems Development* (pp. 113–130). Cham: Springer International Publishing.

102 Hargreaves, N.B., Pantea, S.M., and Taylor, G.A. (2014). Adopting the IEC common information model to enable smart grid interoperability and knowledge representation processes. In: *Large Scale Renewable Power Generation: Advances in Technologies for Generation, Transmission and Storage* (pp. 439–462). Singapore: Springer Singapore.

103 Mutule, A., Antoskova, I., Carroll, P., et al. (2024). Smart grid standardisation: contributions and opportunities of EU Horizon 2020 projects. *Latvian Journal of Physics and Technical Sciences* 61 (4): 17.

104 Papadimitriou, C., Efthymiou, V., and Carroll, P. (2022). Harmonization of IEC 61850 for advanced smart grid control. *Energies* 15 (7): 2539.

105 Zanabria, C., Andrén, F.P., Kathan, J., and Strasser, T.I. (2018). Rapid prototyping of multi-functional battery energy storage system applications. *Applied Sciences* 8 (8): 1326

106 Tada, Y., Tanak, T., and Serizaw, Y. (2012). Status of international standards about CIM on EMS for trunk power network. *IEEJ Transactions on Power and Energy* 132 (10): 829.

107 Chang, K.H., and Mason, B. (2012). The IEEE 802.15. 4g standard for smart metering utility networks. In: *2012 IEEE Third International Conference on Smart Grid Communications (SmartGridComm)*. IEEE.

108 Miao, X., Chen, X., Ma, X.-M., et al. (2012). Comparing smart grid technology standards roadmap of the IEC, NIST and SGCC. In: *2012 China International Conference on Electricity Distribution CICED* (pp. 1–4). IEEE.

109 Venkata Pavan Kumar, Y., Naga Venkata Bramareswara Rao, S., and Kannan, R. (2023). Islanding detection in grid-connected urban community multi-microgrid clusters using decision-tree-based fuzzy logic controller for improved transient response. *Urban Science* 7 (3): 72.

110 Zhou, Z., and Wang, W. (2020). Standardization of smart grid technology for the development of global energy interconnection. *IEEE Student Conference on Electric Machines and Systems (SCEMS)* (pp. 569–573). IEEE.

111 Rana, M.M., Uddin, M., Sarkar, M.R., et al. (2023). Applications of energy storage systems in power grids with and without renewable energy integration—a comprehensive review. *Journal of energy storage* 68: 107811.

112 Wagner, M., Kuba, M., and Oeder, A. (2012). Smart grid cyber security: a German perspective. In: *2012 International Conference on Smart Grid Technology, Economics and Policies (SG-TEP)* (pp. 1–4). IEEE.

113 Yagiz, M.A., MohajerAnsari, P., Pesé, M.D., and Goktas, P. (2024). Transforming in-vehicle network intrusion detection: VAE-based knowledge distillation meets explainable AI. In: *Proceedings of the Sixth Workshop on CPS & IoT Security and Privacy* (pp. 93–103). Association for Computing Machinery.

114 Yagiz, M.A., and Goktas, P. (2025). LENS-XAI: redefining lightweight and explainable network security through knowledge distillation and variational autoencoders for scalable intrusion detection in cybersecurity. *arXiv preprint arXiv:2501.00790*.

115 Sharma, K., and Saini, LM. (2017). Power-line communications for smart grid: progress, challenges, opportunities and status. *Renewable and Sustainable Energy Reviews* 67: 704–751.

116 Samad, T., and Bienert, R. (2020). Smart manufacturing and smart grids. In: *Smart Manufacturing: Applications and Case Studies* (pp. 455–475). Elsevier.

117 Bennett, C., and Highfill, D. (2008). Networking AMI smart meters. In: *2008 IEEE Energy 2030 Conference* (pp. 1–8). IEEE.

118 Chhaya, L., Sharma, P., Kumar, A., and Bhagwatikar, G. (2020). Cybersecurity for smart grid: threats, solutions and standardization. In: *Advances in Greener Energy Technologies* (pp. 17–29). Singapore: Springer Singapore.

119 Bui, N., Castellani, A.P., Casari, P., and Zorzi, M. (2012). The internet of energy: a web-enabled smart grid system. *IEEE Network* 26 (4): 39–45.

120 Kaainoa, C.E., and Jenkins, S.E. (2019). Foundational challenges and opportunities with turning data into information for grid modernization. In: *2019 International Energy and Sustainability Conference (IESC)* (pp. 1–19). IEEE.

121 Meng, Y., Yang, Y., Chung, H., et al. (2018). Enhancing sustainability and energy efficiency in smart factories: a review. *Sustainability* 10 (12): 4779.

122 Cui, Y., Kara, S., and Chan, K.C. (2020). Manufacturing big data ecosystem: a systematic literature review. *Robotics and Computer-Integrated Manufacturing* 62: 101861.

123 Gupta, S., Dangayach, G.S., Singh, A.K., et al. (2018). Implementation of sustainable manufacturing practices in Indian manufacturing companies. *Benchmarking: An International Journal* 25 (7): 2441–2459.

124 Choi, S., Kim, B.H., and Do Noh, S. (2015). A diagnosis and evaluation method for strategic planning and systematic design of a virtual factory in smart manufacturing systems. *International Journal of Precision Engineering and Manufacturing* 16: 1107–1115.

125 Fathy, Y., Jaber, M., and Nadeem, Z. (2021). Digital twin-driven decision making and planning for energy consumption. *Journal of Sensor and Actuator Networks* 10 (2): 37.

126 Wu, Y., Wu, Y., Cimen, H., et al. (2022). Towards collective energy community: potential roles of microgrid and blockchain to go beyond P2P energy trading. *Applied Energy* 314: 119003.

127 Sarkar, B., Mridha, B., and Pareek, S. (2022). A sustainable smart multi-type biofuel manufacturing with the optimum energy utilization under flexible production. *Journal of Cleaner Production* 332: 129869.

128 Sarker, M.AA., Shanmugam, B., Azam, S., and Thennadil, S. (2024). Enhancing smart grid load forecasting: an attention-based deep learning model integrated with federated learning and XAI for security and interpretability. *Intelligent Systems with Applications* 23: 200422.

129 Khalil, M., Esseghir, M., and Boulahia, L.M. (2023). Privacy-preserving federated learning: an application for big data load forecast in buildings. *Computers & Security* 131: 103211.

9

Urban Environmental Sensing for Sustainable Development: A Focus on Air Quality, Temperature, and Humidity

Pankaj P. Tasgaonkar, Siddhant Kamlaskar, Tarun Dhote, and Aditya Chougule*

COEP Technological University, Pune, India

*Corresponding author: ppt.extc@coeptech.ac.in

9.1 Introduction

Urbanization is a global phenomenon transforming cities into hubs of economic, social, and cultural activity. However, the rapid growth of urban populations has also given rise to a multitude of environmental challenges, particularly concerning air quality, temperature fluctuations, and humidity levels. These factors significantly impact human health, ecosystem stability, and the overall sustainability of urban environments. As cities continue to expand, it becomes crucial to monitor and manage these environmental variables to ensure that urban development is aligned with sustainable practices. Environmental sensing technologies have emerged as essential tools in addressing these challenges. The integration of real-time monitoring systems for air quality, temperature, and humidity can provide valuable insights into the state of urban environments, enabling informed decision-making for urban planning and policy development. This research paper explores the role of urban environmental sensing in achieving sustainable development goals, with a particular focus on air quality, temperature, and humidity. By leveraging advanced sensor

AI and ML Techniques in IoT-based Communication: A Path to Sustainable Development Goals, First Edition. Edited by Sumita Mishra, Nishu Gupta and Polat Goktas.

technologies, data analytics, and Internet of Things (IoT) solutions, cities can not only monitor environmental parameters but also implement effective strategies to mitigate pollution, optimize energy use, and improve the overall quality of life for urban residents. Through a comprehensive review of existing literature and case studies, this chapter aims to highlight the importance of urban environmental sensing in shaping resilient, sustainable cities. Additionally, it will examine the challenges, opportunities, and future directions for the integration of these technologies into urban planning frameworks. Ultimately, the research will contribute to a deeper understanding of how environmental sensing can help urban areas move toward more sustainable, healthy, and liveable futures.

9.1.1 Overview

Urbanization has accelerated globally, creating complex environmental pressures such as hazardous air pollution, unregulated temperature shifts, and volatile humidity patterns. These conditions exacerbate public health crises, degrade urban livability, and threaten ecological balance. Growing urban populations strain infrastructure, amplifying pollution and creating heat islands that heighten risks of respiratory diseases, cardiovascular disorders, and heat exhaustion. To combat these challenges, innovative IoT-driven solutions are being deployed to enable precise, real-time environmental monitoring. Such systems leverage interconnected sensor networks to collect hyperlocal data on air composition, thermal gradients, and moisture levels, empowering municipalities and communities to identify risks, optimize resource deployment, and implement adaptive policies. By transforming raw environmental data into actionable insights, IoT technologies bridge the gap between urban growth and ecological stewardship, fostering smarter, more resilient cities.

This study introduces a scalable IoT architecture tailored for multiparameter environmental surveillance in urban settings. The framework employs an ESP32 microcontroller – a cost-effective, energy-efficient device with dual-core processing and integrated Wireless Fidelity (Wi-Fi) – to aggregate inputs from a heterogeneous sensor array. Key components include the SHARP GP2Y1014AU0F sensor for optical-based particulate matter detection, the DHT11 for calibrated thermal and hygrometric readings, and MQ-series gas detectors (MQ-135 for volatile organic compounds, MQ-9 for carbon monoxide, and MQ-2 for hydrocarbon emissions). Data is transmitted to the ThingSpeak IoT platform for cloud-based visualization, enabling remote access via dashboards and APIs [1]. This hybrid design – combining edge computing for real-time alerts and cloud analytics for trend forecasting – supports both immediate interventions and strategic planning. By mapping environmental dynamics across spatial and temporal scales, the system aids in mitigating pollution hotspots, optimizing green infrastructure, and advancing climate-resilient urban policies. The findings highlight IoT's role in democratizing environmental intelligence, fostering equitable health outcomes, and steering urbanization toward sustainable trajectories through data transparency and civic engagement.

The system's core components include the following:

- **ESP32 Microcontroller**: Acts as the central controller, managing sensor data collection and communication.
- **Input System**: Includes sensors such as the SHARP GP2Y1014AU0F dust sensor, DHT11 temperature and humidity sensor, and various MQ sensors for monitoring air quality and harmful gases.
- **Output System**: Data is shown remotely on the ThingSpeak, which provides real-time tracking and alerts.

This setup effectively tracks urban environmental factors, aiding sustainable urban planning and improving public health and safety.

9.2 Flowchart

This flowchart outlines the operational workflow of an IoT-based environmental monitoring system using an ESP32 microcontroller. The process begins with initializing sensors (dust/particulate matter sensor, DHT11 for temperature/humidity, and MQ-series gas sensors MQ2, MQ9, and MQ135 for detecting pollutants like smoke, CO, and VOCs) and establishing Wi-Fi connectivity. The system then checks the Wi-Fi connection status; if unavailable (NO), it retries until a connection is established (YES). Once connected, the ESP32 reads raw data from all sensors, processes and converts it into interpretable formats (e.g. analog-to-digital conversion, calibration). The processed data is then transmitted to ThingSpeak, an IoT cloud platform, for real-time visualization and storage. The cycle concludes (END) after successful data transmission, enabling continuous environmental monitoring. This loop ensures seamless data collection, connectivity resilience, and cloud-based analytics for urban sustainability applications, as shown in Figure 9.1.

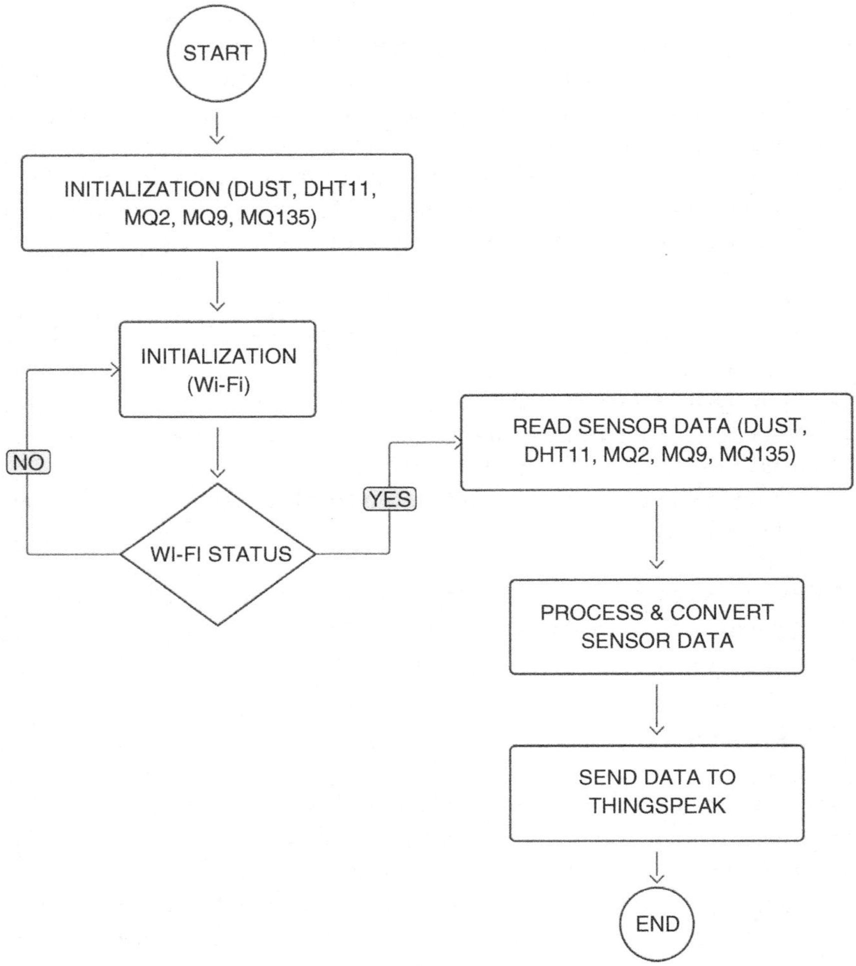

Figure 9.1 Flowchart of IoT-based urban environmental sensing system. *Source*: Authors.

9.3 Methodology

This study adopts a systematic, multiphase methodology to develop an IoT-enabled environmental monitoring platform tailored for urban ecosystems. The framework is organized into five core components: hardware architecture design, multisensor integration and calibration, edge-based data processing, hybrid output interfaces, and cloud-enabled remote analytics. The hardware setup revolves around a modular sensor array comprising the ESP32 microcontroller, chosen for its dual-core processing, low-power operation, and seamless Wi-Fi/Bluetooth connectivity, which serves as the central hub. Sensor integration involves coupling specialized detectors – such as the DHT11 for calibrated temperature/humidity tracking, the SHARP GP2Y1014AU0F for airborne particulate analysis, and MQ-series gas sensors (MQ-135 for NO_x/CO_2, MQ-9 for CO/combustible gases, and MQ-2 for hydrocarbon detection) – to ensure comprehensive environmental profiling [2].

Data acquisition is optimized through real-time signal conditioning, noise reduction algorithms, and on-device preprocessing to minimize latency. The system employs a dual-channel output approach: locally, data is displayed via OLED screens and LED indicators for immediate user feedback, while remotely, encrypted datasets are transmitted to the ThingSpeak IoT platform using MQTT protocols for cloud storage and visualization. Software interfaces include customizable dashboards for trend analysis, threshold-based alert systems, and API integration with urban management tools. Rigorous field testing and cross-validation with reference instruments ensure accuracy, scalability, and interoperability with existing smart city infrastructures. By prioritizing energy efficiency, data security, and user-centric design, this methodology supports adaptive urban planning, pollution mitigation, and climate-resilient policy formulation.

9.4 Pollution and Their Impacts

Once the system is operational, it is tested for accuracy and reliability. The calibration of the sensors is performed to ensure that the readings are within acceptable limits. Additionally, the system's performance is evaluated based on data transmission reliability, sensor response time, and real-time monitoring capabilities [2, 3]. The system's scalability and potential for integration with larger urban sensing networks are also considered [4].

9.4.1 Air Quality Monitoring and Its Impacts

Urban air quality is the most urgent issue concerning city planners, citizens, and environmental scientists. The fast growth of cities has made air pollution a major environmental issue with direct impacts on human health, biodiversity, and the climate. Urban air quality monitoring is the constant tracking of pollutants in the air, including particulate matter (PM2.5 and PM10), nitrogen oxides (NO_X), sulfur dioxide (SO_2), carbon monoxide (CO), ozone (O_3), and volatile organic compounds (VOCs) [2, 5, 6].

Human activities, including industrial manufacturing, transportation, and the combustion of fossil fuels, are the main causes of air pollution. In cities, these pollutants usually surpass the acceptable threshold, causing extreme environmental and health implications. Real-time monitoring of air quality enables governments to issue alerts, implement air quality standards, and make preparations for future city planning that reduce pollution [3].

9.4.1.1 Effects of Poor Air Quality

- **Health Risks**: Poor air quality has been associated with various respiratory and cardiovascular ailments, such as asthma, bronchitis, and heart attacks. Vulnerable groups include children, older people, and those with existing health conditions. Prolonged exposure to air pollutants may result in chronic illnesses and even death [2].
- **Environmental Harm**: Ground-level ozone and nitrogen oxides are among the pollutants that can lead to acid rain, which destroys plants, water bodies, and soil quality. Particulate matter in large quantities can smother plants, limit photosynthesis, and affect urban ecosystem biodiversity [6].
- **Climate Change**: Some of the air pollutants like carbon dioxide (CO_2) and methane (CH_4) are greenhouses as well. They tend to retain heat within the atmosphere, thus leading to global warming. Urban centers, with a large number of industries and cars, are key contributors to climate change [4].

Urban air quality monitoring is essential in the identification of areas with pollution hotspots and how pollutants spread throughout cities. This information can inform policy, encourage green technology, and advocate for behavior that minimizes emissions, thus enhancing urban living standards.

9.4.2 Sound Quality Monitoring and Its Impacts

Urban sound quality refers to the nature and amount of noise pollution present within an area. Urban noise monitoring is interested in monitoring and controlling the impacts of noise pollution, which has become an increasingly significant issue as cities grow. The primary causes of noise pollution within cities are traffic, construction work, industrial processes, public events, and domestic disturbance. Excessive noise can result in damaging effects on physical and mental well-being, and the monitoring of noise levels throughout the urban environment is vital for successful environmental management.

Monitoring the quality of sound in cities is usually done through the application of sound level meters and noise sensors that detect levels of noise in decibels (dB). These devices are installed in several locations, like along roads, industrial areas, or residential districts, to provide information on the intensity and frequency of sound.

It is monitored over a 24-hour period to evaluate noise variations during the day and night so that urban planners can plan measures to reduce noise pollution [7].

9.4.2.1 Effects of Excessive Noise

- **Health Impacts**: Long-term exposure to excessive noise has been associated with various health issues. Repeated exposure to noise can lead to hearing loss, disrupted sleep patterns, hypertension, and heightened levels of stress. The World Health Organization (WHO) has identified noise pollution as a risk factor for cardiovascular disease. Noise can also harm the development of children and result in cognitive impairment [7].
- **Psychological Impact**: Regular exposure to prolonged periods of loud noise may bring about psychological outcomes, including irritability, anxiety, and reduction in quality of life. Concentration may be affected, compromising productivity and learning capacity in workplaces or schools [7].
- **Biodiversity Impact**: Noise pollution can disrupt the communication, navigation, and mating habits of animals in cities. Birds, for instance, can be unable to hear mating calls or predator warnings, which can have a detrimental impact on their populations [7].

Urban sound quality monitoring is a critical component of the mitigation of noise pollution by guiding legislation, zoning regulations, and noise-abatement measures. It informs the authorities on how noise levels are related to human or environmental health, thus promoting improved urban planning.

9.4.3 Water Quality Monitoring and Its Impacts

Urban water quality is usually under the danger of contamination from industrial effluents, agricultural runoff, sewage outlets, and stormwater. With growing cities, there is a growing need to implement effective monitoring of water quality in order to guarantee safe water for drinking, maintain aquatic environments, and stop the spread of waterborne illnesses. Urban aquatic ecosystems like lakes, rivers, and reservoirs have to be constantly monitored for water pollutants including heavy metals, nitrates, phosphates, pesticides, bacteria, and other pollutants that impact human health negatively and the environment.

Water quality monitoring is the process of collecting and analyzing water samples to determine parameters like pH, dissolved oxygen, turbidity, temperature, and the presence of harmful chemicals or pathogens. These parameters give important information about the general health of water bodies. Advanced water quality monitoring systems employ sensors and remote sensing technology to give continuous, real-time information, enabling authorities to act quickly if water quality declines [8, 9].

9.4.3.1 Impacts of Poor Water Quality

- **Human Health**: Polluted water is one of the main causes of waterborne diseases like cholera, dysentery, and typhoid. Polluted water also leads to long-term health problems, such as heavy metal poisoning, which may result in neurological damage and kidney failure. Pesticides and other chemicals in drinking water are also problematic, as they may lead to cancer and reproductive problems [8].
- **Impact on the Environment**: Unhealthy water can ruin local environments. Excessive levels of nutrients (eutrophication) may cause algal blooms, which remove oxygen from the water and kill fish and other aquatic organisms. Plastics and chemicals are also pollutants that destroy food chains and kill sea creatures [9].
- **Economic Impact**: Polluted water sources can result in higher treatment expenses for drinking water. Furthermore, the tourism and recreation sectors can be affected as dirty rivers, lakes, and beaches are no longer attractive to the public.

Water quality monitoring is critical to guaranteeing that urban water resources are safe and sustainable. Proper monitoring enables cities to identify sources of pollution, monitor trends over time, and apply the right policies to protect water supplies.

9.5 Hardware Setup

The IoT-based environmental sensing system relies on the ESP32 microcontroller, which is equipped with Wi-Fi connectivity to facilitate data transmission to remote platforms for real-time monitoring. The ESP32 is chosen for its processing power, low energy consumption, and seamless integration with Wi-Fi-based communication.

9.5.1 Sensors

- **SHARP GP2Y1014AU0F Dust Sensor**: This sensor measures the concentration of particulate matter in the air, particularly PM2.5, which is a significant indicator of air quality. It operates on an infrared light-scattering principle to detect dust particles.
- **DHT11 Temperature and Humidity Sensor**: This sensor measures both ambient temperature and relative humidity, which are critical factors in assessing urban climate and comfort levels. The DHT11 provides accurate measurements within the range of 20–90% humidity and temperatures from 0 °C to 50 °C.

- **MQ-135 Air Quality Sensor**: The MQ-135 sensor is used to measure a range of gases including ammonia (NH_3), nitrogen oxide (NO_x), alcohol, benzene, and smoke. It provides real-time data on the general air quality, crucial for understanding the pollution levels in urban environments.
- **MQ-9 Carbon Monoxide and Flammable Gases Sensor**: This sensor detects carbon monoxide (CO) and other flammable gases, making it essential for monitoring hazardous gas levels that may be present in urban environments, particularly from vehicles and industrial activities.
- **MQ-2 Methane, Butane, LPG, and Smoke Sensor**: The MQ-2 sensor detects gases like methane, butane, liquefied petroleum gas (LPG), and smoke. These are crucial indicators of potential fire hazards and indoor air quality concerns in urban areas.

9.5.1.1 Sensor Integration and Data Acquisition

Each sensor is interfaced with the ESP32 through analog or digital pins, depending on the sensor's output type. The data from each sensor is continuously collected by the microcontroller. The readings are then processed to obtain meaningful environmental data, such as particulate matter concentration, air quality index, temperature, and humidity levels. The sensors' data is sampled at regular intervals to ensure real-time monitoring.

9.5.1.2 Data Processing and Transmission

The ESP32 microcontroller performs basic data processing, including filtering and averaging of sensor readings, to improve the accuracy and reliability of the data. Once the data is processed, it is transmitted over Wi-Fi to an online platform for remote monitoring and visualization. The communication protocol used is MQTT (Message Queuing Telemetry Transport), which allows efficient data exchange between the microcontroller and cloud services.

9.5.1.3 Output Systems

The system provides multiple output options for real-time monitoring:

ThingSpeak: It is an open-source IoT analytics platform that enables real-time data collection, visualization, and analysis. It allows devices such as ESP32, Arduino, and other microcontrollers to send sensor data to the cloud using HTTP requests or MQTT. The platform provides a user-friendly dashboard with live graphs, data storage, MATLAB-based analytics, and alert notifications. ThingSpeak is widely used for applications like environmental monitoring, smart agriculture, and industrial IoT solutions.

9.5.1.4 Software Setup

The software for the system is designed using the Arduino Integrated Development Environment (IDE) ThingSpeak, which is compatible with the ESP32.

The following steps are involved in the software development process:

- **Sensor Data Collection**: The code interfaces with the sensors and collects data at set intervals.
- **Data Processing**: The collected raw data is processed, including conversion from analog values to meaningful environmental parameters.
- **Cloud Communication**: Using MQTT and the ThingSpeak API, the processed data is sent to a remote cloud server for storage and access. The ThingSpeak app is configured to receive and visualize data from the cloud.
- **Threshold-based Notifications**: The software includes logic for triggering alerts in case the environmental data exceeds predefined thresholds, such as a high level of particulate **matter or gas concentrations**.

9.5.1.5 Data Visualization and Analysis

The collected data is analyzed and presented in an intuitive format through the ThingSpeak Cloud platform. Users can visualize trends in environmental conditions over time, making it easier to assess the urban environment's status. The system allows for both real-time monitoring and historical data analysis to support long-term sustainability efforts.

9.6 Different Types of Communication Interfaces

Communication interfaces allow data to be exchanged between systems or devices. They are often classified by communication method, protocol, or medium. Below are some of the most important communication interfaces:

9.6.1 Serial Communication Interfaces

Serial communication interfaces provide the transfer of data between devices in a serial fashion, a bit at a time, with simplicity and reliability. Microcontrollers utilize the Universal Asynchronous Receiver/Transmitter (UART) interface for straightforward and efficient serial communication. The Serial Peripheral Interface (SPI) is a synchronous protocol that performs well for high-speed data exchange between microcontrollers and peripherals and is suitable for applications demanding fast communication [3, 10]. The Inter-Integrated Circuit (I2C) interface is especially convenient for the interconnection of several devices using just two wires (SDA and SCL), which makes circuit design easy and wiring less complex. RS-232 and RS-485 are standard serial communications widely applied in industrial and legacy environments because they are strong and support long-distance communications. The CAN interface is crucial in automotive and industrial applications, delivering dependable and robust communication between multiple microcontrollers, particularly in environments where data integrity and speed are paramount. In combination, these serial communication interfaces provide flexible solutions for a variety of applications, ranging from straightforward device communication to sophisticated industrial networks.

9.6.2 Parallel Communication Interfaces

Parallel communication interfaces such as Peripheral Component Interconnect (PCI), Small Computer System Interface (SCSI), and Integrated Drive Electronics/Advanced Technology Attachment (IDE/ATA) have important roles to play in interfacing internal hardware and peripherals on computer systems. PCI is widely used on computer motherboards in interfacing internal hardware components such as graphics cards, network cards, and storage devices and achieves parallel data transfer at high speeds. SCSI is utilized mainly for connecting peripherals and storage devices and provides benefits in multidevice environments and supporting features such as daisy-chaining. IDE/ATA interfaces are mainly employed in hard disk connections for transferring data between the motherboard and storage devices. Parallel interfaces provide efficient data exchange with support for both legacy and contemporary computing environments.

9.6.3 Wireless Communication Interfaces

Wireless communication interfaces such as Wi-Fi, Bluetooth, ZigBee, Near Field Communication (NFC), and cellular (4G/5G) technology offer flexible and convenient connectivity in the absence of physical cables. Wi-Fi is commonly used for connecting devices to networks wirelessly, allowing for Internet access and data sharing between homes, offices, and public areas. Bluetooth provides short-range communication for the connection of devices like headphones, speakers, and IoT devices with easy device pairing and data transfer. ZigBee is a low-data-rate, low-power wireless protocol widely used in smart-home applications and automation systems because it is energy efficient and supports mesh networking. NFC enables communication between devices over a few centimeters and is thus suitable for contactless payments and secure data exchanges. Cellular networks, such as 4G and 5G, provide reliable wireless communication over extended distances, providing high-speed Internet access and reliable connections for mobile networks and smart devices. Collectively, these wireless interfaces enable a broad variety of applications, from individual device connectivity to advanced industrial and smart city networks [3, 9].

9.6.4 Network Communication Interfaces

Network communication interfaces like Ethernet, Universal Serial Bus (USB), High-Definition Multimedia Interface (HDMI), and Musical Instrument Digital Interface (MIDI) enable effective data transfer and connectivity between devices. Ethernet is a high-speed communication standard that is commonly used in local area networks (LANs) for wired networking, offering secure and fast Internet and data connections in homes, offices, and data centers. The USB interface is a general-purpose and widely used standard for peripheral connection of keyboards, mice, printers, and external storage devices to computers and other devices with both data transfer and power supply functionality. HDMI is mainly employed for video and audio data transmission between devices, like from computers or media players to displays and TVs, with high-definition and ultra-high-definition content support. MIDI is a dedicated communication protocol that enables electronic musical instruments, computers, and other digital audio devices to talk to each other, and facilitate music creation, digital sound manipulation, and synchronization between hardware and software. Network communication interfaces like these are central to contemporary digital ecosystems, providing greater connectivity, media streaming, and creative processes.

9.6.5 Analog and Digital Communication Interfaces

Analog and digital interfaces for communication, such as General-Purpose Input/Output (GPIO), analog interfaces such as Analog-to-Digital Converters (ADC) and Digital-to-Analog Converters (DAC), and Modbus, are all important for connectivity and control across a range of applications. GPIO provides simple on/off control of external devices and is suitable for straightforward tasks such as switching LEDs on and off or sensing the state of switches in embedded systems. Analog interfaces, such as ADCs and DACs, provide the analog to digital and digital to analog conversion, allowing analog sensors and audio devices to be integrated with digital systems. Analog interfaces are essential in applications that need accurate measurement and signal processing, for example, in environmental sensors and audio devices. Modbus is a very strong communication protocol applied in industrial environments that can be used for connecting and controlling devices in serial or Ethernet networks. It is extensively used in automation and industrial control systems due to its ease of use, reliability, and compatibility with a wide variety of devices. Combined, these interfaces accommodate varied communication requirements in consumer electronics and industrial settings [2, 3, 7].

9.6.6 Human–Machine Interfaces

Human–machine interfaces (HMIs) include an assortment of technologies that allow the user to interact with electronic devices. Touchscreens are an interactive dynamic graphical interface that enables users to interact with devices using intuitive touch motions, widely present in smartphones, tablets, and new-generation industrial control panels. Keypads and buttons are simple input interfaces, providing straightforward and dependable ways of data entry and control for use in devices like home appliances and security systems. Voice recognition systems make hands-free communication possible by processing voice commands, increasing convenience and accessibility, especially in smart home appliances, virtual assistants, and car systems. All these HMI technologies make the user experience richer by offering different and flexible means of controlling and communicating with devices.

9.6.7 Specialized Communication Interfaces

Specialized interface communications, for example, Mobile Industry Processor Interface (MIPI), Serial Digital Interface (SDI), and Thunderbolt, are designed to meet particular high-performance applications. MIPI finds extensive application in mobile devices for interconnecting internal components like display, camera, and sensors in order to offer efficient data communication with low power consumption. SDI is utilized mainly in professional video equipment and offers strong and reliable transmission of high-definition video signals in broadcast and production environments. Thunderbolt provides a high-speed interface for peripherals to be connected to computers, which is a blend of PCI Express and DisplayPort protocols. It provides fast data transfer, video output, and power delivery in a single cable, which makes it suitable for high-performance computing and creative workstations. Such specialized interfaces provide advanced features and high data throughput, which suits the needs of particular industries and applications.

9.7 Ways to Encounter the Problem

To properly face urban environmental issues determined by comparative data analysis, a strategic and multidimensional approach is required. The following are some efficient means of addressing these issues:

9.7.1 Measures to Improve Air Quality

Air quality enhancement efforts encompass a mix of emission reduction, green infrastructure, and technology applications. Strategies for emission reductions include implementing tighter emission standards and frequent monitoring of industries, particularly in highly polluted areas such as Chandrapur and Ghuggus. Strengthening vehicle emission policies through incentives for the use of electric vehicles (EVs) and increasing emission standards for both public and private transport can go a long way in curbing air pollution. Green infrastructure plays a vital role in improving air quality, with urban greening initiatives aimed at increasing green cover in high-pollution areas such as Mumbai's Sion region, which has recorded high PM10 levels. Additionally, creating green belts around industrial zones by planting trees can serve as natural air filters. Integration of technology is also an important element, with the proliferation of low-cost air quality sensors, for example, those tested by IIT Kanpur in Mumbai, aiding in increased monitoring capacity. In addition, the use of digital platforms to share real-time data can increase public awareness and drive behavior change, supporting healthier and more sustainable cities [1–3].

9.7.2 Water Quality Management

Water quality management consists of an integrated approach that entails wastewater treatment, pollution source management, and restoration programs. Enhanced efficiency of sewage treatment plants (STPs) is essential, especially in the case of urban areas with high levels of water pollution. Better performance of common effluent treatment plants (CETPs) in industrial areas, as reported by the Maharashtra Pollution Control Board (MPCB), will assist in treating industrial wastewater better. It is also important to control pollution at the source level with the enforcement of rigorous guidelines for the treatment of effluents before discharge into water bodies so that industrial discharges are minimized. Encouraging environmentally sustainable agriculture can also reduce chemical runoff into surface and groundwater, safeguarding these precious resources. Restoration efforts, like river and lake restoration projects under the "Majhi Vasundhara" program in Maharashtra, are crucial for reviving natural water bodies and sustaining ecological balance. These holistic approaches can greatly improve water quality and contribute to long-term environmental sustainability [8, 9].

9.7.3 Noise Pollution Control

Mitigation of noise pollution is a blend of strategic planning, infrastructure development, and community participation to achieve a quieter and healthier city. The enforcement of zoning regulations is an effective measure, allocating certain areas for commercial, industrial, and residential use to reduce noise disturbance in sensitive areas. Constructing physical sound barriers like walls and sound-reducing fences, and green buffers by way of tree planting, will effectively absorb and lower the levels of noise. Also, ensuring the use of low-noise equipment and machines in industrial processes and construction operations can control the noise at its source. Support for the application of noise-minimizing technologies like silent generators and electric cars can also keep ambient noise at lower levels. Strict enforcement of noise laws, particularly during festivals, construction activities, and late-night events, through continuous monitoring and stringent penalties is essential. Public campaigns can raise awareness among communities about the effects of noise pollution and encourage responsible practices, like curbing honking and keeping moderate sound levels during public events. These combined strategies can contribute to a balanced soundscape, improving urban liveability and encouraging public health [7].

9.7.4 Community Engagement and Awareness

Community awareness and involvement are vital aspects of effective environmental management. Mobilizing people to participate by initiating campaigns towards sustainable practices like the reduction of plastic usage and organizing green festivities can contribute majorly to environmental improvements. School and community education programs are the backbone of the movement by tying up with educational institutions, nongovernmental agencies (NGOs), and other community organizations to make people aware of pollution mitigation practices. These initiatives may comprise workshops, training sessions, and interactive events that emphasize the value of sustainable activities and empower citizens to take initiative in preserving the environment. Getting communities involved in activities such as tree planting, clean-up initiatives, and conservation efforts can also create a sense of responsibility and ownership towards a cleaner and greener environment [4, 6].

9.7.5 Policy and Governance

Sound policy and good governance are needed for the realization of sustainable environmental management. Ensuring strong legal regimes through higher punishments for failure to comply with norms of pollution control can serve as a deterrent for environmental crime. Offering incentives to industries for their adoption of cleaner technologies and greener practices can prompt enterprises to make positive contributions toward environmental sustainability. In addition, regular environmental audits and making data publicly available can increase transparency and accountability. Such openness not only fosters public confidence but also enables stakeholders to track progress, detect gaps, and push for action to enhance environmental health. A balanced system that encourages compliance and sustainable development can be created through an integrated policy and governance approach that balances strict regulations with supportive policies [5].

9.7.6 Technological and Innovative Solutions

Technology and innovation are fundamental in improving environmental management and sustainability. The application of the IoT-enabled smart sensors in real-time tracking of air and water quality is possible, providing timely and correct data. Sophisticated sensors can identify contaminants, give instantaneous feedback, and connect to digital platforms to forward information to governments and the people. Besides that, the utilization of data analytics and predictive modelling can also facilitate the prediction of pollution patterns in order to facilitate proactive steps for preventing likely environmental hazards. Decision-making processes, resource allocation optimization, and efficient pollution control intervention can be assisted by predictive analytics. With a combination of intelligent technology and higher level analytics, cities can optimize their reaction rates to environmental concerns, facilitate compliance with regulations, and help promote healthier and more sustainable urban habitats.

9.7.7 Sustainable Development Practices

Sustainable development practices are vital in the creation of long-term economic and environmental stability. One of the key strategies is advocating for green building practices, supporting construction techniques that reduce environmental effects through energy-saving designs, renewable materials, and waste management measures. Green buildings tend to integrate elements such as natural lighting, rainwater harvesting, and energy-saving systems that minimize carbon footprints. Moreover, facilitating the transition to renewable energy sources, including solar, wind, and hydropower, can lower fossil fuel dependence considerably. Increasing the application of renewables not only lowers greenhouse gas emissions but also enhances energy security and minimizes pollution. Through the incorporation of green building and renewable energy programs, cities can enhance their sustainability objectives and help create a healthier and more resilient urban landscape.

9.8 Summary

This chapter presents the development of an IoT-based environmental monitoring system designed to track urban air quality, temperature, and humidity. The integration of the ESP32 microcontroller with advanced sensors, including SHARP GP2Y1014AU0F, DHT11, MQ-135, MQ-9, and MQ-2, ensures accurate and efficient data acquisition. The system supports real-time data visualization through an LCD interface, remote monitoring via the ThingSpeak Cloud platform, and seamless communication using the MQTT protocol. Its capabilities extend to threshold-based notifications and historical data analysis, facilitating informed decision-making for sustainable urban development. Through extensive testing, the system has demonstrated high accuracy, reliability, and scalability, making it suitable for incorporation into broader urban sensing frameworks. By providing real-time environmental insights, this work supports smart city initiatives and promotes data-driven approaches to urban sustainability. The findings of this study contribute to the advancement of IoT applications in environmental monitoring, enhancing the efficiency of urban ecosystem management.

References

1 Guvvala, M.V., and Ch, D. (2023). ThingSpeak based air pollution monitoring system using Raspberry Pi. In: *2023 3rd Asian Conference on Innovation in Technology (ASIANCON)* (pp. 1–6). doi: 10.1109/ASIANCON58793.2023.10269833.

2 Datta, A., Islam, M.M., Hassan, M.S., et al. (2023). IoT based air quality and noise pollution monitoring system. In: *2023 3rd International Conference on Robotics, Electrical and Signal Processing Techniques (ICREST)* (pp. 202–206). doi: 10.1109/ICREST57604.2023.10070039.

3 Saha, A.K., Sircar, S., Chatterjee, P., et al. (2018). A Raspberry Pi controlled cloud-based air and sound pollution monitoring system with temperature and humidity sensing. In: *2018 IEEE 8th Annual Computing and Communication Workshop and Conference (CCWC)* (pp. 607–611). doi: 10.1109/CCWC.2018.8301660.

4 Zeng, F., Pang, C., and Tang, H. (2024). Sensors on Internet of Things systems for the sustainable development of smart cities: a systematic literature review. *Sensors* 24 (7): 2074. https://www.mdpi.com/1424-8220/24/7/2074.

5 Chekuri, R.K., and Lekshmi, R.R. (2024). IoT based air quality monitoring system for apartment building. In: *2024 International Conference on Advancements in Power, Communication and Intelligent Systems (APCI)* (pp. 1–5). doi: 10.1109/APCI61480.2024.10617328.

6 Kumar, A., Kumari, M., and Gupta, H. (2020). Design and analysis of IoT based air quality monitoring system. In: *2020 International Conference on Power Electronics & IoT Applications in Renewable Energy and its Control (PARC)* (pp. 242–245). doi: 10.1109/PARC49193.2020.236600.

7 AbeBer, J., Gotze, M., Kuhnlenz, S., et al. (2018). A distributed sensor network for monitoring noise level and noise sources in urban environments. In: *2018 IEEE 6th International Conference on Future Internet of Things and Cloud (FiCloud)* (pp. 318–324).

8 Kumar, T.A., Ramapraba, J., Kanimozhi, P., and Ajagbe, S.A. (2023). IoT based smart water quality monitoring system. In: *7th IET Smart Cities Symposium (SCS 2023)* (pp. 313–322). doi: 10.1049/icp.2024.0944.

9 Hemdan, E.E., Essa, Y.M., Shouman, M., et al. (2023). An efficient IoT based smart water quality monitoring system. *Multimedia Tools Applications* 82 (19): 28827–28851.

10 Ibrahim, M., Elgamri, A., Babiker, S., and Mohamed, A. (2015). Internet of Things based smart environmental monitoring using the Raspberry-Pi computer. In: *2015 Fifth International Conference on Digital Information Processing and Communications (ICDIPC)* (pp. 159–164). doi: 10.1109/ICDIPC.2015.7323023.

Part III

Security, Privacy, and Ethical Aspects

10

IoT Security and Privacy in Sustainable Communication

Polat Goktas[1,2,] and Sumita Mishra[3]*

[1] *UCD School of Computer Science, University College Dublin, Dublin, Ireland*
[2] *Faculty of Engineering and Natural Sciences, Sabanci University, Istanbul, Turkey*
[3] *Amity School of Engineering and Technology, Amity University Uttar Pradesh, Lucknow Campus, India*

*Corresponding author: polatgoktas@gmail.com

10.1 Introduction: The Intersection of Security, Privacy, and Sustainable IoT Communication

The Internet of Things (IoT) has become a foundational pillar of digital transformation, connecting billions of devices and enabling intelligent, data-driven applications across domains such as energy, healthcare, manufacturing, transportation, agriculture, and urban development. As an essential enabler of the United Nations Sustainable Development Goals (SDGs), IoT technologies are instrumental in advancing environmental sustainability, operational efficiency, and social inclusion [1, 2]. From smart grids that optimize energy usage to remote healthcare monitoring and precision farming, IoT systems support scalable, context-aware solutions that align with global sustainability agendas. However, this unprecedented proliferation of IoT devices has also given rise to a sophisticated range of problems, especially in terms of data safety, privacy protection, energy frugality, and ethical AI integration. The ubiquitous setting of IoT means that vulnerabilities have far-reaching, indeed critical implications. IoT-infrastructure attacks pose not only data confidentiality but also physical security, public trust, and service availability [3, 4].

AI and ML Techniques in IoT-based Communication: A Path to Sustainable Development Goals, First Edition. Edited by Sumita Mishra, Nishu Gupta and Polat Goktas.

In the context of sustainability, ensuring the resilience and trustworthiness of IoT ecosystems requires security mechanisms that are not only effective but also energy-aware and contextually adaptable [5]. Furthermore, the integration of advanced technologies such as artificial intelligence (AI), machine learning (ML), federated learning (FL), blockchain, and quantum-inspired encryption into IoT infrastructures presents both opportunities and ethical dilemmas. These technologies offer powerful tools for intrusion detection, autonomous threat mitigation, and privacy preservation; yet, their deployment also raises critical concerns regarding algorithmic bias, explainability, accountability, and compliance with global regulatory frameworks [6, 7]. Consequently, a balanced approach that harmonizes innovation with responsibility is essential for the sustainable evolution of secure IoT systems.

Therefore, this chapter aims to present an in-depth exploration of the convergence between security, privacy, and sustainability in IoT ecosystems. It brings together cutting-edge technological developments, identifies critical challenges, and outlines practical strategies for implementing ethical, energy-conscious, and regulation-compliant IoT systems. With a focus on building resilience against evolving cyber threats, the discussion emphasizes the importance of aligning technical innovation with societal values and environmental responsibility. By examining AI-enabled threat detection, privacy-preserving frameworks, and resource-efficient security protocols, alongside emerging regulatory and ethical considerations, the chapter offers a forward-looking, multidisciplinary roadmap for designing trustworthy and SDG-aligned IoT infrastructures. Collectively, these sections offer a multidisciplinary perspective on building resilient, ethical, and sustainable IoT infrastructures.

10.2 Artificial Intelligence and Machine Learning-enabled Security Mechanisms in IoT Networks

To support scalable and privacy-preserving IoT deployments, a multilayered architecture that integrates privacy-enhancing technologies and energy-efficient security mechanisms is essential. As illustrated in Figure 10.1, such a system can include FL and lightweight AI at the edge, ensuring real-time data inference without exposing sensitive information to centralized servers. Simultaneously, blockchain-based identity management, coupled with homomorphic encryption and differential privacy techniques, secures data authenticity and confidentiality across distributed nodes. These capabilities, when designed with energy-aware protocols, support sustainable operation and compliance with global sustainability targets like SDG 3, SDG 7, SDG 11, and SDG 12.

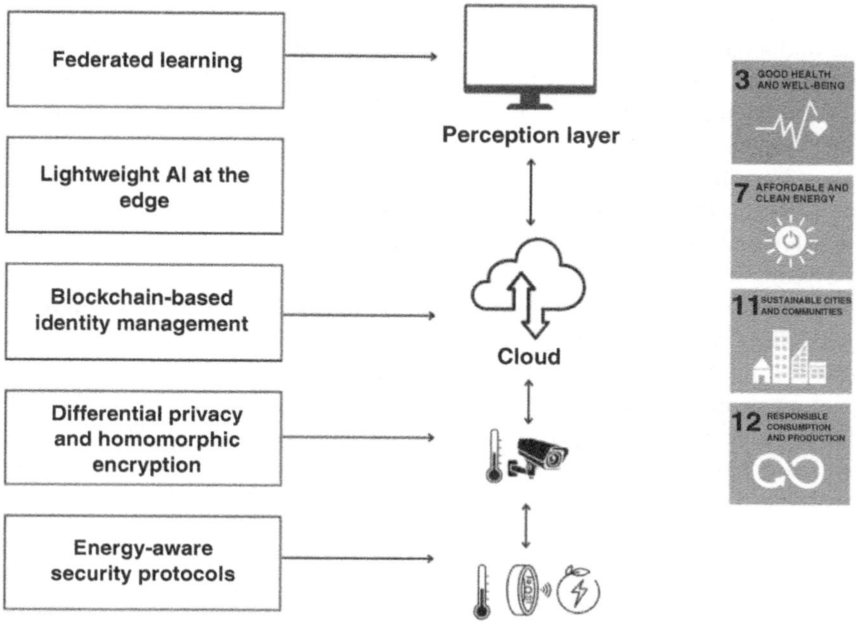

Figure 10.1 Integrated security and privacy architecture for sustainable IoT systems.

10.2.1 Machine Learning Algorithms for Anomaly and Intrusion Detection

ML refers to the capability of systems to learn from data, identify patterns, and make decisions with minimal human intervention [8]. In the context of IoT networks, this learning ability becomes indispensable due to the vast and heterogeneous nature of data generated by interconnected sensors and devices. Anomaly and intrusion detection systems (IDS) that leverage ML offer a dynamic defence mechanism by enabling the identification of irregularities and unauthorized access attempts in real time [9, 10]. As the IoT ecosystem expands into sectors like healthcare, smart grids, and environmental monitoring, sustainable communication becomes both a necessity and a vulnerability. ML-based IDS are emerging as sustainable solutions that provide adaptive protection without the need for constant manual updates, which is resource-intensive and unsustainable in the long run [11, 12].

Supervised learning techniques, such as Support Vector Machines (SVM), Decision Trees (DT), and K-Nearest Neighbors (KNN), are widely used in classifying known types of intrusions. These models depend on labeled datasets and are known for high detection accuracy [13, 14]. However, their reliance on annotated data limits their effectiveness in detecting zero-day attacks. On the other hand, unsupervised learning approaches, including clustering and autoencoders, are suited for identifying previously unseen threats by learning the normal behavior of IoT systems and flagging deviations [15, 16].

Deep learning (DL) models, especially Convolutional Neural Networks (CNNs) and Long Short-Term Memory (LSTM) networks, provide advanced feature extraction and sequential data modelling capabilities, which are critical in analyzing complex IoT traffic data [9, 17]. Reinforcement learning (RL) methods are also gaining traction in intrusion detection by enabling agents to learn optimal policies through interaction with network environments [18].

Despite their potential, these systems face challenges related to computational overhead, model explainability, and the imbalance of datasets [8, 19]. To enhance sustainable deployment, recent approaches advocate for FL and edge computing-based architectures that allow for distributed model training while preserving data privacy and reducing latency [20, 21]. Table 10.1 presents a comparative overview of prominent ML techniques for intrusion and anomaly

Table 10.1 ML techniques for anomaly and intrusion detection in IoT environments.

ML technique	Functionality	Application area	Strengths	Limitation	Key references
SVM, KNN, DT	Supervised classification	Known attack detection	High accuracy	Requires labeled data	[13, 14]
Autoencoders, Clustering	Unsupervised anomaly detection	Unknown/zero-day threat detection	No prior labels needed	Difficult validation	[15, 16]
CNN, LSTM	Deep learning for temporal/spatial features	Complex traffic analysis	Automatic feature extraction	Computationally intensive	[9, 17]
Federated Learning	Distributed model training	Privacy-preserving IDS	Data remains local, lower latency	Communication overhead, convergence	[20, 21]
Reinforcement Learning	Policy learning via trial-and-error	Adaptive intrusion response systems	Suited for dynamic environments	Slow learning rate	[18]

Note: The content of this table is synthesized from key insights and findings reported in the current literature.

detection in IoT networks, highlighting their strengths and limitations within sustainable communication infrastructures.

10.2.2 DL for Advanced Threat Intelligence

DL, a subset of ML, employs neural networks with multiple layers to model complex patterns in data, enabling systems to make highly accurate predictions and decisions. In the context of IoT security, DL has emerged as a transformative force, offering advanced threat intelligence capabilities that surpass traditional signature-based intrusion detection systems [22, 23]. IoT environments, characterized by heterogeneous devices and vast data streams, are particularly vulnerable to sophisticated cyber threats. Conventional methods often struggle to keep up with zero-day attacks, dynamic topologies, and real-time detection requirements. DL addresses these challenges through automatic feature extraction, scalability, and adaptability, thus enabling sustainable, intelligent intrusion detection systems [24, 25].

One of the most widely adopted DL techniques in threat intelligence is the CNN, often combined with LSTM or BiLSTM architectures to detect temporal and spatial dependencies in IoT traffic data [26, 27]. Hybrid models such as CNN–BiLSTM and deep autoencoders have shown promise in identifying previously unseen threats while maintaining low false-positive rates [25, 28]. Recent advances in DL-based threat intelligence also integrate swarm optimization and evolutionary computation for improved feature selection and model convergence [24, 29]. Moreover, graph neural networks (GNNs) and explainable AI (XAI) have been utilized to improve interpretability and trust in automated decision systems, which is critical in regulated environments such as healthcare and industrial IoT [23, 30].

Furthermore, DL-driven threat detection frameworks are increasingly deployed in fog-enabled and edge-computing environments, ensuring low-latency responses and energy efficiency, key factors in sustainable communication infrastructures [31, 32]. FL and blockchain-based deep models are also gaining traction for privacy-preserving distributed threat detection [28]. Despite these advancements, challenges persist. DL models require vast annotated datasets, and their computational demands can be prohibitive in resource-constrained IoT environments [33, 34]. Nonetheless, the growing integration of DL with SDN and adaptive frameworks has the potential to evolve threat intelligence into a self-healing and sustainable paradigm [35, 36].

10.2.3 FL for Privacy-preserving IoT Security

The increasing ubiquity of IoT devices in sectors such as smart cities, healthcare, industrial control, and autonomous vehicles has brought significant advancements to data-driven systems. However, this interconnected ecosystem also introduces unprecedented security and privacy risks due to its decentralized and heterogeneous architecture. Traditional ML methods, which require centralized data collection for model training, are no longer viable in such environments because of data sensitivity, communication bottlenecks, and sustainability concerns. In this context, FL has emerged as a cutting-edge privacy-preserving framework that supports secure, distributed intelligence across IoT networks [37, 38].

FL shifts the paradigm by enabling edge devices to collaboratively train a global model without exchanging raw data. Only model updates or gradients are transmitted, thus reducing the attack surface and preserving user data privacy [39, 40]. This decentralized approach aligns well with the design philosophy of IoT systems and offers a scalable, communication-efficient, and privacy-conscious alternative to conventional centralized learning architectures [41].

10.2.3.1 FL Architectures in IoT

There are several FL architectures tailored to the IoT landscape. Horizontal FL is commonly used when devices share the same feature space but differ in data samples, while vertical FL suits systems with heterogeneous feature sets [42]. In advanced IoT deployments, asynchronous FL frameworks have been adopted to accommodate varied computation and communication capacities across devices [40]. These frameworks reduce synchronization delays and support energy efficiency, key goals in sustainable communication systems [43, 44]. FL frameworks have also integrated secure multiparty computation (SMPC) and differential privacy (DP) to reinforce confidentiality and prevent information leakage through model updates [45, 46]. These privacy-enhancing techniques are particularly essential in sensitive sectors such as smart healthcare, where patient data must remain confidential [47].

10.2.3.2 FL in Intrusion Detection and Threat Intelligence

One of the most impactful applications of FL in IoT is intrusion detection. Federated Intrusion Detection Systems (FIDS) leverage local anomaly detection capabilities across devices to detect malicious behavior patterns without relying on a centralized threat database [48, 49]. For instance, the FRHIDS framework combines FL with hybrid recommender systems and DL layers to enhance software-defined networking security [39]. Similarly, knowledge distillation-based FL models have been proposed to enable lightweight yet effective threat detection suitable for resource-constrained IoT nodes [44]. Emerging architectures also explore the combination of FL with GNNs and autoencoders for uncovering complex spatiotemporal patterns in network traffic [50]. These integrations enable dynamic, real-time adaptation to new threats, thereby enhancing resilience and sustainability.

10.2.3.3 FL with Edge Intelligence and Blockchain

In tandem with edge computing, FL empowers IoT devices to perform local computations, reducing reliance on cloud resources and minimizing latency – a crucial factor for time-sensitive applications like autonomous vehicles and industrial control systems [37, 51]. Edge-intelligence-based FL systems also improve energy efficiency and reduce carbon footprints, which are pivotal for environmentally sustainable communication infrastructures [41, 52]. The fusion of FL with blockchain technology further strengthens trust and auditability by enabling decentralized identity management and tamper-proof logs of model updates [53]. Blockchain-enhanced FL platforms are especially beneficial in adversarial environments where model poisoning and update manipulation attacks are prevalent.

10.2.3.4 Key Challenges and Sustainability Dimensions

While FL offers significant advantages, challenges such as data heterogeneity, communication overhead, non-IID data distribution, and model convergence delay remain substantial [48, 54]. To address these issues, recent research explores model pruning, selective client participation, and adaptive learning rates to optimize performance and energy consumption [43, 50]. From a sustainability lens, FL minimizes the need for large-scale cloud data centers by distributing the computational burden across edge devices. This distribution reduces energy-intensive data transfers and fosters localized intelligence, both of which are essential for achieving the SDGs, particularly SDG 9 (Industry, Innovation, and Infrastructure) and SDG 12 (Responsible Consumption and Production).

10.2.4 Lightweight AI Models for Resource-constrained Devices

The proliferation of IoT devices has enabled unusual levels of automation, sensing, and communication across domains such as healthcare, agriculture, environmental monitoring, and smart infrastructure. However, the majority of IoT devices are highly resource-constrained, operating under tight restrictions on processing power, memory, and energy consumption. These limitations result in significant challenges to the deployment of conventional AI and ML models, which are often computationally intensive and storage-heavy. This has led to a growing interest in lightweight AI models, particularly in the form of TinyML and edge-optimized architectures, designed specifically for low-power environments [55–57]. TinyML refers to the deployment of ML algorithms on ultra-low-power microcontrollers and edge devices, often consuming less than a few milliwatts of energy [58]. These models are typically designed for tasks such as anomaly detection, speech recognition, or sensor data classification, enabling real-time processing without reliance on cloud infrastructure. This local inferencing capability not only reduces latency and bandwidth usage but also enhances data privacy and security – key priorities in sustainable IoT systems [57].

Several strategies have been employed to enable DL on constrained devices. These include model quantization, pruning, and knowledge distillation, which aim to reduce the computational complexity and memory footprint of models without significantly compromising their performance [59, 60]. For example, quantized neural networks transform 32-bit floating-point weights into 8-bit integers, resulting in faster inference and reduced energy usage [55]. Moreover, lightweight CNNs and recurrent models have been adapted for embedded systems, allowing tasks like intrusion detection and acoustic signal analysis to be performed locally [58].

In the realm of IoT security, lightweight AI models are proving essential for detecting Distributed Denial of Service (DDoS) attacks and other malicious behaviors in Software-defined IoT (SD-IoT) environments [61]. By leveraging tailored ML architectures optimized for edge deployment, such systems can identify threats in real time without the need for data offloading to centralized servers, mitigating the risks associated with communication bottlenecks and data breaches. Furthermore, TinyML models contribute to the sustainability of IoT systems by extending battery life, lowering carbon emissions from data transmission, and enabling devices to operate autonomously in remote or off-grid locations. This decentralized intelligence supports environmentally conscious design, particularly in applications like wildlife tracking, soil monitoring, and smart farming [57].

Despite the progress, challenges persist. These include ensuring model generalizability across diverse environments, managing overfitting due to limited onboard training capabilities, and standardizing deployment frameworks for edge ML [61]. However, ongoing advancements in hardware accelerators, such as AI-specific microcontrollers and neural processing units (NPUs), promise to further bridge the gap between performance and efficiency.

10.2.5 Real-world Case Studies and Implementations

The application of IoT technologies in real-world systems has introduced significant security and privacy concerns that go beyond theoretical discussions and simulation environments. While proof-of-concept models often present idealized solutions, actual implementations show complex challenges including device heterogeneity, scalability, legacy integration, physical tampering risks, and the trade-off between security robustness and energy efficiency. The following real-world case studies illustrate how different sectors, ranging from smart homes to defence systems, are addressing IoT security and how these implementations contribute to sustainable, secure communication infrastructures. A summary of

Table 10.2 Selected real-world implementations of IoT security strategies.

Use case/sector	Implementation focus	Key technology	Key reference
Smart Home	Security analysis in trigger-action environments	TAESim TestbedTAESim Testbed	[62]
Smart Home (Z-Wave)	Protocol vulnerability assessment	Model-based Security Model	[63]
Healthcare Monitoring Healthcare	Authentication protocol for diabetic patient data	Custom Auth Protocol	[67]
Biosensor Networks	Secure wearable health-monitoring systems	Layered Security Framework	[66]
National Defense	Secure IoT in battlefield and logistics	Federated and Encrypted IoT	[68]
Smart City Infrastructure	ML-enhanced blockchain for authentication	Decentralized Ledger + AI	[73]
Firmware Attack Defense	Malware behavior simulation and defence	Epidemiology Modelling	[70]
Surveillance Devices	Penetration testing of IoT-enabled IP cameras	Forensics & Patch Evaluation	[71]
Supply Chain Risk	Framework for evaluating IoT supply vulnerabilities	I-SCRAM Decision Framework	[74]

selected real-world implementations and their security strategies is provided in Table 10.2.

10.2.5.1 Smart Homes and Trigger-action Automation

The TAESim testbed developed by Ban et al. [62] provides a dedicated environment for analyzing the security of trigger-action programming models in smart homes. These models, used by systems like Alexa routines, are vulnerable to chain-reaction attacks where one device's compromised behavior can trigger multiple actions across the network. TAESim enables the simulation of such attack chains and evaluation of mitigation strategies. Its modularity and extensibility also allow testing of smart home setups in a controlled, repeatable manner – an essential component for sustainable development of secure smart environments.

10.2.5.2 Z-wave Protocol in Consumer IoT

In another significant deployment-focused study, Braghin et al. [63] conducted a model-based security analysis on Z-wave, a widely adopted wireless communication protocol used in home automation systems. Their formal vulnerability analysis demonstrated that even mature protocols could harbor critical flaws if not continually tested against modern threat models. The Z-wave case exemplifies the need for ongoing evaluation of embedded systems, especially when used in energy-efficient buildings and IoT-enabled housing, where reliability and low power consumption are critical.

10.2.5.3 IoT in Healthcare: Securing Biosensors

Security in healthcare is critical not only for patient safety but also for compliance with regulatory standards [64, 65]. Chaki et al. [66] and Mekki et al. [67] provided comprehensive security frameworks for biosensor networks and diabetic patient-monitoring systems, respectively. Their case studies emphasized the need for robust authentication protocols, secure device pairing, and privacy-preserving communication. These implementations illustrate how the integration of security-by-design principles can promote trust in health-monitoring technologies while reducing reliance on centralized healthcare infrastructure, aligning well with sustainability and access goals in remote care scenarios.

10.2.5.4 National Defense and Mission-critical IoT Systems

The defence sector has been an early adopter of IoT, deploying it for logistics, surveillance, and threat detection. Douglass et al. [68] documented use cases where IoT supports real-time decision-making in military logistics and battlefield monitoring. The demand for ultra-reliable communication, rapid authentication, and tamper-proof data is exceptionally high. Thus, the adoption of FL, lightweight encryption, and redundancy in these networks sets an example for securing critical infrastructure in other civilian domains such as disaster response and energy grids.

10.2.5.5 Security Threat Simulation and Malware Defense

Mukhtar et al. [69] studied the SILEX malware, which specifically targets IoT firmware, rendering devices unusable by corrupting memory storage. Their case study illustrated the consequences of firmware-level vulnerabilities and the importance of physical hardening. In parallel, Casado-Vara et al. [70] applied mathematical epidemiology models to simulate the spread and mitigation of malware in IoT networks. These simulations have real-world relevance for sustainable cities where a compromised network of sensors can lead to public safety and environmental monitoring failures.

10.2.5.6 IoT Device Penetration Testing and Digital Forensics

Abdalla and Varol [71] conducted penetration testing on commercial IP cameras, revealing weak authentication defaults and insecure firmware update mechanisms. Their findings highlight the importance of digital forensics and vulnerability disclosure in the consumer IoT space. Similarly, Xu et al. [72] discussed design-level vulnerabilities, such as insecure boot processes and poor key management, which remain persistent in many off-the-shelf IoT devices.

10.2.5.7 Blockchain and ML Integration in Smart City Security

In smart city deployments, where thousands of interconnected devices monitor traffic, lighting, and energy systems, Hemashree et al. [73] implemented blockchain-integrated ML systems to ensure authentication and anomaly detection in decentralized architectures. These systems reduce energy overhead from centralized servers and promote sustainable network scaling by localizing decision-making processes.

10.3 Privacy-preserving Frameworks and Data Governance in Sustainable IoT

10.3.1 Privacy-by-design Principles for IoT Architectures

The fast growth of the IoT has brought serious data privacy issues, especially in the systems that constantly gather, process, and communicate sensitive data through distributed networks. To manage such issues, the Privacy-by-design (PbD) approach has surfaced as a leading strategy, which is interested in the incorporation of privacy into the fundamental architecture of IoT systems right from the start and not as an afterthought [75]. PbD supports sustainable and responsible communication with user-focused, open, and context-aware handling of data behavior.

10.3.1.1 Core Principles and Architectural Implications

The *seven foundational principles of PbD* – proactive not reactive, privacy as the default, privacy embedded into design, full functionality, end-to-end security, visibility and transparency, and respect for user privacy, serve as a guiding framework for secure IoT architecture development [76]. Applying these principles requires careful architectural choices at every layer of the IoT stack, including data minimization strategies at the edge, secure device pairing, identity management, and encryption protocols embedded within firmware. In practice, these principles can be operationalized through architectural elements such as fog computing, blockchain, edge-based access control, and privacy-preserving data analytics. For example, Mayer et al. [77] proposed a hybrid fog-blockchain IoT architecture, FogChain, to manage personal health records. The decentralized design minimizes reliance on central servers, reducing single points of failure and exposure.

10.3.1.2 Regulatory Compliance and Developer Responsibility

Recent legal frameworks such as the General Data Protection Regulation (GDPR) and the California Consumer Privacy Act (CCPA) demand stricter compliance from IoT service providers. Aljeraisy et al. [75] emphasized that developers need tools and design templates aligned with these regulations to integrate PbD into practical software engineering workflows. The SAFETHINGS initiative by Barbosa et al. [78] exemplifies such efforts, offering a comprehensive security-by-design approach tailored for embedded IoT systems.

10.3.1.3 Design Patterns and Standardization

Pape and Rannenberg [76] introduced privacy design patterns, reusable architectural templates that embed privacy safeguards into the design phase. Examples include patterns for data minimization, unlinkability, and user-controlled data retention. These patterns help system architects implement compliant and verifiable privacy features in resource-constrained IoT devices without compromising performance or functionality. Furthermore, recent work explores middleware approaches that enforce privacy policies across devices and services. Moldovan et al. [79] presented the RERUM middleware, which uses data-centric access control policies and trusted data provenance models to maintain data integrity and confidentiality in smart city deployments.

10.3.1.4 Edge-based and Distributed Solutions

To align with sustainability goals, modern PbD solutions advocate for edge and fog computing to process data closer to its source. This not only reduces network load and energy consumption but also limits the data's exposure to external networks. Pavithran et al. [80] introduced an edge-based blockchain architecture that uses hierarchical identity-based encryption to secure event-driven IoT environments, supporting both privacy and low-latency operations. Moreover, blockchain integration enhances trust and auditability by ensuring immutability of transactions and transparency of data flows. As shown by Chhabra et al. [81], the convergence of IoT and blockchain can fulfil both privacy and security requirements while enabling scalable and decentralized systems for smart cities and critical infrastructure.

10.3.2 Homomorphic Encryption and Secure Multiparty Computation in IoT

As IoT networks become ubiquitous in healthcare, smart cities, and industries increasingly, they produce and send vast quantities of sensitive information. Privacy, as well as the integrity of that data, particularly while in an untrusted or distributed setting as it is being processed, is paramount. Of those two top-priority technologies in such a setting, Homomorphic Encryption (HE) and SMPC were among them. These cryptographic methods enable data processing in its encrypted state or collaborative analysis without exposing the raw content, thus facilitating PbD in green IoT communications.

10.3.2.1 Homomorphic Encryption in IoT

HE enables computations on ciphertexts such that the decrypted result of the computation matches the output had it been performed on the plaintext. This property is particularly beneficial in IoT systems, where sensor data is often offloaded to fog or cloud layers for analysis. With HE, sensitive health records, location data, or industrial metrics can be securely processed in untrusted domains without exposure [82]. For instance, Yoon et al. [82] presented a low-power HE crypto-engine optimized for IoT nodes, demonstrating that secure computation on encrypted data can be feasible even on resource-constrained devices. Meanwhile, Song et al. [83] developed a blockchain-based secure outsourcing framework for HE polynomial multiplication, emphasizing scalability and verifiability in decentralized networks. In a recent deployment scenario, López et al. [84] applied HE in conjunction with FL for predicting the remaining useful life (RUL) of lithium-ion batteries. This method ensured model accuracy while protecting sensitive operational data from exposure during training.

10.3.2.2 Secure Multiparty Computation

SMPC complements HE by enabling multiple parties to jointly compute a function over their private inputs without revealing them to each other. This is particularly valuable in collaborative IoT scenarios, such as traffic coordination among autonomous vehicles or cooperative energy grid optimization. Vashishth et al. [85] explored privacy-preserving ML techniques using SMPC to securely train models on distributed IoT datasets hosted in cloud environments, thereby mitigating the risks of data leakage or centralized compromise. SMPC-based frameworks are also becoming essential in edge computing scenarios, where fog nodes coordinate to perform privacy-sensitive tasks such as anomaly detection or smart grid load balancing [86]. These systems align with sustainability goals by reducing the need to transmit large volumes of raw data and by distributing computation closer to data sources.

10.3.2.3 Integration with Broader Security Architectures

Both HE and SMPC are being integrated into broader privacy-preserving architectures involving blockchain, FL, and secure outsourcing mechanisms. Hlaing and Asaeda [87] proposed ShieldDINC, a privacy-preserving framework for distributed in-network computation using HE. Their work demonstrated that integrating HE into in-network aggregation processes enhances both privacy and efficiency – essential for delay-sensitive IoT applications. Moreover, the versatility of these cryptographic primitives allows them to be applied across diverse domains, from secure health informatics to encrypted industrial automation [88]. Their integration into IoT platforms enhances trust, supports compliance with privacy regulations, and contributes to the development of secure-by-design and sustainable communication systems.

10.3.3 Differential Privacy Techniques in Data Sharing and Analytics

The IoT environment constantly collects, transmits, and elaborates large quantities of sensitive information, including personal, behavioral, and contextual information. While facts analytics are more critical to enable insight extraction from data, it also brings important privacy threats. DP is a growing mathematically rooted technique to make information-preserving facts elaboration so that systems may extract knowledge from datasets without divulging information about any individual information point. DP adds noise to datasets or calculations in a managed way so that the contribution of an individual is masked statistically. Such a procedure becomes most applicable to IoT systems where data frequently gets relayed to third parties or calculated in cloud or federated platforms [89]. With DP implemented, IoT networks can maintain utility in data while meeting strong privacy requirements – a crucial issue for sustainable and ethical communication design.

10.3.3.1 Applications in Federated and Blockchain-based Learning

With the rise of FL in IoT, where models are trained across distributed devices without centralizing raw data, differential privacy serves as an added layer of protection. Zhao et al. [90] introduced a multimodal FL approach incorporating DP for IoT data. This architecture prevents information leakage from gradient updates during collaborative training, which is particularly important in smart city and industrial IoT scenarios. In blockchain-based systems, DP complements encryption and consensus mechanisms to protect data integrity and user identity. Shen et al. [91] implemented a DP-enhanced SVM training algorithm over blockchain-encrypted smart city data, achieving accurate predictions without compromising user privacy. Similarly, Ul Haque et al. [92] demonstrated a privacy-preserving KNN model for encrypted health data, incorporating DP techniques to resist membership inference attacks.

10.3.3.2 Industrial IoT and Sparse Data Analysis

DP has also proven effective in addressing challenges of sparse or high-dimensional IoT data. Wang et al. [93] applied multiple DP strategies to tensor factorization methods used for 5G network traffic analysis, optimizing the balance between model accuracy and data privacy. These methods are crucial for large-scale industrial IoT networks where anomaly detection and traffic analysis are continuously performed. Feng et al. [94] extended this paradigm by integrating DP with blockchain-based Tucker decomposition, enabling privacy-preserving data analytics in encrypted industrial environments. Their approach highlights how combining DP with secure data storage (e.g. blockchain) and decomposition algorithms ensures robustness and transparency in data-driven decision systems.

10.3.3.3 Toward Privacy-accountable Data Sharing

Beyond analytics, DP is also being embedded into data-sharing frameworks to enforce accountability. Li et al. [95] proposed an attribute-based encryption model incorporating DP to manage access control and privacy policies across Cloud-IoT architectures. Such frameworks are increasingly necessary in compliance-driven domains such as healthcare, finance, and public infrastructure. In another application, Ma et al. [96] utilized blockchain-driven DP mechanisms to create trusted IoT data-sharing models, focusing on real-time sensor data streams. Their system supports both scalability and strong privacy guarantees, addressing critical gaps in conventional edge-cloud architectures.

10.3.4 Blockchain and Decentralized Identity Management in IoT

With the growing number of IoT devices globally, the conventional identity management models are challenged by scalability, security, and privacy. The conventional identity systems, centralized server-based, are prone to single points of failure, data breaches, and cross-platform compatibility challenges. These drawbacks have prompted blockchain-based Decentralized Identity Management (DID) systems as possible substitutes to secure, interoperable, and privacy-protecting user and device authentication in IoT environments [97, 98].

10.3.4.1 The Role of Blockchain in Decentralized Identity

Blockchain offers a distributed, tamper-proof ledger that inherently supports transparency and auditability without relying on central authorities. This architecture is particularly advantageous for IoT security, as it enables identity credentials to be verified and recorded immutably while ensuring that ownership and control remain with the data subject [99]. In one implementation, Sadique et al. [98] introduced IMSC-EIOTD, a blockchain-integrated identity management framework for edge IoT devices. Their model supports scalable and secure communication among heterogeneous devices by eliminating reliance on central identity providers. Similarly, Tu et al. [100] proposed a decentralized identity authentication and key management scheme based on blockchain that minimizes latency and enhances identity privacy in distributed energy systems.

10.3.4.2 Federated Identity and Cross-domain Integration

Cross-domain interoperability is a pressing issue in large-scale IoT networks, especially when integrating industrial, healthcare, or vehicular IoT systems. Kim et al. [101] posed a critical question about building federated identity models without reliance on centralized giants like Google. Their study suggested that blockchain-based systems can facilitate cross-domain federated identities by enabling secure interactions between entities under different administrative domains without compromising privacy. Moreover, multifactor authentication (MFA) frameworks that use blockchain as a trust anchor have been applied in smart homes and healthcare [102]. These systems combine biometrics, visual cryptography, and digital memories to build decentralized identity profiles that are difficult to spoof or tamper with.

10.3.4.3 Fog Computing and Lightweight Blockchain Authentication

To meet IoT's energy and latency requirements, lightweight blockchain-based authentication mechanisms have been put forward that are complementary to fog computing environments. Ahmed et al. [103] developed a hybrid framework in which fog nodes enable decentralized authentication with minimized end device computational loads. The edge-centric model is conducive to sustainability while providing strong security and low latencies. A comparative view of governance approaches across sectors is summarized in Table 10.3.

Table 10.3 Domain-specific data governance applications and features.

Sector	Governance focus	Key approach	Key reference
Consumer IoT	Transparency and user control	Consent, access logs Management of real time data permissions, Implementation of well-defined privacy policies	[105]
National Policy	Legal framework implementation	GDPR, DSA, AI Act	[124]
Industry 4.0	Factory data pipelines and analytics	Machine trust models	[126]
Healthcare	Interoperability and privacy	Federated learning	[130]
Agriculture	Environmental data governance	Real-time data traceability	[127]
Urban systems	Smart city interoperability	Governance-aware middleware	[128]
Cloud IoT	Security and recovery strategies	Policy-based recovery	[129]

10.3.4.4 Performance and Protocol Evaluation

Blockchain performance remains a topic of debate in the IoT context due to the limited processing capabilities of most devices. Sharma et al. [104] compared various communication protocols under decentralized identity management systems for IoT. Their results demonstrated that protocol optimization and selection directly affect throughput, latency, and energy consumption, impacting the feasibility of real-time DID solutions in large-scale networks. Trust is another central pillar in decentralized identity models. El Majdoubi et al. [99] proposed a trust establishment protocol leveraging blockchain for smart IoT systems, aiming to evaluate the reputation and behavior of participating devices before granting access. This approach helps create resilient systems that can dynamically respond to identity-based threats without human intervention.

10.3.5 Data Governance, Policy, and Ethical Standards for Sustainable IoT

The rapidly accelerating growth of the IoT has spawned data governance as the pre-eminent cornerstone of sustainable, secure, and privacy-friendly digital worlds. Given that IoT things continuously interact to produce real-time personal, environmental, and operating data, ethics-driven data stewardship becomes the solution – both for compliance and societal trustworthiness and long-term system sustainability [122].

10.3.5.1 The Foundations of IoT Data Governance

In the IoT context, data governance extends beyond classical information management. It encompasses the real-time, distributed, and often autonomous nature of data generation and decision-making across smart devices, vehicles, homes, factories, and even agricultural fields. Traditional top-down governance models are insufficient to cope with the dynamic and federated nature of IoT, especially when ethical, legal, and environmental responsibilities are involved [123, 124]. Emerging governance models prioritize user-centricity, transparency, interoperability, and accountability. These principles are vital in building public trust and ensuring long-term sustainability. As emphasized by Tapsell et al. [125], placing the user at the center of governance efforts through mechanisms such as data portability, access control, and real-time auditing can mitigate exploitation and foster ethical innovation.

10.3.5.2 Policy Evolution and Regulatory Trends

The regulatory environment for IoT data has rapidly evolved, particularly in Europe. The GDPR established the groundwork for privacy and data protection, introducing key rights such as data minimization, purpose limitation, and user consent. However, recent legislative developments – including the Digital Services

Act (DSA), Digital Markets Act (DMA), Data Governance Act (DGA), and AI Act – highlight a shift from isolated privacy enforcement to a holistic digital governance model [124]. These newer regulations aim to address the economic, ethical, and power asymmetries in data ecosystems. For instance, the Europe Union (EU) Data Act encourages data interoperability and fair access between businesses and governments, while the AI Act targets algorithmic accountability and risk classification. At the national level, Cheryl and Ng [122] examined how Malaysia is adapting global best practices to local contexts, showing that harmonizing technical design and legal mandates is a complex but crucial task for inclusive digital governance. Together, these frameworks promote sustainable governance by design, where data sharing and innovation do not compromise rights or security.

10.3.5.3 Ethical Dimensions and Sectoral Applications

Data governance is not purely a technical or legal matter, it is deeply ethical. It raises questions about power dynamics, surveillance, inequality, and digital rights. In the healthcare domain, where privacy is paramount, Verma and Kishor (2024) [130] proposed FL models integrated with AR (Augmented Reality) to preserve data locality while still enabling advanced analytics. This design embeds privacy and fairness directly into the computational model, avoiding centralized data collection. In Industry 4.0, the sheer volume and velocity of data necessitate robust governance layers. Yebenes Serrano and Zorrilla [126] proposed a factory-centric model to handle the automated exchange of information between machines, emphasizing transparency in algorithmic decision-making and traceability of production data.

Agriculture provides a unique governance challenge due to its environmental and socio-economic implications. Amir et al. [127] introduced the Horti-IoT framework to manage horticultural data streams with traceability and accountability in mind. Their work highlights how data frameworks must be aligned with sustainable agricultural practices, ensuring not just efficiency but environmental stewardship. Urban systems also demand multi-layered governance strategies. Kazmi et al. [128] presented a governance-aware middleware for smart cities, enabling dynamic policy enforcement across heterogeneous IoT devices. The model accounts for data provenance, cross-domain identity validation, and real-time compliance monitoring, all of which are critical for ethical urban innovation.

10.4 Energy-aware and Resource-efficient Security Protocols

Since the number of low-power IoT networks in industrial automation, sensor networks, and smart homes is increasing, there is a requirement for strong security without compromising energy efficiency. Current cryptography and authentication methods are usually too heavy for resource-constrained IoT devices under highly constrained computation, memory, and battery resources. Thus, energy-efficient and resource-conscious security protocols are needed to provide sustainable, secure, and durable IoT network deployments [106, 107].

10.4.1 Energy Constraints in IoT Security Design

In IoT-enabled wireless sensor networks (WSNs), energy is a finite and often non-replenishable resource. Security mechanisms that introduce excess overhead in computation or communication can drastically shorten device lifespans or create system bottlenecks. Recognizing this, researchers have begun optimizing security protocols by minimizing transmission payloads, reducing cryptographic complexity, and implementing dynamic trust-based mechanisms [108]. Verma et al. [109] proposed Li-MSD, a lightweight mitigation strategy for Routing Protocol for Low-power and Lossy (RPL)-based IoT networks under Decentralized Autonomous Organizations (DAO) insider attack scenarios. Their solution not only enhances resilience to internal routing attacks but also maintains a minimal memory footprint, making it suitable for real-world embedded systems. Similarly, Srivastava and Paulus [108] developed an optimization-driven hybrid trust algorithm for energy-efficient cluster head selection in IoT WSNs, integrating security features without degrading network performance.

10.4.2 Protocol Innovations in the Perception Layer

At the perception layer, where physical sensing and data acquisition occur, energy-sensitive routing and cryptographic protocols have been tailored for sustainable deployments. Hasan and Patle [107] showed a JSO-enhanced LEACH protocol that incorporates energy-aware decision-making into the routing process, supporting secure and efficient cluster formation in low-power IoT networks. This is particularly impactful for healthcare and agricultural IoT use cases, where continuous operation is required over extended periods with limited battery replacement opportunities. Mazunga and Nechibvute [110] emphasized the ongoing challenges in low-power WSNs and provided a systematic review of energy-aware communication frameworks.

10.4.3 Adaptive and Learning-based Security Mechanisms

Recent innovations have also introduced RL and graph-based intelligence into energy-aware IoT security. Kumar et al. [111] implemented a hardware-level RL-based energy-efficient protocol for WSNs that dynamically adapts security measures based on environmental conditions. This form of intelligent decision-making allows systems to selectively invoke security protocols, conserving resources while responding effectively to threats. Shiny et al. [112] demonstrated how GNNs, when adapted with domain-specific knowledge, can enable secure and energy-efficient data dissemination in WSN-enabled IoT. Their domain-adaptive message passing approach balances latency, energy, and security, exemplifying a new class of context-aware, lightweight frameworks.

10.4.4 Edge-centric Optimization and Multilevel Strategies

IoT systems are increasingly integrating edge computing to offload data processing and enhance decision latency. In this context, Agrawal et al. [56] proposed a gaming theory approach tackling key challenges such as minimizing latency, improving energy efficiency, and balancing load in IoT-based service allocation via edge computing. Jagho Mdemaya et al. [113] introduced ATENA, an adaptive technique for network area coverage and routing in edge-based IoT environments. By optimizing the interplay between energy constraints and routing intelligence, their protocol enhances both performance and system security in resource-limited scenarios. Additionally, Mann and Singh [106] proposed a multilevel hierarchical routing protocol that adapts to green IoT conditions by balancing energy-aware security with node role rotation, reducing individual device burden while maintaining secure communications throughout the network.

10.5 Future Directions and Ethical Considerations in IoT Security for Sustainability

As IoT technologies continue to advance, security frameworks need to be forward-looking in anticipating changing threats and aligned with ethical, environmental, and societal requirements. Beyond ensuring basic confidentiality, integrity, and availability, next-generation IoT security solutions need to enable sustainability through energy efficiency, ethical alignment, and the ability to adapt with dynamic, decentralized architectures. The adoption of new technologies like quantum computing, AI, 6G, and blockchain into IoT demands not just new security measures but also doubled scrutiny of the values and meanings that are the basis for their implementation [114, 115].

10.5.1 Emerging Technologies and the Future of Secure IoT

The shift toward quantum-enabled FL and 6G wireless systems marks a turning point in scalable and privacy-preserving IoT security. Javeed et al. [114] proposed quantum-empowered models that enhance federated security while reducing attack surfaces across distributed environments. As the number of interconnected devices multiplies, lightweight cryptographic schemes remain crucial. Alauthman et al. [116] emphasized the need to optimize cryptographic protocols for resource-constrained IoT ecosystems, promoting real-world adoption without performance trade-offs. Meanwhile, blockchain continues to be a promising framework for decentralized authentication and trust. Maurya et al. [115] highlighted blockchain-driven security architectures as central to future IoT ecosystems, especially in handling autonomous identity and secure consensus without centralized control.

10.5.2 Ethical Imperatives in Security Design

Ethics must be embedded in the early stages of IoT system development. This includes concerns related to data transparency, bias in AI models, user consent, algorithmic accountability, and digital sovereignty. The use of AI-based anomaly detection in IoT systems, as shown by Yagiz et al. [59] and Yagiz and Goktas [60], raises issues around explainability and false positives – both of which affect user trust and societal acceptance. Furthermore, ethical concerns also emerge in relation to mass surveillance, behavioral prediction, and data monetization, particularly in smart city deployments. Hu et al. [117] emphasized the importance of aligning intrusion detection and defence strategies with fairness, privacy, and societal norms, proposing the adoption of nature-inspired and metaheuristic algorithms to reduce bias and optimize energy efficiency.

10.5.3 Sustainability and Environmental Responsibility

Sustainability in IoT security extends to green computing, where energy usage, electronic waste, and processing overhead must be optimized. Alauthman et al. [116] and Tiwari et al. [118] pointed out how deploying security at the firmware and microcontroller levels – such as on ESP32 – requires novel hardware–software co-design strategies to ensure both security and efficiency. In addition, Prajapati et al. [119] identified RPL-based routing as a hotspot for attacks in low-power networks and called for self-healing, adaptive protocols that reduce communication costs and carbon footprints. These solutions must also support lifecycle-aware designs, managing security updates, deprecation, and end-of-life policies without human intervention.

10.5.4 Research Trends and Gaps

As shown by bibliometric and systematic reviews [120, 121], IoT security research is shifting toward automated, explainable, and ethically grounded AI solutions, while prioritizing adaptability to heterogeneous networks, including those

Table 10.4 Key future directions and ethical considerations in sustainable IoT security.

Domain	Future trend/ innovation	Ethical or sustainability concern	Key reference
Federated Learning + 6G	Quantum-enhanced decentralized training	Bias, E-waste, Carbon footprint, Data sovereignty, fairness	[114]
Lightweight Cryptography	Protocol optimization for constrained devices	Access equity, resource efficiency	[116]
Blockchain	Distributed identity & trust	Transparency, decentralization ethics	[115]
AI-Based Anomaly Detection	Deep learning for behavioral analysis	Bias, explainability, false alarms	[59, 60]
Metaheuristic Defences	Adaptive intrusion detection using AI	Energy waste, algorithmic opacity	[117]
Routing Protocols	RPL attack mitigation strategies	Network sustainability, low-power compliance	119]
Microcontroller Security	Firmware-layer security for ESP32 & Wi-Fi modules	Energy optimization, firmware accountability	[118]

operating in terahertz bands and beyond. Future research must also explore policy-based governance integration, quantum-safe encryption standards, and bio-inspired decentralized security mechanisms. A summary of key trends, future research avenues, and ethical concerns is outlined in Table 10.4.

As shown in Figure 10.2, each AI and privacy-preserving technique contributes uniquely to the overall sustainability and security of IoT systems. While FL scores high on privacy and scalability, homomorphic encryption excels in privacy but lags in energy efficiency. Supervised and DL models, despite their strong performance in scalability and effectiveness, fall short in explainability and privacy. Differential privacy, although well-balanced in privacy and energy use, requires further tuning to match the security strength of encryption-based methods. This comparative analysis serves as a valuable reference for selecting appropriate technologies tailored to specific industrial needs and regulatory expectations in sustainable IoT ecosystems.

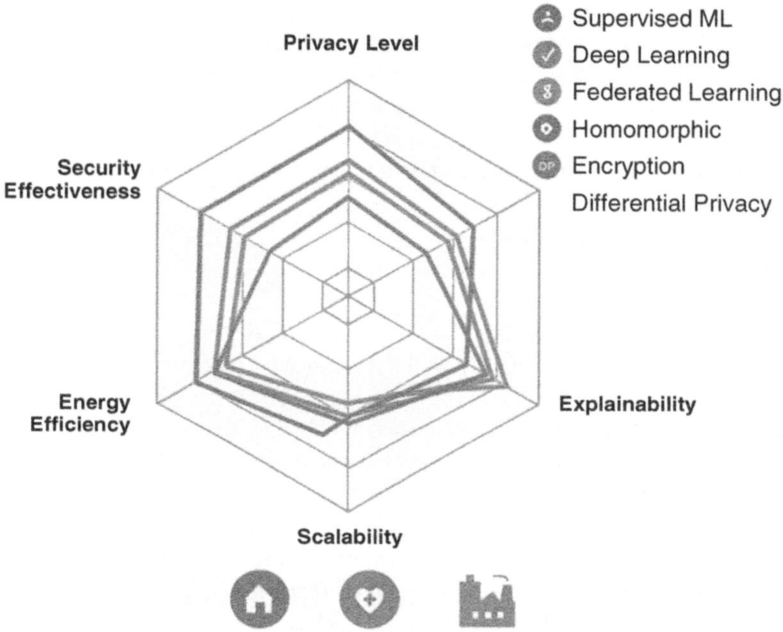

Figure 10.2 Comparative view of AI and privacy-preserving technologies for secure and sustainable IoT.

10.6 Conclusion

The convergence of security, privacy, and sustainability has become imperative in the design and deployment of modern IoT ecosystems. As this chapter has demonstrated, achieving resilience in IoT systems extends beyond technical robustness; it necessitates ethical foresight, regulatory alignment, and resource-conscious innovation. AI and ML techniques, including anomaly detection, DL, and federated models, play a crucial role in enabling intelligent and autonomous security mechanisms customized to IoT's decentralized nature. Meanwhile, privacy-preserving techniques such as homomorphic encryption, differential privacy, and decentralized identity systems foster trust and compliance in sensitive applications like healthcare and smart cities. Equally critical is the shift toward energy-aware security protocols and lightweight AI models that ensure operability within constrained environments without compromising performance. As real-world case studies have illustrated, sustainable IoT security must be grounded in practical, adaptable, and forward-thinking strategies. Emerging frameworks that combine edge computing, blockchain, and explainable AI are poised to transform threat intelligence and governance practices. As a future-ready mindset, the path to sustainable IoT security could require cross-disciplinary collaboration, continuous technological adaptation, and a strong commitment to ethical principles. By aligning innovation with the broader goals of environmental stewardship and social responsibility, the future of IoT can be both secure and sustainable.

References

1 Sicari, S., Rizzardi, A., Grieco, L.A., and Coen-Porisini, A. (2015). Security, privacy and trust in Internet of Things: the road ahead. *Computer Networks* 76: 146–164.

2 Sadique, K.M., Rahmani, R., and Johannesson, P. (2018). Towards security on internet of things: applications and challenges in technology. *Procedia Computer Science* 141: 199–206.

3 Rizvi, S., Kurtz, A., Pfeffer, J., and Rizvi, M. (2018). Securing the internet of things (IoT): a security taxonomy for IoT. In: *2018 17th IEEE International Conference on Trust, Security and Privacy in Computing and Communications/12th IEEE International Conference on Big Data Science and Engineering (TrustCom/BigDataSE)* (pp. 163–168). IEEE.

4 Rey, V., Sánchez, P.M.S., Celdrán, A.H., and Bovet, G. (2022). Federated learning for malware detection in IoT devices. *Computer Networks* 204: 108693.

5 Sadio, O., Ngom, I., and Lishou, C. (2019). Lightweight security scheme for mqtt/mqtt-sn protocol. In: *2019 Sixth International Conference on Internet of Things: Systems, Management and Security (IOTSMS)* (pp. 119–123). IEEE.

6 Attkan, A., and Ranga, V. (2022). Cyber-physical security for IoT networks: a comprehensive review on traditional, blockchain and artificial intelligence based key-security. *Complex & Intelligent Systems* 8 (4): 3559–3591.

7 Nayak, J., Meher, S.K., Souri, A., et al. (2022). Extreme learning machine and bayesian optimization-driven intelligent framework for IoMT cyber-attack detection. *The Journal of Supercomputing* 78 (13): 14866–14891.

8 Khraisat, A., and Alazab, A. (2021). A critical review of intrusion detection systems in the internet of things: techniques, deployment strategy, validation strategy, attacks, public datasets and challenges. *Cybersecurity* 4: 1–27.

9 Wei, C., Xie, G., and Diao, Z. (2023). A lightweight deep learning framework for botnet detecting at the IoT edge. *Computers & Security* 129: 103195.

10 Mazhar, N., Salleh, R., Zeeshan, M., et al. (2021). R-IDPS: real time SDN-based IDPS system for IoT security. In: *2021 IEEE 18th International Conference on Smart Communities: Improving Quality of Life Using ICT, IoT and AI (HONET)* (pp. 71–76). IEEE.

11 El Hajla, S., Ennaji, E.M., Maleh, Y., and Mounir, S. (2024). Enhancing internet of things network security through an ensemble-learning approach. In: *Proceedings of the 7th International Conference on Networking, Intelligent Systems and Security* (pp. 1–7).

12 Ramaiah, N.S., Andrews, S.K., Shenbagharaman, A., et al. (2023). Enhancing IoT security through AI-based anomaly detection and intrusion prevention. In: *2023 6th International Conference on Contemporary Computing and Informatics (IC3I)* (Vol. 6, pp. 1786–1790). IEEE.

13 Abinaya, A.B., Preethi, K., Janani, S., and Hymlin Rose, S.G. (2024). Secure IoT: fortifying IoT security with support vector machines. *Proceedings of NKCon 2024 – 3rd Edition of IEEE NKSS's Flagship International Conference: Digital Transformation: Unleashing the Power of Information* (pp. 1–7). IEEE.

14 Alsalman, D. (2024). A comparative study of anomaly detection techniques for IoT security using adaptive machine learning for IoT threats. *IEEE Access* 12: 14719–14730.

15 Tyagi, H., and Kumar, R. (2021). Attack and anomaly detection in IoT networks using supervised machine learning approaches. *Revue d'Intelligence Artificielle* 35 (1): 11–21.

16 Sharma, T., and Prasad, S.K. (2024). Enhancing cybersecurity in IoT networks: SLSTM-WCO algorithm for anomaly detection. *Peer-to-Peer Networking and Applications* 17 (4): 2237–2258.

17 Sampath, A.K. (2024). A survey paper on device type identification and intrusion detection using deep learning techniques in securing IoT environments. *Grenze International Journal of Engineering & Technology (GIJET)* 10 (2): 602.

18 Manzano, R., Goel, N., Zaman, M., et al. (2022). Design of a machine learning based intrusion detection framework and methodology for IoT networks. In: *2022 IEEE 12th Annual Computing and Communication Workshop and Conference (CCWC)* (pp. 0191–0198). IEEE.

19 Boyanapalli, A., and Shanthini, A. (2021). A comparative study of techniques, datasets and performances for intrusion detection systems in IoT. In: *Artificial Intelligence Techniques for Advanced Computing Applications: Proceedings of ICACT 2020* (pp. 225–236). Springer Singapore.

20 Sáez-de-Cámara, X., Flores, J.L., Arellano, C., et al. (2023). Federated explainability for network anomaly characterization. In: *Proceedings of the 26th International Symposium on Research in Attacks, Intrusions and Defenses* (pp. 346–365).

21 Abdulla, H., Al-Raweshidy, H., and Awad, W. (2023). The era of internet of things: towards better security using machine learning. In: *2023 International Conference on IT Innovation and Knowledge Discovery, ITIKD 2023* (pp. 1–5). IEEE.

22 Kharoubi, K., Cherbal, S., Mechta, D., and Gawanmeh, A. (2025). Network intrusion detection system using convolutional neural networks: NIDS-DL-CNN for IoT security. *Cluster Computing* 28 (4): 219.

23 Rbah, Y., Mahfoudi, M., Fattah, M., et al. (2024). Deep learning for enhanced IoMT security: a GNN-BiLSTM intrusion detection system. In: *2024 International Conference on Circuit, Systems and Communication (ICCSC)* (pp. 1–6). IEEE.

24 Alabduallah, B., Maray, M., Alruwais, N., et al. (2024). Class imbalanced data handling with cyberattack classification using Hybrid Salp Swarm Algorithm with deep learning approach. *Alexandria Engineering Journal* 106: 654–663.

25 Gupta, G.P., Tripathi, R., Gupta, B.B., and Chui, K.T. (eds.). (2023). *Big Data Analytics in Fog-Enabled IoT Networks: Towards a Privacy and Security Perspective*. CRC Press.

26 Agbor, B.A., Stephen, B.U.A., Asuquo, P., et al. (2025). Hybrid CNN–BiLSTM–DNN approach for detecting cybersecurity threats in IoT networks. *Computers* 14 (2): 58.

27 Alqahtani, A. (2025). Optimized deep autoencoder and BiLSTM for intrusion detection in IoTs-Fog computing. *Multimedia Tools and Applications* 84: 4907–4943.

28 Ali, S., Li, Q., and Yousafzai, A. (2024). Blockchain and federated learning-based intrusion detection approaches for edge-enabled industrial IoT networks: a survey. *Ad Hoc Networks* 152: 103320.

29 Lohitha, N.S., and Pounambal, M. (2022). A novel hybrid spotted hyena-swarm optimization (HS-FFO) framework for effective feature selection in IOT based cloud security data. *IJRITCC* 10 (12): 290–303.

30 Al Hammadi, A.Y., Yeun, C.Y., Damiani, E., et al. (2021). Explainable artificial intelligence to evaluate industrial internal security using EEG signals in IoT framework. *Ad Hoc Networks* 123: 102641.

31 Tang, W., Li, D., Fan, W., et al. (2023). An intrusion detection system empowered by deep learning algorithms. In: *2023 IEEE Intl Conf on Dependable, Autonomic and Secure Computing, Intl Conf on Pervasive Intelligence and Computing, Intl Conf on Cloud and Big Data Computing, Intl Conf on Cyber Science and Technology Congress (DASC/PiCom/CBDCom/CyberSciTech)* (pp. 1137–1142). IEEE.

32 Rawat, R., Chakrawarti, R.K., Vyas, P., et al. (2023). Intelligent fog computing surveillance system for crime and vulnerability identification and tracing. *International Journal of Information Security and Privacy (IJISP)* 17 (1): 1–25.

33 Muthanna, M.S.A., Alkanhel, R., Muthanna, A., et al. (2022). Towards SDN-enabled, intelligent intrusion detection system for internet of things (IoT). *IEEE Access* 10: 22756–22768.

34 Nicho, M., Cusack, B., Girija, S., and Arachchilage, N. (2024). Evaluating the cost of classifier discrimination choices for IoT sensor attack detection. *International Journal of Computers and Applications* 46 (11): 945–956.

35 Kumar, P., Jolfaei, A., and Islam, A.N. (2025). An enhanced deep-learning empowered threat-hunting framework for software-defined internet of things. *Computers & Security* 148: 104109.

36 Atheeq, C., Sultana, R., Sabahath, S.A., and Mohammed, M.A.K. (2024). Advancing IoT cybersecurity: adaptive threat identification with deep learning in cyber-physical systems. *Engineering, Technology & Applied Science Research* 14 (2): 13559–13566.

37 Mao, S., Liu, L., Zhang, N., et al. (2022). Intelligent reflecting surface-assisted low-latency federated learning over wireless networks. *IEEE Internet of Things Journal* 10 (2): 1223–1235.

38 Wang, J., Liu, Z., Yang, X., et al. (2025). The internet of things under federated learning: a review of the latest advances and applications. *Computers, Materials and Continua* 82 (1): 1–39.

39 Babbar, H., and Rani, S. (2023). FRHIDS: Federated learning recommender hybrid intrusion detection system model in software-defined networking for consumer devices. *IEEE Transactions on Consumer Electronics* 70 (1): 2492–2499.

40 Yan, X., Miao, Y., Li, X., et al. (2023). Privacy-preserving asynchronous federated learning framework in distributed IoT. *IEEE Internet of Things Journal* 10 (15): 13281–13291.

41 Garro, E., Lacalle, I., Bogacka, K., et al. (2024). Decentralized strategy for artificial intelligence in distributed IoT ecosystems: federation in ASSIST-IoT. *Shaping the Future of IoT with Edge Intelligence: How Edge Computing Enables the Next Generation of IoT Applications* (p. 231). Abingdon (UK): River Publishers.

42 Li, H., Ge, L., and Tian, L. (2024). Survey: federated learning data security and privacy-preserving in edge-Internet of Things. *Artificial Intelligence Review* 57 (5): 130.

43 Ellaji, C.H., Ponmagal, R.S., and Saritha, V. (2024). Study on federated learning model with pruning, optimal quantization and selective updating for IOT and big data analytics. In: *AIP Conference Proceedings* (Vol. 3075, No. 1). AIP Publishing.

44 Shen, J., Yang, W., Chu, Z., et al. (2024). Effective intrusion detection in heterogeneous Internet-of-Things networks via ensemble knowledge distillation-based federated learning. In: *ICC 2024-IEEE International Conference on Communications* (pp. 2034–2039). IEEE.

45 Geng, T., Liu, J., and Huang, C.T. (2024). A privacy-preserving federated learning framework for IoT environment based on secure multi-party computation. In: *2024 IEEE Annual Congress on Artificial Intelligence of Things (AIoT)* (pp. 117–122). IEEE.

46 Chen, Y.C., Hsu, S.Y., Xie, X., et al. (2024). Privacy preserving support vector machine based on federated learning for distributed IoT-enabled data analysis. *Computational Intelligence*, 40 (2): e12636.

47 Lim, W.Y.B., Garg, S., Xiong, Z., et al. (2020). Dynamic contract design for federated learning in smart healthcare applications. *IEEE Internet of Things Journal* 8 (23): 16853–16862.

48 Belarbi, O., Spyridopoulos, T., Anthi, E., et al. (2023). Federated deep learning for intrusion detection in IoT networks. In: *GLOBECOM 2023-2023 IEEE Global Communications Conference* (pp. 237–242). IEEE.

49 Hamid, S., and Bawany, N.Z. (2024). Federated learning for enhanced intrusion detection in smart city environments. In: *2024 18th International Conference on Open Source Systems and Technologies (ICOSST)* (pp. 1–6). IEEE.

50 Di, Y., Shi, H., Wang, Q., et al. (2025). Federated cross-domain recommendation system based on bias eliminator and personalized extractor. *Knowledge and Information Systems* 67 (3): 2935–2965.

51 Wang, W., Zhang, Y., Liu, Q., et al. (2024). Edge-intelligence-based computation offloading technology for distributed internet of unmanned aerial vehicles. *IEEE Internet of Things Journal* 11 (12): 20948–20957.

52 Alsharif, M.H., Kannadasan, R., Wei, W., et al. (2024). A contemporary survey of recent advances in federated learning: taxonomies, applications, and challenges. *Internet of Things* 27: 101251.

53 Al Asqah, M., and Moulahi, T. (2023). Federated learning and blockchain integration for privacy protection in the internet of things: challenges and solutions. *Future Internet* 15 (6): 203.

54 Shang, B., Liu, S., Lu, S., et al. (2020). A cross-layer optimization framework for distributed computing in IoT networks. In: *2020 IEEE/ACM Symposium on Edge Computing (SEC)* (pp. 440–444). IEEE.

55 Alajlan, N.N., and Ibrahim, D.M. (2022). TinyML: enabling of inference deep learning models on ultra-low-power IoT edge devices for AI applications. *Micromachines* 13 (6): 851.

56 Agrawal, K., Goktas, P., Sahoo, B., et al. (2024). IoT-based service allocation in edge computing using game theory. In: *International Conference on Distributed Computing and Intelligent Technology* (pp. 45–60). Cham: Springer Nature Switzerland.

57 Chaudhari, B.S., Ghorpade, S.N., Zennaro, M., and Paškauskas, R. (2024). TinyML for low-power internet of things. In: *TinyML for Edge Intelligence in IoT and LPWAN Networks* (pp. 1–12). Academic Press.

58 Muthumala, U., Zhang, Y., Martinez-Rau, L.S., and Bader, S. (2024). Comparison of tiny machine learning techniques for embedded acoustic emission analysis. In: *2024 IEEE 10th World Forum on Internet of Things (WF-IoT)* (pp. 444–449). IEEE.

59 Yagiz, M.A., MohajerAnsari, P., Pesé, M.D., and Goktas, P. (2024). Transforming in-vehicle network intrusion detection: VAE-based knowledge distillation meets explainable AI. In: *Proceedings of the Sixth Workshop on CPS&IoT Security and Privacy* (pp. 93–103).

60 Yagiz, M.A., and Goktas, P. (2025). LENS-XAI: redefining lightweight and explainable network security through knowledge distillation and variational autoencoders for scalable intrusion detection in cybersecurity. *arXiv preprint* arXiv:2501.00790.

61 Ali, J., Song, H.H., Sharma, V., and Al-Khasawneh, M.A. (2025). DDoS intrusions detection in low power SD-IoT devices leveraging effective machine learning. *IEEE Transactions on Consumer Electronics* 71 (1): 343–351.

62 Ban, X., Ding, M., Liu, S., et al. (2021). TAESim: a testbed for IoT security analysis of trigger-action environment. In: *European Symposium on Research in Computer Security* (pp. 218–237). Cham: Springer International Publishing.

63 Braghin, C., Lilli, M., and Riccobene, E. (2023). A model-based approach for vulnerability analysis of IoT security protocols: the Z-wave case study. *Computers & Security* 127: 103037.

64 Goktas, P. (2024). Ethics, transparency, and explainability in generative AI decision-making systems: a comprehensive bibliometric study. *Journal of Decision Systems* 1–29.

65 Goktas, P., and Grzybowski, A. (2025). Shaping the future of healthcare: ethical clinical challenges and pathways to trustworthy AI. *Journal of Clinical Medicine* 14 (5): 1605.

66 Chaki, J., Dey, N., and De, D. (eds.). (2020). *Smart Biosensors in Medical Care.* Academic Press.

67 Mekki, N., Hamdi, M., Aguili, T., and Kim, T.H. (2023). An authentication protocol for healthcare application: a case study of a diabetic patient. In: *International Conference on Advanced Information Networking and Applications* (pp. 434–445). Cham: Springer International Publishing.

68 Douglass, R., Gremban, K., Swami, A., and Gerali, S. (eds.). (2023). *IoT for Defense and National Security.* John Wiley & Sons.

69 Mukhtar, B.I., Elsayed, M.S., Jurcut, A.D., and Azer, M.A. (2023). IoT vulnerabilities and attacks: SILEX malware case study. *Symmetry* 15 (11): 1978.

70 Casado-Vara, R., Severt, M., Díaz-Longueira, A., et al. (2024). Dynamic malware mitigation strategies for IoT networks: a mathematical epidemiology approach. *Mathematics* 12 (2): 250.

71 Abdalla, P.A., and Varol, C. (2020). Testing IoT security: the case study of an IP camera. In: *2020 8th International Symposium on Digital Forensics and Security (ISDFS)* (pp. 1–5). IEEE.

72 Xu, T., Wendt, J.B., and Potkonjak, M. (2014). Security of IoT systems: design challenges and opportunities. In: *2014 IEEE/ACM International Conference on Computer-Aided Design (ICCAD)* (pp. 417–423). IEEE.

73 Hemashree, P., Kavitha, V., Mahalakshmi, S.B., et al. (2024). Machine learning approaches in blockchain technology-based IoT security: an investigation on current developments and open challenges. In: *Blockchain Transformations: Navigating the Decentralized Protocols Era* (pp. 107–130). Cham: Springer Nature Switzerland.

74 Kieras, T., Farooq, J., and Zhu, Q. (2021). I-SCRAM: a framework for IoT supply chain risk analysis and mitigation decisions. *IEEE Access* 9: 29827–29840.

75 Aljeraisy, A., Barati, M., Rana, O., and Perera, C. (2021). Privacy laws and privacy by design schemes for the internet of things: a developer's perspective. *ACM Computing Surveys (Csur)* 54 (5): 1–38.

76 Pape, S., and Rannenberg, K. (2019). Applying privacy patterns to the internet of things' (IoT) architecture. *Mobile Networks and Applications* 24: 925–933.

77 Mayer, A.H., Rodrigues, V.F., da Costa, C.A., et al. (2021). Fogchain: a fog computing architecture integrating blockchain and internet of things for personal health records. *IEEE Access* 9: 122723–122737.

78 Barbosa, M., Mokhtar, S.B., Felber, P., et al. (2017). SAFETHINGS: data security by design in the IoT. In: *2017 13th European Dependable Computing Conference (EDCC)* (pp. 117–120). IEEE.

79 Moldovan, G., Tragos, E.Z., Fragkiadakis, A., et al. (2016). An IoT middleware for enhanced security and privacy: the RERUM approach. In: *2016 8th IFIP International Conference on New Technologies, Mobility and Security (NTMS)* (pp. 1–5). IEEE.

80 Pavithran, D., Al-Karaki, J.N., and Shaalan, K. (2021). Edge-based blockchain architecture for event-driven IoT using hierarchical identity based encryption. *Information Processing & Management* 58 (3): 102528.

81 Chhabra, S., Mor, P., Mahdi, H.F., and Choudhury, T. (2021). Block chain and IoT architecture. In: *Blockchain Applications in IoT Ecosystem* (pp. 15–27). Cham: Springer International Publishing.

82 Yoon, I., Cao, N., Amaravati, A., and Raychowdhury, A. (2019). A 55nm 50nj/encode 13nj/decode homomorphic encryption crypto-engine for IoT nodes to enable secure computation on encrypted data. In: *2019 IEEE Custom Integrated Circuits Conference (CICC)* (pp. 1–4). IEEE.

83 Song, M., Sang, Y., Zeng, Y., and Luo, S. (2021). Blockchain-based secure outsourcing of polynomial multiplication and its application in fully homomorphic encryption. *Security and Communication Networks* 2021 (1): 9962575.

84 López, V., Fontenla-Romero, O., Hernández-Pereira, E., et al. (2024). RUL prediction of lithium-ion batteries using a federated and homomorphically encrypted learning method. In: *Proceedings of the 39th ACM/SIGAPP Symposium on Applied Computing* (pp. 565–571).

85 Vashishth, T.K., Sharma, V., Sharma, K.K., et al. (2024). Privacy-preserving machine learning techniques for IoT data in cloud environments. In: *Emerging Technologies for Securing the Cloud and IoT* (pp. 144–173). IGI Global.

86 Zhao, Y., Barnaghi, P., and Haddadi, H. (2022). Multimodal federated learning on IoT data. In: *2022 IEEE/ACM Seventh International Conference on Internet-of-Things Design and Implementation (IoTDI)* (pp. 43–54). IEEE.

87 Hlaing, H.H., and Asaeda, H. (2024). ShieldDINC: privacy-preserving distributed In-network computations with efficient homomorphic encryption. In: *2024 IEEE 49th Conference on Local Computer Networks (LCN)* (pp. 1–6). IEEE.

88 Sharma, S., Agrawal, S.S., and Kumar, S.A. (2024). Unlocking cybersecurity horizons: exploring cutting-edge technologies, strategies, and trends in the dynamic cyber threat landscape. In: *2024 International Conference on Intelligent Computing and Emerging Communication Technologies (ICEC)* (pp. 1–6). IEEE.

89 Hu, F. (2016). *Security and privacy in Internet of things (IoTs): Models, Algorithms, and Implementations*. CRC Press.

90 Zhao, S., Li, F., Li, H., et al. (2020). Smart and practical privacy-preserving data aggregation for fog-based smart grids. *IEEE Transactions on Information Forensics and Security* 16: 521–536.

91 Shen, M., Tang, X., Zhu, L., et al. (2019). Privacy-preserving support vector machine training over blockchain-based encrypted IoT data in smart cities. *IEEE Internet of Things Journal* 6 (5): 7702–7712.

92 Haque, R.U., Hasan, A.T., Jiang, Q., and Qu, Q. (2020). Privacy-preserving K-nearest neighbors training over blockchain-based encrypted health data. *Electronics* 9 (12): 2096.

93 Wang, J., Han, H., Li, H., et al. (2021). Multiple strategies differential privacy on sparse tensor factorization for network traffic analysis in 5G. *IEEE Transactions on Industrial Informatics* 18 (3): 1939–1948.

94 Feng, J., Yang, L.T., Zhang, R., and Gavuna, B.S. (2020). Privacy-preserving tucker train decomposition over blockchain-based encrypted industrial IoT data. *IEEE Transactions on Industrial Informatics* 17 (7): 4904–4913.

95 Li, J., Zhang, Y., Ning, J., et al. (2020). Attribute based encryption with privacy protection and accountability for Cloud IoT. *IEEE Transactions on Cloud Computing* 10 (2): 762–773.

96 Ma, Z., Wang, L., and Zhao, W. (2020). Blockchain-driven trusted data sharing with privacy protection in IoT sensor network. *IEEE Sensors Journal* 21 (22): 25472–25479.

97 Gupta, S., Bairwa, A.K., Kushwaha, S.S., and Joshi, S. (2023). Decentralized identity management system using the amalgamation of blockchain technology.

In: *2023 3rd International Conference on Intelligent Communication and Computational Techniques (ICCT)* (pp. 1–6). IEEE.

98 Sadique, K.M., Rahmani, R., and Johannesson, P. (2020). IMSC-EIoTD: Identity management and secure communication for edge IoT devices. *Sensors* 20 (22): 6546.

99 El Majdoubi, D., El Bakkali, H., Bensaih, M., and Sadki, S. (2022). A decentralized trust establishment protocol for smart IoT systems. *Internet of Things* 20: 100634.

100 Tu, Y., Gan, J., Hu, Y., et al. (2019). Decentralized identity authentication and key management scheme. In: *2019 IEEE 3rd Conference on Energy Internet and Energy System Integration (EI2)* (pp. 2697–2702). IEEE.

101 Kim, E., Cho, Y.S., Kim, B., et al. (2021). Can we create a cross-domain federated identity for the industrial internet of things without Google? *IEEE Internet of Things Magazine* 3 (4): 82–87.

102 Thompson, A., Abayomi, A., and Gabriel, A.J. (2022). Multifactor IoT authentication system for smart homes using visual cryptography, digital memory, and blockchain technologies. In: *Blockchain Applications in the Smart Era* (pp. 273–290). Cham: Springer International Publishing.

103 Ahmed, N.A., Ammar, M., and Hefny, H.A. (2020). Proposed authentication protocol for IoT using blockchain and fog nodes. *International Journal of Advanced Computer Science and Applications* 11 (4): 710–716.

104 Sharma, P., Wilfred Godfrey, W., and Trivedi, A. (2024). When blockchain meets IoT: a comparison of the performance of communication protocols in a decentralized identity solution for IoT using blockchain. *Cluster Computing* 27 (1): 269–284.

105 Pathmabandu, C., Grundy, J., Chhetri, M.B., and Baig, Z. (2023). Privacy for IoT: informed consent management in smart buildings. *Future Generation Computer Systems* 145: 367–383.

106 Mann, G., and Singh, R. (2023). Energy efficient multilevel hierarchical routing protocol in green Internet of Things. In: *Congress on Intelligent Systems* (pp. 325–340). Singapore: Springer Nature Singapore.

107 Hasan, A., and Patle, V.K. (2024). JSO-based enhanced energy efficient LEACH protocol for IoT-perception layer. *International Journal of Information Technology* 16 (2): 979–991.

108 Srivastava, A., and Paulus, R. (2024). Optimization-driven hybrid trust aware algorithm for energy efficient cluster head selection and routing in IoT enabled WSN. In: *2024 International Conference on IoT Based Control Networks and Intelligent Systems (ICICNIS)* (pp. 532–537). IEEE.

109 Verma, A., Verma, S.K., Pandey, A.C., et al. (2024). Li-MSD: a lightweight mitigation solution for DAO insider attack in RPL-based IoT. *Future Generation Computer Systems* 159: 327–339.

110 Mazunga, F., and Nechibvute, A. (2021). Ultra-low power techniques in energy harvesting wireless sensor networks: recent advances and issues. *Scientific African* 11: e00720.

111 Kumar, A., Gupta, N., Derawi, M., and Pal, R. (2024). Hardware implementation of a reinforcement learning based energy efficient protocol for wireless sensor network. *2024 Second International Conference on Microwave, Antenna and Communication (MAC)* 1–5.

112 Shiny, G.S., Ram, R.S., and Muthalakshmi, M. (2024). Energy efficient data dissemination in wireless sensor network enabled IoT using domain-adaptive message passing graph neural network. *International Journal of Communication Systems* 37 (13): e5825.

113 Jagho Mdemaya, G.B., Kengne Tchendji, V., Velempini, M., and Atchaze, A. (2024). ATENA: adaptive TEchniques for Network Area coverage and routing in IoT-based edge computing. *Journal of Network and Systems Management* 32 (4): 83.

114 Javeed, D., Saeed, M.S., Ahmad, I., et al. (2024). Quantum-empowered federated learning and 6G wireless networks for IoT security: concept, challenges and future directions. *Future Generation Computer Systems* 160: 577–597.

115 Maurya, V., Rishiwal, V., Yadav, M., et al. (2025). Blockchain-driven security for IoT networks: state-of-the-art, challenges and future directions. *Peer-to-Peer Networking and Applications* 18 (1): 1–35.

116 Alauthman, M., Aldweesh, A., Al-Qerem, A., et al. (2024). Cryptographic protocols for internet of things (IoT) security lightweight schemes and practical deployment. *Innovations in Modern Cryptography* 431–448.

117 Hu, W., Cao, Q., Darbandi, M., and Jafari Navimipour, N. (2024). A deep analysis of nature-inspired and meta-heuristic algorithms for designing intrusion detection systems in cloud/edge and IoT: state-of-the-art techniques, challenges, and future directions. *Cluster Computing* 27 (7): 8789–8815.

118 Tiwari, H., Tomar, A., Patil, S., et al. (2024). Slipper zero: exploring Wi-Fi security vulnerabilities and attack implementations on ESP32 microcontrollers. In: *2024 Global Conference on Wireless and Optical Technologies (GCWOT)* (pp. 1–7). IEEE.

119 Prajapati, A.K., Pilli, E.S., Battula, R.B., et al. (2025). A comprehensive survey on RPL routing-based attacks, defences and future directions in Internet of Things. *Computers and Electrical Engineering* 123: 110071.

120 Hajoary, D., Basumatary, R., and Narzary, R. (2024). Bibliometric analysis of IoT application research: unveiling trends and future directions for terahertz communication in wireless systems. In: *2024 4th International Conference on Technological Advancements in Computational Sciences (ICTACS)* (pp. 328–332). IEEE.

121 Nadimi-Shahraki, M.H., Zamani, H., Varzaneh, Z.A., et al. (2024). A systematic review of applying grey wolf optimizer, its variants, and its developments in different internet of things applications. *Internet of Things* 26: 101135.

122 Cheryl, B.-K., and Ng, B.-K. (2022). Protecting the unprotected consumer data in Internet of Things: Current scenario of data governance in Malaysia. *Sustainability* 14 (16): 9893.

123 Dasgupta, A., Gill, A., and Hussain, F. (2019). A conceptual framework for data governance in IoT-enabled digital IS ecosystems. *DATA 2019 – Proceedings of the 8th International Conference on Data Science, Technology and Applications* 209–216.

124 Pathak, M. (2024). Data governance redefined: The evolution of EU data regulations from the GDPR to the DMA, DSA, DGA, Data Act and AI Act. *European Data Protection Law Review* 10 (1): 43–56.

125 Tapsell, J., Akram, R. N., and Markantonakis, K. (2018). Consumer centric data control, tracking and transparency – A position paper. *Proceedings – 17th IEEE International Conference on Trust, Security and Privacy in Computing and Communications and 12th IEEE International Conference on Big Data Science and Engineering (TrustCom/BigDataSE 2018)* 1380–1385.

126 Yebenes Serrano, J., and Zorrilla, M. (2021). A data governance framework for Industry 4.0. *IEEE Latin America Transactions* 19 (12): 2130–2138.

127 Amir, A., Vinke, C., and van Heijningen, R. (2024). Horti-IoT: A new data framework for Dutch horticulture. *International Food and Agribusiness Management Review* 27 (5): 858–881.

128 Kazmi, A., Serrano, M., and Lenis, A. (2018). Smart governance of heterogeneous internet of things for smart cities. *Proceedings of the International Conference on Sensing Technology (ICST 2018)* 58–64.

129 Ahanger, A. S., Masoodi, F. S., Khanam, A., and Ashraf, W. (2024). Managing and securing information storage in the Internet of Things. *Internet of Things Vulnerabilities and Recovery Strategies* 102–151. CRC Press.

130 Verma R.K., and Kishor K. (2024). Innovative solutions: Exploring federated learning-based resource virtualization with AR integration in healthcare environments. *Federated Learning for Smart Communication using IoT Application*. pp. 155–177. CRC Press, Taylor & Francis. doi: 10.1201/9781003489368-8

11

Machine Learning-empowered Physical Layer Security Techniques Toward 6G Wireless Communication

Aditya Verma[1], Sumita Mishra[1], Sachin Kumar[1], and Jyoti Grover[2,]*

[1] *Amity School of Engineering and Technology, Amity University, Lucknow Campus, Lucknow, Uttar Pradesh, India*
[2] *Malaviya National Institute of Technology, Jaipur, Rajasthan, India*

*Corresponding author: jgrover.cse@mnit.ac.in

11.1 Introduction

The move to 6G networks marks a major progression towards tremendous information speeds, and much more diverse application regions. Whereas the fifth era (5G) is already changing businesses with its ultra-low latency and high throughput, 6G is expected to further push these boundaries, offering ground-breaking advancements in applications like autonomous vehicles, holographic communications, counterfeit insights, and immersive virtual environments as depicted in Figure 11.1. In any case, the scale and complexity of these advancements bring with them noteworthy security challenges that must be handled to ensure reliable information exchange, guarantee client protection, and keep up in general the dependability of the network [1].

Customary security measures, like encryption and confirmation, ordinarily work at the upper layers of the arrangement and frequently depend on resource-intensive forms or preconfigured security keys. Whereas these strategies have been compelling in 5G, they may not be adequate to address the requests

AI and ML Techniques in IoT-based Communication: A Path to Sustainable Development Goals, First Edition. Edited by Sumita Mishra, Nishu Gupta and Polat Goktas.

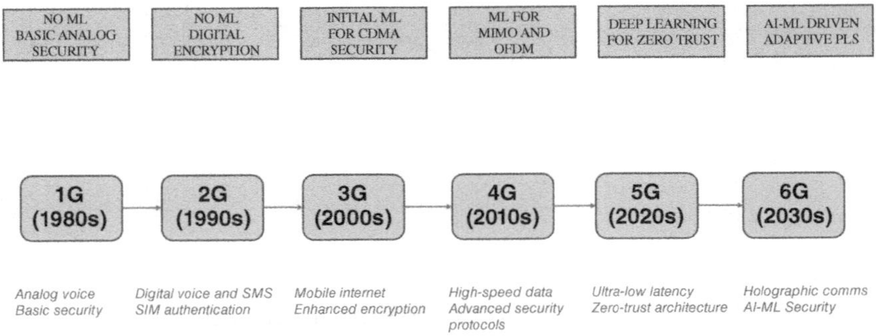

Figure 11.1 Evolution of wireless technologies:machine learning-empowered physical layer security (PLS).

of 6G, given the network's extended scale, and the assortment of connected devices, coupled with threats such as quantum-based attacks and sophisticated jamming or eavesdropping methods, which necessitate robust security measures. Subsequently, guaranteeing strong security across all layers of the communication stack has become a key area of concern for the next generation of wireless systems.

Physical layer security (PLS) has risen as an viable procedure to handle these challenges. Not at all like conventional cryptographic strategies that depend on keys and calculations to ensure information, PLS centers on securing communication specifically at the physical layer – the center level of the communication framework. It takes advantage of the common properties of remote channels, such as blurring, clamor, and obstructions, to make conditions that make it troublesome for busybodies to catch or translate the transmitted signals. By consolidating PLS into 6G systems, it becomes conceivable to safeguard sensitive data, even in the presence of powerful adversaries attempting to spy or disrupt the transmission. In recent years, the combination of machine learning (ML) and PLS has gained considerable interest because of its potential to improve the flexibility, efficiency, and robustness of security frameworks. ML techniques, including deep learning and reinforcement learning, enable systems to learn from data, adapt to changing conditions, and continuously improve their capability to detect and counter security threats. This integration of ML with PLS clears the way for the creation of brilliantly, independent, and context-aware security arrangements that are in a perfect world suited to the energetic and assorted environment of 6G networks [2].

With 6G systems expected to associate billions of gadgets, from smartphones and wearables to independent vehicles and mechanical Internet of Things (IoT) frameworks, the potential applications of ML-enhanced PLS are broad. These applications incorporate defending the communications of basic framework in smart cities, securing vehicle-to-vehicle (V2V) and vehicle-to-infrastructure (V2I) communications in independent vehicles, and guaranteeing the security of information traded among different IoE gadgets. The integration of ML into PLS can moreover play a key part in identifying and relieving developing dangers, such as sticking, spoofing, and information infusion, by persistently adjusting to the dynamic nature of the network.

In spite of the promising potential of coordination ML with PLS in 6G, a few challenges stay. These incorporate the prerequisite for huge datasets, the computational requests of ML calculations, and the helplessness of ML models to ill-disposed assaults. Also, incorporating ML-based PLS arrangements over the wide extent of differing and resource-constrained gadgets in a 6G environment requires the development of adaptable approaches [3].

This chapter analyses the merging of ML and PLS in context of 6G as shown in Figure 11.2, emphasizing how ML-driven methods can fortify security at the physical layer. It investigates key applications over businesses such as smart cities, independent vehicles, and the Internet of Everything (IoE), whereas too talking about the progressing challenges and potential future inquire about bearings in this field. Eventually, the integration of ML into PLS is pivotal for creating secure, versatile, and versatile systems that can meet the requests of 6G and past.

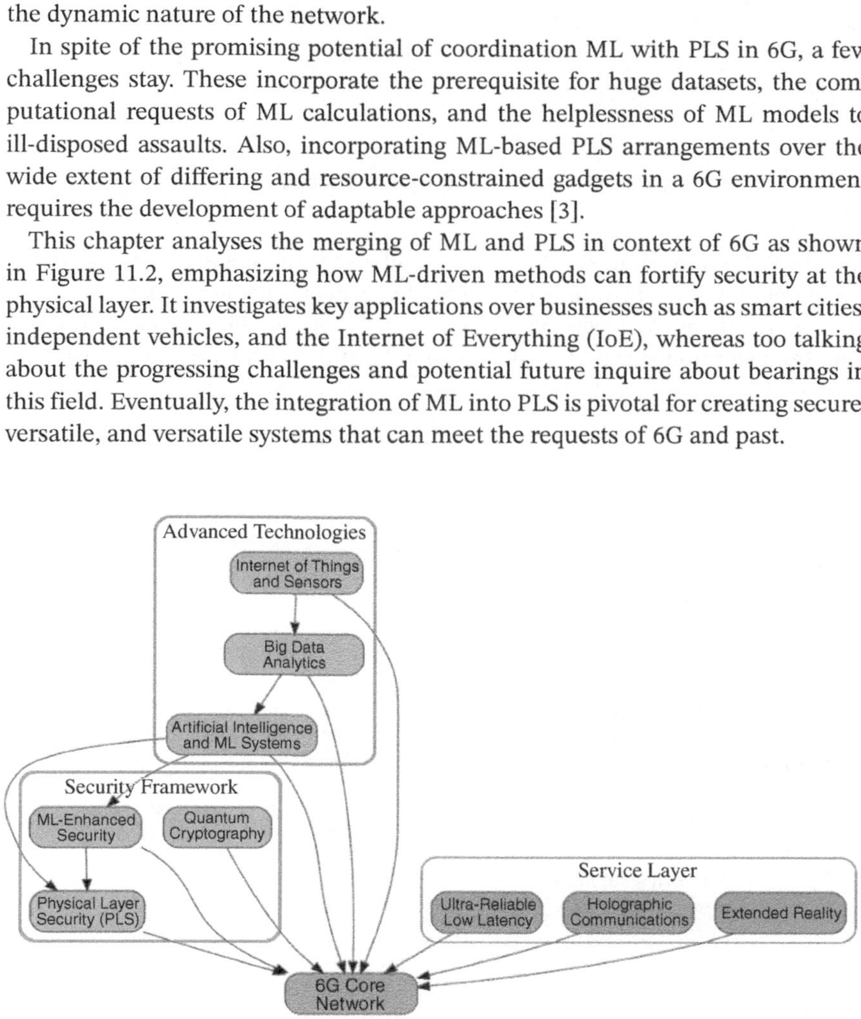

Figure 11.2 A high-level diagram of 6G network architecture.

11.2 6G Technologies: Security and Privacy Issues

The transition to 6G systems is expected to revolutionize worldwide network, empowering ultra-high-speed communications and seamless support to billions of interconnected devices over diverse use cases. It simultaneously presents a wave of security challenges as depicted in Figure 11.3. These challenges must be handled to guarantee secure, flexible, and reliable network operations. These issues are not essentially an expansion of those experienced in past eras, but many of them arise due to the tremendous scale, complexity, and huge data volume characteristic to 6G networks [4].

11.2.1 Expanding Attack Surface and Rising Threat Landscape in 6G

One of the foremost noteworthy concerns related with 6G systems is the sheer development within the number of associated gadgets. Not at all like 5G, which underpins millions of gadgets, 6G is anticipated to bolster billions, counting everything from smartphones and wearables to independent vehicles and mechanical

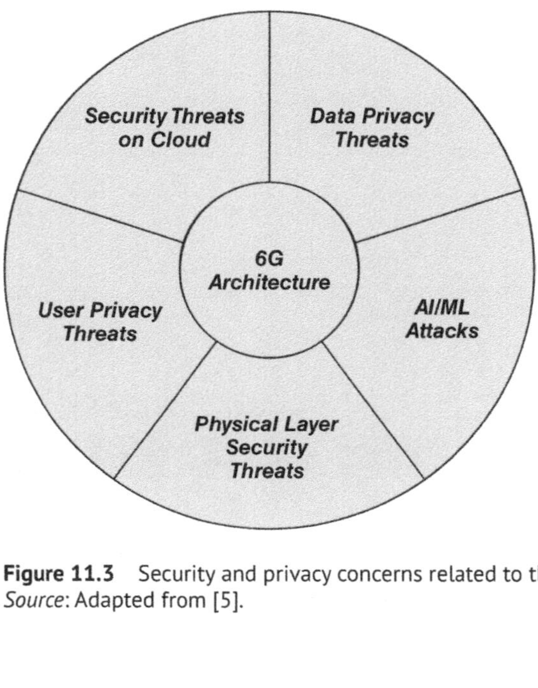

Figure 11.3 Security and privacy concerns related to the 6G architecture. *Source*: Adapted from [5].

IoT frameworks. This proliferation of connected devices increases the attack surface making more opening for cyber threats.

11.2.1.1 Modern Risk Vectors

As 6G systems will depend on a wide blend of technologies – including AI, quantum computing, and independent systems – they will be vulnerable to novel assault methods. For illustration, assailants might use AI-powered assaults to consequently bypass security measures, or misuse quantum computing capabilities to break routine encryption strategies. In expansion to conventional dangers like spoofing, listening stealthily, and sticking, progressed cyberattacks will require unused strategies of protection, counting real-time location and adjustment to advancing strategies.

11.2.1.2 IoT and Gadget Vulnerabilities

The integration of billions of IoT gadgets over basic infrastructure – such as control lattices, healthcare frameworks, and transportation networks – will make openings for gadget seizing, information control, and other malicious interferences.

11.2.2 Security Dangers in a Hyperconnected Environment

As 6G systems will encourage the trade of enormous volumes of information, much of it delicate, security concerns are increased. The consistent interaction between keen homes, wearable gadgets, independent vehicles, and other associated frameworks will result in a ceaseless stream of information being exchanged – often containing individual and private data.

11.2.2.1 Information Capture Attempts

With so much individual data being transmitted over the network, information breaches and capture attempts pose noteworthy dangers. Malicious actors may obtain delicate information from individual gadgets or frameworks, such as health data, location history, and private communications, in the event that the network's security is not robust enough to ensure it.

11.2.2.2 Information Proprietorship and Control

In a world where information will stream openly over different gadgets and frameworks, deciding who claims the information, who can get to it, and how it can be shared gets to be a complex issue. With dispersed cloud models and edge computing at the heart of 6G, the purview and proprietorship of information will obscure, particularly with regard to worldwide security laws just like the GDPR. Guaranteeing that clients have control over their information and their privacy is maintained may be a developing concern.

11.2.3 The Danger of Quantum Computing to Cryptography

One of the greatest potential dangers to the security of 6G systems lies in the advancement of quantum computing. Quantum computers have the potential to break many of the currently used encryption algorithms in wireless networks.

11.2.3.1 Affect on Current Encryption

Advanced encryption strategies like RSA and ECC (Elliptic Bend Cryptography) are broadly utilized to defend information in travel. In any case, quantum calculations like Shor's calculation may possibly break these encryption schemes much more productively than classical computers. Therefore, cryptographic algorithms, designed to resist attacks from quantum adversaries, are required to ensure data confidentiality and integrity.

11.2.3.2 Move to Post-quantum Cryptography

Transitioning to post-quantum cryptographic guidelines is not a basic errand. It requires upgrading the fundamental framework of 6G systems and guaranteeing in reverse compatibility with existing frameworks. Furthermore, quantum-resistant calculations must be adequately effective to handle the high-speed, low-latency requests of 6G applications.

11.2.4 Guaranteeing Security in Dynamic and Heterogeneous Systems

6G systems will be dynamic and heterogeneous, meaning they will comprise of an assortment of gadgets, each with distinctive security capabilities and authorization mechanism. Guaranteeing reliable security over such a differing environment could be a complex task.

11.2.4.1 Edge Computing and Transmission Risks

A key highlight of 6G is edge computing, where information is prepared closer to the source, such as in associated gadgets or at neighborhood edge hubs. Whereas this decreases idleness and improves execution, it moreover presents new security dangers. For example, edge nodes may be compromised, or sensitive information may be leaked during transmission between the edge and core network. Securing these decentralized frameworks and guaranteeing the system integrity may be a growing challenge.

11.2.4.2 Cross-slice Security in 6G

6G will utilize network slicing, where the network is separated into smaller, customized virtual networks. These virtual networks suited for specific applications will have special prerequisites for execution and security. Securing communications over diverse slices while avoiding cross-slice assaults requires careful isolation of network resources.

11.2.5 Insights and Threats That AI Technology Could Form

AI has progressed to a stage where it is necessary to the functionality of the 6G systems, but it will also create new risks that will require advanced security protocols as well as improving existing ones [6].

11.2.5.1 AI-driven Assaults

There are, however, some negative ramifications to the use of AI. The benefits that proprietary systems provide will make cyber assaults more common. Ill-disposed ML or AI driven cyberattacks greatly enhances the ability of their creators to harness the intelligence behind the AI systems and will be able to bypass existing security measures. AI technology based attacks will become agile and sophisticated which will force us to entirely rethink how we counter these AI based threats. They will be harder to prevent from and traditional security elements will not be able to counter them effectively.

11.2.6 The Requirement for Robust and Adaptive Security System

The inherent features of 6G systems including phenomenal scale and complexity renders the conventional security measures inadequate in presence of threats discussed in above section. Therefore, a versatile and robust security system is required that supports adaptive, real-time response mechanisms to safeguard user privacy, confidentiality in presence of quantum-based attacks, advanced jamming, and malicious eavesdropping techniques.

11.3 The Role of ML in PLS for 6G

In 6G systems, PLS integrated with ML leverages the special characteristics of the physical communication medium itself. Not at all like conventional security conventions that work at higher layers (like encryption at the application or transport layers), PLS centers on securing the information straightforwardly at the physical layer – using variables such as flag impedances, channel clamor, and blurring to guarantee secrecy and integrity [7].

As 6G systems guarantee altogether higher information rates, ultra-low latencies, and gigantic network (with billions of gadgets), the complexity of keeping up secure communication increments exponentially. In this environment, ML can

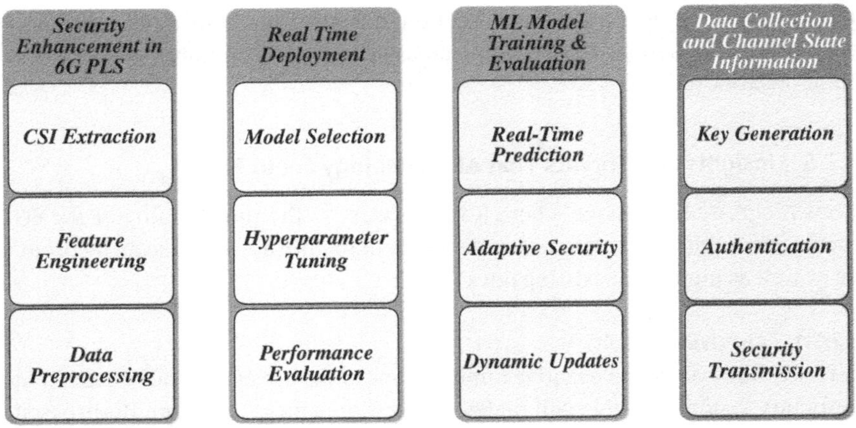

Figure 11.4 ML integration in PLS for 6G.

be integrated with PLS as shown in Figure 11.4 to address the security concerns. Underneath is an investigation of how ML can boost PLS within the setting of 6G.

11.3.1 Real-time Risk Location and Anomaly Identification

Within the fast-paced, exceedingly variable situations of 6G systems, dangers may advance quickly. ML can be a game changer in recognizing security breaches in genuine time by examining expansive volumes of information from the physical layer, counting channel conditions, flag quality, and impedances designs. Here's how:

11.3.1.1 Spotting Unpredictable Conduct

ML models can be prepared to recognize anomalous communication patterns, such as irregular signal-to-noise proportions or sudden changes that conform with malicious interference like sticking or listening in.

11.3.1.2 Recognizing Flag Contrasts

By utilizing ML methods, systems can separate between true blue client signals and potential dangers by learning the one of a kind characteristics of signals from authorized clients versus those created by aggressors.

11.3.1.3 Higher Layer Security Breech Detection

In addition to identifying physical-layer assaults, ML calculations can dissect the stream of information over the network to distinguish signs of higher layer breaches, making a difference to anticipate both conventional and unusual assaults.

Table 11.1 Comparison of traditional vs. ML-enhanced security.

Ref.	Metric	Traditional PLS	ML-enhanced PLS	Comments
[8]	Security Level	Fixed cryptographic techniques	Adaptive threat detection	ML-enhanced PLS adapts to threats better than traditional methods
[8]	Energy Efficiency	Moderate power consumption	Optimized power use	ML-enhanced PLS optimizes power use more effectively
[9]	Robustness (to Attacks)	Vulnerable to evolving attack strategies	Learns and adapts to new attack patterns	ML-enhanced PLS resists attacks better than fixed models
[9]	Computational Efficiency	Computationally lightweight but less adaptive	High processing power, optimized use	ML-enhanced PLS needs more processing but optimizes use
[10]	Adaptability to Network Changes	Slow adjustments	Real-time adaptation	ML-enhanced PLS adjusts faster to changes
[11]	Scalability	Limited scalability	Handles large and dynamic networks efficiently	ML-enhanced PLS handles larger networks more efficiently
[11]	Real-time Performance	Slower response time due to predefined rule-based security	Fast anomaly detection	ML-enhanced PLS detects threats faster than traditional methods

11.3.2 Efficient Key Administration

Rather than depending on settled plans, ML can be utilized to oversee the secure dissemination and turn of keys, altering strategies based on past execution, channel inconstancy, or threat perceptions. A comparison of ML-enhanced PLS systems and traditional security systems is shown in Table 11.1.

11.3.3 Adaptive Threat Response and Secured Communication Through ML

PLS points to guarantee that transmitted information remains secret and tamper-free. ML can be significant for:

11.3.3.1 Confirming Signals

Each flag transmitted in a remote organize has special physical characteristics due to defects in equipment, natural conditions, and indeed the development of the sender. ML models can examine these "physical layer fingerprints" to confirm the personality of the signal's root and guarantee it hasn't been capturing or changed.

11.3.3.2 Versatile Mistake Rectification

With ML, mistake location and adjustment instruments can adjust powerfully to the particular conditions of the communication channel. This versatility upgrades both the unwavering quality of transmission and the security of the information against any malicious disturbances.

11.3.4 Jamming Location and Moderation

One of the foremost predominant assaults on remote systems is jamming, where an assailant surges the organize with commotion to disturb communication. In 6G, where the number of gadgets and information streams will skyrocket, sticking discovery and moderation will be critical [12]. Here's where ML can sparkle:

11.3.4.1 Identifying Sticking Assaults

ML can recognize the signs of sticking by ceaselessly observing obstructions designs, flag quality, and range utilization. Once a potential sticking assault is recognized, it can trigger programmed reactions to ensure the organize.

11.3.4.2 Versatile Reactions

ML can moreover offer assistance plan brilliantly countermeasures. For case, once sticking is recognized, ML may consequently alter communication parameters, such as exchanging to a diverse recurrence band or tweaking transmission control, to bypass the obstructions and keep up secure communication.

11.3.5 Optimized Channel Estimation for Secure Communication

The quality of the communication channel is constantly changing in wireless environments, making it difficult to maintain consistent security. ML can enhance channel estimation – the process of understanding the state of the communication channel – by predicting future conditions and optimizing parameters to maintain secure transmission. For example:

11.3.5.1 Predictive Channel Estimation

ML models can analyze past channel data to predict future changes, allowing the communication system to pre-emptively adjust encryption schemes, power levels, or modulation techniques to stay secure in fluctuating environments.

11.3.5.2 MIMO Systems

With the use of Multiple Input Multiple Output (MIMO) technology, which will be critical for 6G's performance, ML can be applied to improve beamforming techniques, helping to reduce eavesdropping and enhance the secrecy of transmitted data.

11.3.6 Secrecy and Transmission Rate Enhancement

In 6G, ensuring the secrecy capacity (the ability to securely transmit information even in the presence of eavesdroppers) will be fundamental to maintaining confidentiality [13]. ML can enhance this by dynamically adjusting system parameters to optimize the balance between communication speed and security:

11.3.6.1 Parameter Optimization

ML can optimize parameters like transmission power, modulation techniques, and beamforming strategies to maximize the rate of secure communication while minimizing the risk of interception by unauthorized users.

11.3.6.2 Adaptive Modulation and Coding

Through ML, communication systems can continuously adapt modulation schemes and coding techniques in response to varying channel conditions, ensuring that security is maintained even as the physical layer's characteristics change.

11.3.7 Collaborative Learning for Distributed Security

6G networks will involve an unprecedented number of connected devices, base stations, and IoT devices, making it challenging to ensure security across such a vast network. Federated learning (FL), a type of distributed ML, can enable devices and network nodes to work together in detecting and mitigating threats while maintaining data privacy:

11.3.7.1 Cooperative Threat Detection

Instead of each device working in isolation, FL enables devices across the network to share models (not raw data), improving overall detection capabilities without compromising user privacy.

11.3.7.2 Decentralized Security

ML can help coordinate distributed encryption strategies across different network nodes. By dynamically adapting security measures based on localized data, each node can enhance the overall security of the network without introducing a single point of failure.

11.4 AI-driven Extensions for PLS in 6G

Apart from traditional Machine Learning approaches, other advanced intelligent techniques further enhance PLS in 6G by offering adaptable, robust, and scalable privacy preserving solutions. In essence, deep learning (DL), reinforcement learning (RL), and FL are indispensable and critical toward solving security predicaments for next-generation networks [14].

11.4.1 DL for PLS

It is the use of neural networks that are deep and numerous to understand complex data that defines DL. For instance, DL methods are applied when performing a number of crucial tasks that improve the performance and dependability of communications in real time in a certain context of PLS [15].

11.4.1.1 Estimation of the Channel, as well as the Detection of the Signal

Detecting patterns using DL models in vast and complex datasets is phenomenally Out of this world. In PLS, it is possible to more accurately estimate the wireless channel using DL methods like convolutional neural networks (CNN) or recurrent neural networks (RNN) and likewise. With these models, it is possible to identify and adjust to the changes in the conditions of the channel which provides the ability to improve signal detection and interference. This provides security by diminishing the chances of interception or eavesdropping.

11.4.1.2 Secret Key Generation

A key aspect of PLS is generating secure keys for communication. DL can enhance this process by learning the statistical properties of the channel noise, making it possible to generate more secure, unpredictable keys. Models like autoencoders or generative adversarial networks (GANs) can be trained to extract randomness from the environment, further strengthening the security of shared keys between legitimate users while minimizing the chances of a malicious entity acquiring the same key.

11.4.1.3 Eavesdropper Detection and Signal Classification

DL can also be used to differentiate between legitimate communication and potential eavesdropping. By training a deep neural network with data from various wireless environments, the system can automatically identify and flag unusual behavior indicative of eavesdropping or malicious interference. This proactive detection mechanism makes it more difficult for unauthorized users to access the network.

11.4.1.4 Optimizing Beamforming for Security

DL methods, particularly those based on RL, can optimize beamforming techniques. Beamforming is a method used to direct the wireless signal toward the legitimate receiver while minimizing leakage to unintended recipients (like eavesdroppers). By training DL models to adapt beamforming patterns in real time, we can ensure that the signal is always directed in the most secure manner possible.

11.4.2 RL for PLS

RL is an area of ML where an agent learns to make decisions by interacting with its environment. RL is particularly useful for dynamic, decision-making processes where feedback is critical. When applied to PLS, RL helps improve security by continuously adapting to changing network conditions [16].

11.4.2.1 Adaptive Power Control

One of the primary applications of RL in PLS is power control. In traditional networks, power levels are often fixed, but RL can adjust power dynamically to reduce the risk of eavesdropping. An RL agent, by observing feedback about the current communication environment, learns to select the optimal transmission power, balancing secure communication with energy efficiency.

11.4.2.2 Jamming-aware Transmission

Jamming attacks, where malicious entities intentionally disrupt the signal, pose a significant threat to secure communication. RL can be employed to detect and respond to jamming in real time. By learning from previous experiences, an RL agent can adjust the transmission parameters or switch to alternative frequencies to maintain secure communication, minimizing the effects of jamming on legitimate data transfer.

11.4.2.3 Secure Multiuser Communication

In modern networks, multiple users may share the same resources. RL is used to optimize the allocation of network resources, such as bandwidth and power, in a way that maximizes security for all users. An RL agent learns to make decisions that protect the entire network from potential eavesdropping and interference, even in dense, multiuser environments.

11.4.2.4 Beamforming and Scheduling for Security

Just like DL, RL can be applied to optimize beamforming strategies. By selecting the right beam direction and power allocation, RL agents can maximize the signal strength toward the legitimate receiver while minimizing the risk of interception by eavesdroppers. RL is also used to adjust scheduling algorithms, ensuring that transmission times are optimized for security without compromising throughput.

11.4.3 FL for PLS

With FL, it is possible for numerous units to work within a central model without having to share any raw data. This method of ML is optimal for preserving privacy in wireless networks where data cannot be individually collected due to privacy issues.

11.4.3.1 Collaborative, Privacy-preserving Learning

In the context of PLS, FL enables multiple devices to collaboratively improve security models while keeping their data local. For example, each mobile device or base station in the network can train a model on local data (such as signal characteristics, interference, or noise) and share only the model updates (not the raw data). This allows for the development of more robust security models that can detect eavesdropping or other malicious activities without compromising user privacy [17].

11.4.3.2 Distributed Eavesdropper Detection

Using FL, devices across different locations can work together to detect eavesdropping without sharing sensitive data. Each device analyses its local environment and provides feedback, helping to build a global model that detects unusual patterns of behavior associated with unauthorized access or malicious interference. Since raw data is not shared, privacy is maintained, but the network benefits from collective learning.

11.4.3.3 Collaborative Beamforming and Security Strategy

In a large-scale network, multiple base stations or devices can use FL to optimize beamforming strategies. By collaborating, they can refine their security strategies in a decentralized manner without sharing their data. This is particularly useful in multiuser or multicell environments where privacy is a concern, and centralized training would be inefficient or infeasible.

11.4.3.4 Network-wide Secure Resource Allocation

FL allows for decentralized learning in network-wide resource allocation decisions. Devices collaborate to determine optimal power, frequency, and beamforming strategies while ensuring that the shared model respects the privacy of individual nodes. By doing so, FL enables a more efficient and secure management of resources across large, distributed networks. Table 11.2 lists some applications of DL, RL, and FL in PLS.

Table 11.2 Applications of DL, RL and FL in PLS.

Reference	Machine learning approach	Applications in PLS	Advantages
[18]	Deep learning	Accurate channel estimation Eavesdropper detection Secure key generation	Handles complex data High accuracy Learns patterns from large datasets
[19]	Reinforcement learning	Adaptive power control Jamming-aware transmission Secure beamforming	Real-time adaptation Long-term optimization Learns optimal security strategies
[20]	Federated learning	Privacy-preserving collaboration Distributed eavesdropper detection Collaborative beamforming	Protects privacy Scalable and efficient Decentralized learning

11.5 Applications of ML-powered PLS in 6G

With the onset of 6G, ML has become imperative as it merges into PLS on the path to securing communication networks with various industries. The amount of data and intricacies of the systems in next-generation networks require dynamic and intelligent solutions that can adapt to threats in real time. Moving on, we will discuss how ML will enhance PLS to secure domains such as Smart Cities, Autonomous Vehicles, the Internet of Everything (IoE), and other developing applications.

11.5.1 Secure Smart Cities

Securing Critical Infrastructure and Privacy: Smart cities are built on a network of interconnected devices and systems-from energy grids, traffic management systems, surveillance cameras, and smart streetlights. The networks that allow the cities to run efficiently create enormous amounts of data that could be targeted by malicious actors. The entire aim is to secure all critical infrastructure while maintaining the privacy of the citizens.

In a world of PLS powered by ML enhancement security, many things could go right. Communication patterns can be analyzed through DL frameworks to detect irregularities possibly signifying security breaches, such as a hacker intercepting data from smart sensors. In co-occurrence, adaptive beamforming and secure key generation strategies will ensure prevention from unauthorized access to sensitive information intended for legitimate users to access or control city infrastructure.

Also, here ML techniques can be used to secure city residents' privacy. Smart devices like wearables, home assistants, or even vehicles tend to generate a plethora of personal data. PLS encrypted with ML models would ensure that the data were secured before the transmission, in adherence to security mechanisms relevant to that specific context. With ML models dynamically changing communication parameters like signal strength or frequency, the real key data becomes exceptionally private in highly connected urban zones.

11.5.2 Autonomous Vehicles: Protecting V2V and V2I Communications

Autonomous vehicles (AVs) rely heavily on communication technologies to navigate safely. These vehicles need to exchange data not only with each other (Vehicle-to-Vehicle or V2V) but also with surrounding infrastructure (Vehicle-to-Infrastructure or V2I), like traffic lights and road sensors. This communication ensures smooth operations, safety, and the efficient flow of traffic, but it also introduces potential security vulnerabilities, such as eavesdropping, jamming, and spoofing attacks [21]. In an AV ecosystem, secure and low-latency communication crucial to guarantee safety in real-time operations. ML techniques can optimize both the security and efficiency of communication as depicted in Figure 11.5. ML-powered PLS techniques can help

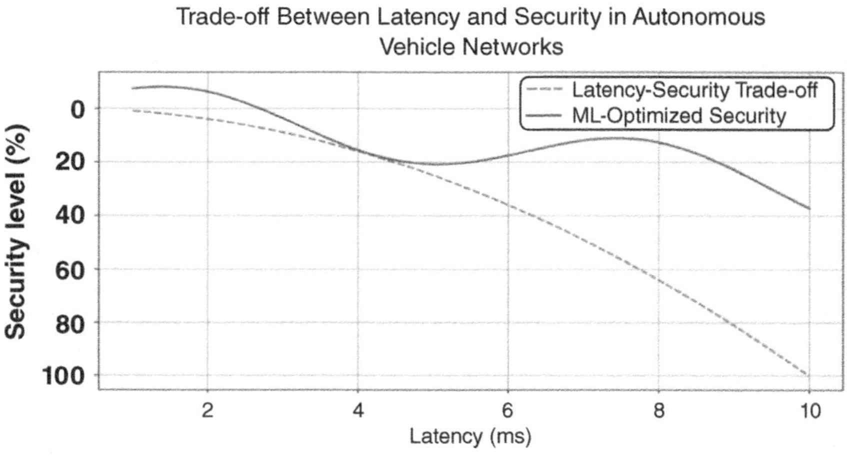

Figure 11.5 A graph showing the trade-off between latency and security in autonomous vehicle networks, and how ML can optimize both aspects for real-time security.

in detecting malicious devices trying to gain access to the vehicle network. For instance, ML models can keep a constant check on the communication channels, examining any signs of abnormal behavior or interference to catch malign actors in the act. The RL algorithms could be leveraged for dynamic adjustment of the vehicle's communication parameters, such as transmission power, frequency, or encryption, in accordance with the real-time conditions and identified threats.

Moreover, they can also address jamming attacks, where malicious actions would block communication between vehicles or between a vehicle and infrastructure. ML algorithms would assume this process, continually learning from the environment to help in developing signal pathways and frequency allocations that avoid areas of interference for secure and effective communication by AVs.

11.5.3 The Internet of Everything

Security of Data Exchange and Privacy: The IoE generally refers to a more extensive network of connected devices going beyond standard smartphones and computers: home appliances, industrial, or building automation; wearable technologies, and medical devices. Billions of these devices communicate simultaneously with one another, and any assurance of the security and privacy of the data exchanged among them is a considerable challenge [22].

ML-based PLS can secure IoE communications with an adaptive system that provides real-time solutions for protecting data integrity and privacy. For example, ML models can assess communication patterns to prevent and detect eavesdropping. Further, the system can create a behavioral pattern of normal device activity and flag any abnormal activity, such as unauthorized devices trying to access the network or tamper with data.

Besides, PLS will empower privacy-preserving techniques. Many IoE devices generate sensitive private data such as health metrics, location information, or even financial transactions. On the one hand, ML could encrypt the data in motion across networks. On the other hand, FL allows for the model updates only to be communicated across the network while keeping the data safe and residing on local devices. In this way, sensitive private data is never communicated away from the device, protecting privacy while still allowing for continuous improvement of the system.

11.5.4 Other Applications

Apart from smart cities, autonomous vehicles, and IoE, ML-powered PLS has a transformational capability in many other fields, like healthcare, smart grids, and industrial applications. Each of them is dependent on secure data transfer and infrastructure protection.

11.5.4.1 Healthcare

The ML-based PLS could, in the context of health, ensure confidential transfer of healthcare information between gadgets and hospitals with providers. In an example scenario, wearable monitors which track important parameters such as heart rate or blood sugar are forwarded to remote data centers where data is extracted. If someone listens in during such a process, it risks violation of personal healthcare information about a patient. By using the technique of ML for PLS, this could potentially be assured without eavesdropping of information at the physical layer of transmission.

Additionally, intrusion detection systems (IDS) powered by ML can continuously monitor the network for any signs of malicious activity. These systems can detect and neutralize attacks in real time, ensuring that health-related data remains protected.

11.5.4.2 Smart Grids

The smart grids that manage and optimize the flow of electricity need secure communication against cyberattacks. ML-based PLS will be able to protect the communication between smart meters, substations, and control centers, thereby making it hard for hackers to manipulate grid operations or steal sensitive energy usage data.

In a smart grid, ML can be used to identify any unusual pattern of communication that may indicate a security breach. For instance, RL could be used dynamically to adjust the security parameters of the network in order to achieve optimal protection against potential attacks.

11.5.4.3 Industrial Systems

The industrial IoT is revolutionizing the manufacturing, energy production, and other industrial processes by allowing machines and sensors to communicate with one another in real time. These networks are however susceptible to cyberattacks that may disrupt operations or compromise sensitive data.

PLS techniques will ensure that the communication within the industrial systems is secure. ML algorithms can learn from the previous attacks and are continuously monitoring data exchanges among the machines and sensors for any type of interference or tampering. Secure key generation may protect encryption keys used during communication, thereby not allowing unauthorized parties to decrypt the data. In addition, the dynamic, real-time nature of ML allows industrial systems to adapt their security measures as the environment changes, thus making them more resilient to attacks [23].

11.6 Open Challenges and Future Directions in ML-empowered PLS for 6G

As 6G networks take shape, integrating ML into PLS techniques holds great promise for enhancing security. However, several obstacles must be addressed to make these ML-powered PLS solutions practical and effective in the rapidly evolving 6G ecosystem. In this section, we will explore the key challenges associated with using ML in PLS and the exciting research directions that could shape the future of secure 6G networks [24].

11.6.1 Computational Complexity of ML Calculations

The foremost noteworthy challenge within the sending of ML for PLS in 6G systems is the computational complexity of these calculations. Millions of gadgets and gigantic information streams in 6G systems request real-time, adaptable security arrangements. In any case, numerous ML models, particularly DL models, are computationally seriously and require a parcel of preparing control, which may not be attainable for littler, resource-constrained gadgets.

11.6.1.1 Challenges

- **Tall Preparing Costs**: Preparing modern ML models requires preparing endless sums of information, regularly requiring significant computational assets and time. In real-time security scenarios, this may well be as well moderate to successfully secure the arrange.
- **Real-time Deduction Idleness**: Even when the show is prepared, utilizing it to create real-time choices around organize security might present delays, which is risky in 6G situations that request ultra-low idleness.
- **Vitality Utilization**: Numerous ML models, especially profound learning models, are energy-hungry. For gadgets such as IoT sensors or portable gadgets, which frequently depend on battery control, this gets to be a concern for vitality productivity.

11.6.1.2 Future Headings

- **Edge Computing and Combined Learning**: A arrangement may lie in edge computing, where information handling happens closer to where it is produced, decreasing the ought to transfer huge sums of information to central servers. This could offer assistance overcome idleness and diminish computational burden on person gadgets. Essentially, combined learning, which permits gadgets to prepare ML models locally without sharing delicate information, may empower security arrangements that protect both security and computational effectiveness.

- **Lightweight ML Models**: Another promising approach is creating lighter ML models that are less resource-intensive but still competent of conveying solid security. Strategies like show pruning, quantization, or information refining seem offer assistance decrease the complexity of ML calculations, making them more doable for real-time PLS.
- **Equipment Optimization**: Specialized equipment like ASICs (Application-Specific Coordinates Circuits) or FPGAs (Field-Programmable Entryway Clusters) outlined for ML workloads might too play a key part in boosting computational productivity, lessening control utilization, and empowering speedier handling speeds for PLS applications.

11.6.2 Require for Expansive Preparing Datasets

For ML models to viably secure 6G systems, they ought to be prepared on expansive, shifted datasets that speak to the complete extend of conceivable arrange conditions and assault scenarios. Be that as it may, getting such datasets presents a few challenges.

11.6.2.1 Challenges

- **Differing Qualities of Information**: 6G systems will be exceedingly heterogeneous, with millions of different gadgets working in different situations. Gathering enough data that precisely speaks to the wide assortment of organize conditions and assault sorts could be a critical challenge.
- **Naming Information**: Administered ML models require named datasets to train viably. Be that as it may, collecting named information for real-world assaults on 6G systems, such as listening in or sticking, can be troublesome, because it requires both critical assets and skill.
- **Protection Concerns**: Numerous applications in 6G, such as healthcare or keen homes, include delicate information. Collecting huge datasets for ML preparing may raise genuine security concerns, especially when individual data is included.

11.6.2.2 Future Directions

- **Synthetic Data Generation**: One potential solution is to generate synthetic data using techniques like GANs. These models can simulate various network conditions and attack scenarios, helping to overcome the shortage of real-world data for training purposes.
- **Transfer Learning**: Transfer learning offers another avenue for addressing data limitations. In transfer learning, a model that has already been trained on one task or dataset can be fine-tuned with a smaller dataset for a related

task. This could be particularly useful when labeled data for specific 6G attack scenarios is scarce.

- **FL for Privacy Preservation**: FL allows training across distributed devices while keeping the data localized and private. This approach can enable secure, decentralized data training, mitigating privacy concerns while still benefiting from large-scale model training.

11.6.3 Adversarial Attacks on ML Models

While ML has great potential for enhancing security, ML models themselves are vulnerable to adversarial attacks. In these attacks, small, carefully crafted perturbations are made to the input data that cause the model to make incorrect predictions. If attackers can manipulate the data fed into a security system, they may bypass security measures, making this a critical challenge for PLS applications in 6G [25, 26].

11.6.3.1 Challenges

- **Manipulating Training Data**: Adversaries could influence the training process by introducing misleading or malicious data, which would make the ML model less effective or even vulnerable to specific attacks.
- **Real-time Attacks on Inference**: Even after training, attackers could introduce small changes in the data sent to the ML model during inference (real-time use), leading the system to make erroneous security decisions.
- **Lack of Robustness**: Many ML models are inherently vulnerable to these types of attacks, which could undermine the effectiveness of PLS techniques, particularly in high-stakes environments like autonomous vehicles or healthcare applications.

11.6.3.2 Future Directions

- **Adversarial Training**: One approach to improving the robustness of ML models is adversarial training, where models are trained on both clean data and intentionally modified adversarial data. This helps the model learn to detect and handle such attacks effectively.
- **Certified Robustness**: Another promising area of research is certified robustness, which aims to develop methods for mathematically proving that a model will not be easily fooled by adversarial inputs. This can provide a stronger guarantee of security in PLS systems.
- **Explainable AI (XAI)**: Making ML models more explainable can help security experts understand how the model arrived at its decisions. By being able to interpret the model's decision-making process, it is easier to spot any weaknesses or vulnerabilities in its defence mechanisms.

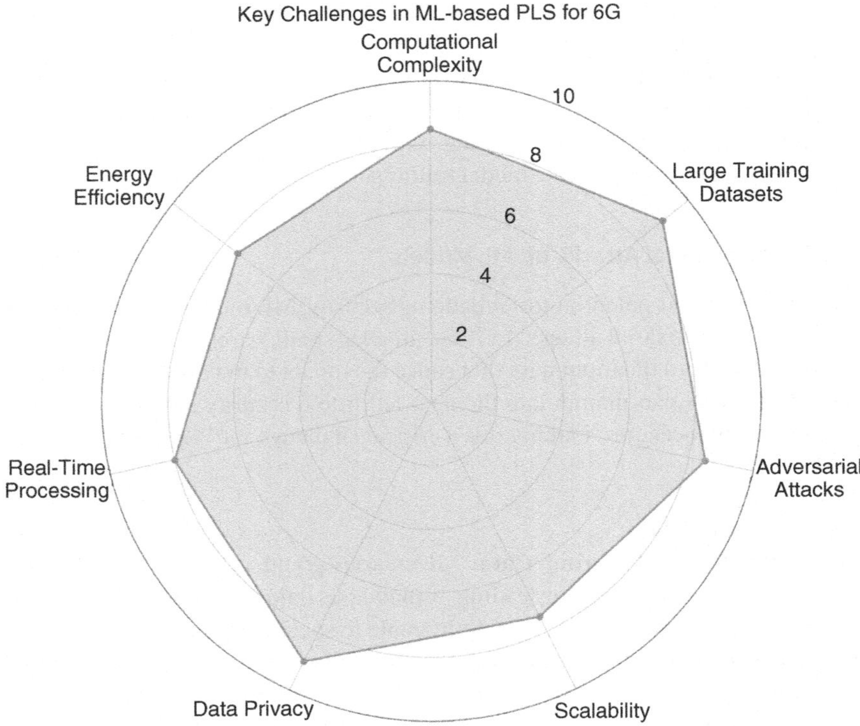

Figure 11.6 Relative challenges in implementing ML for PLS in 6G networks (0: Least Challenging, 10: Most Challenging).

11.6.4 Future Research Directions

Beyond the challenges mentioned, there are several exciting areas of research that could shape the future of ML-powered PLS in 6G as depicted in Figure 11.6.

11.6.4.1 Quantum-resistant Encryption

With the future threat of quantum computing potentially breaking existing encryption methods, there is growing interest in quantum-resistant cryptography. Quantum computers could theoretically solve complex mathematical problems that underpin many of today's encryption techniques (e.g. RSA or ECC), putting 6G networks at risk. Research is underway into lattice-based cryptography, code-based encryption, and other quantum-resistant methods that could secure 6G communications in a post-quantum world. ML can assist in optimizing these cryptographic techniques, ensuring they are both secure and efficient.

11.6.4.2 Adaptive Security Frameworks

As 6G networks become more complex, static security measures will no longer be enough. Adaptive security frameworks powered by ML can respond to dynamic network conditions and emerging threats in real time. These frameworks can continuously learn from the network environment, adjusting security parameters like encryption strength, anomaly detection, and access control to maintain robust protection against evolving threats.

11.6.4.3 Next-generation PLS Protocols

As 6G evolves, so will the need for new PLS protocols that are specifically designed for the unique demands of 6G networks. Research will focus on developing enhanced physical layer authentication techniques, secure beamforming methods, and network slicing strategies tailored to 6G environments. These protocols, powered by ML, will enable better security with minimal overhead and enhanced reliability [27].

11.7 Conclusion

ML is expected to greatly improve PLS in 6G networks by making security smarter, faster, and more adaptable. Traditional security methods may not be enough to handle the new challenges of 6G, such as high-speed communication, billions of connected devices, and AI-driven systems. ML-powered PLS helps networks detect threats, improve encryption, and protect wireless signals from cyberattacks.

Some important ML-based techniques, like secure beamforming, power control, and channel estimation, help keep data safe and make networks more reliable. These solutions are especially useful in smart cities, self-driving cars, and the IoE, where secure communication is crucial.

However, ML-powered PLS still faces challenges, such as high processing needs, data privacy risks, and possible cyberattacks on ML models. Future research in edge computing, FL, and stronger encryption methods will help solve these problems.

By constantly learning and adapting, ML-powered PLS will ensure that 6G networks stay secure, efficient, and resistant to new cyber threats in the future.

References

1 Zhang, S., Zhu, D., and Liu, Y. (2024). Artificial intelligence empowered physical layer security for 6G: state-of-the-art, challenges, and opportunities. *Computer Networks* 242: 110255.

2 Puspitasari, A.A., An, T.T., Alsharif, M.H., and Lee, B.M. (2023). Emerging technologies for 6G communication networks: machine learning approaches. *Sensors* 23 (18): 7709.

3 Sejan, M.A.S., Rahman, M.H., Shin, B.S., et al. (2022). Machine learning for intelligent-reflecting-surface-based wireless communication towards 6G: a review. *Sensors* 22 (14): 5405.

4 Nguyen, V.L., Lin, P.C., Cheng, B.C., et al. (2021). Security and privacy for 6G: a survey on prospective technologies and challenges. *IEEE Communications Surveys & Tutorials* 23 (4): 2384–2428.

5 Porambage, P., Gür, G., Osorio, D.P.M., et al. (2021). The roadmap to 6G security and privacy. *IEEE Open Journal of the Communications Society* 2: 1094–1122.

6 Arastouei, N. (2022). 6G technologies: key features, challenges, security and privacy issues. In: *International Workshop on Digital Sovereignty in Cyber Security: New Challenges in Future Vision* (pp. 94–109). Springer.

7 Chorti, A., Barreto, A.N., Köpsell, S., et al. (2022). Context-aware security for 6G wireless: the role of physical layer security. *IEEE Communications Standards Magazine* 6 (1): 102–108.

8 Ara, I., and Kelley, B. (2024). Physical layer security for 6G: toward achieving intelligent native security at layer-1. *IEEE Access* 12: 1–1.

9 Poor, H.V., and Schaefer, R.F. (2017). Wireless physical layer security: principles, perspectives, and new directions. *IEEE Journal on Selected Areas in Communications* 34 (6): 1276–1290.

10 Oleksandr, T., Oleksandra, D., and Anatolii, M. (2023). Basic technologies and techniques ml/ai for improving physical layer security for 5G/6G communications systems. In: *The 17th International Scientific and Practical Conference "System Analysis and Intelligent Systems for Management" (May 02–05, 2023)* (pp. 403). Ankara, Turkey: International Science Group.

11 Liu, Y., Wang, J., Dai, H., et al. (2020). Machine learning for security enhancement in wireless networks. *IEEE Network* 34 (3): 6–14.

12 Ara, I., and Kelley, B. (2023). 6G physical layer security. In: *Deep Learning—Recent Findings and Researches*. IntechOpen.

13 Mitev, M., Chorti, A., Poor, H.V., and Fettweis, G.P. (2023). What physical layer security can do for 6G security. *IEEE Open Journal of Vehicular Technology* 4: 375–388.

14 Shaik, R.B., Nagaradjane, P., Ioannou, I., et al. (2024). AI/ML-aided capacity maximization strategies for URLLC in 5G/6G wireless systems: a survey. *Computer Networks* 249: 110506.

15 Saleem, R., Ni, W., Ikram, M., and Jamalipour, A. (2022). Deep-reinforcement-learning-driven secrecy design for intelligent-reflecting-surface-based 6G-IoT networks. *IEEE Internet of Things Journal* 10 (10): 8812–8824.

16 Hassan, S.S., Park, Y.M., Tun, Y.K., et al. (2024). Enhancing spectrum efficiency in 6G satellite networks: a GAIL-powered policy learning via asynchronous federated inverse reinforcement learning. *arXiv preprint* arXiv:2409.18718.

17 Shvetsov, A.V., Alsamhi, S.H., Hawbani, A., et al. (2023). Federated learning meets intelligence reflection surface in drones for enabling 6G networks: challenges and opportunities. *IEEE Access* 11: 130860–130887.

18 Rahman, M.A., and Hossain, M.S. (2022). A deep learning assisted software defined security architecture for 6G wireless networks: IIoT perspective. *IEEE Wireless Communications* 29 (2): 52–59.

19 Iqbal, M.S., Farhan, M., Ullah, F., et al. (2024). Intelligent multimedia content delivery in 5G/6G networks: a reinforcement learning approach. *Transactions on Emerging Telecommunications Technologies* 35 (4): e4842.

20 Kazmi, S.H.A., Qamar, F., Hassan, R., et al. (2024). Security of federated learning in 6G era: a review on conceptual techniques and software platforms used for research and analysis. *Computer Networks* 245: 110358.

21 Gupta, M., Benson, J., Patwa, F., and Sandhu, R. (2020). Secure V2V and V2I communication in intelligent transportation using cloudlets. *IEEE Transactions on Services Computing* 15 (4): 1912–1925.

22 Gupta, N., and Mishra, S. (Eds.). (2024). *Internet of Everything for Smart City and Smart Healthcare Applications*. Springer.

23 Sangeetha, S.K.B., and Dhaya, R. (2022). Deep learning era for future 6G wireless communications – theory, applications, and challenges. In: *Artificial Intelligent Techniques for Wireless Communication and Networking* (pp. 105–119). Scrivener Publishing LLC.

24 Okere, E.E., and Balyan, V. (2025). Sixth generation enabling technologies and machine learning intersection: a performance optimization perspective. *Future Internet* 17 (2): 50.

25 Irram, F., Ali, M., Naeem, M., and Mumtaz, S. (2022). Physical layer security for beyond 5G/6G networks: emerging technologies and future directions. *Journal of Network and Computer Applications* 206: 103431.

26 Ziegler, V., Schneider, P., Viswanathan, H., et al. (2021). Security and trust in the 6G era. *IEEE Access* 9: 142314–142327.

27 Alsharif, M.H., Hossain, M.S., Jahid, A., et al. (2022). Milestones of wireless communication networks and technology prospect of next generation (6G). *Computers, Materials & Continua* 71 (3): 4803–4818.

12

Justice in IoT and AI for Sustainable Development

George Belsian[1], Hrishikesh Tamhane[2], Binesh Asok Kumar[3], and Geraldine Shirley Nicholas[4,]*

[1] *Senior Architect, AWS, Charlotte, NC, USA*
[2] *Firmware Engineer, Atom Power, Charlotte, NC, USA*
[3] *CTO, Atom Power, Charlotte, NC, USA*
[4] *Assistant Professor, High Point University, High Point, NC, USA*

*Corresponding author: geraldine.shirley31@gmail.com

12.1 Introduction

The convergence of artificial intelligence (AI) and the Internet of Things (IoT) is transforming industries by enhancing automation, optimizing resource allocation, and enabling predictive analytics. These technologies play a crucial role in achieving sustainability by improving infrastructure efficiency, enhancing energy management, and enabling real-time monitoring in sectors such as healthcare, climate action, and smart governance. AI, powered by machine learning and deep neural networks, processes vast amounts of data collected by IoT systems, generating insights that contribute to smart cities, climate resilience, and precision agriculture. However, despite these advancements, the deployment of AI and IoT presents ethical, security, and justice-related challenges that must be addressed to ensure their fair and responsible adoption. Issues such as lack of transparency in AI decision-making, algorithmic bias, data privacy concerns, and cybersecurity vulnerabilities can lead to social inequities, privacy breaches, and systemic failures, potentially limiting the widespread benefits of these technologies. Addressing these challenges requires interdisciplinary approaches, strong regulatory frameworks, and ethical governance mechanisms to ensure that AI

AI and ML Techniques in IoT-based Communication: A Path to Sustainable Development Goals, First Edition. Edited by Sumita Mishra, Nishu Gupta and Polat Goktas.

and IoT align with justice, equity, and sustainability goals rather than exacerbating existing disparities [1].

Global efforts, including the European Union's AI Act, UNESCO's AI Ethics Guidelines, and ISO/IEC cybersecurity frameworks, highlight the increasing need for accountability, transparency, and ethical AI–IoT deployment [2, 3]. However, the rapid evolution of AI technologies and decentralized IoT architecture poses challenges to effective regulation and enforcement, requiring collaboration between governments, industries, and research institutions. As AI and IoT systems become more embedded in decision-making processes, critical infrastructure, and environmental monitoring, it is essential to develop privacy-preserving models, equitable access strategies, and sustainable governance frameworks that mitigate the risks associated with bias, security breaches, and resource exploitation. Research in AI fairness and responsible innovation has emphasized the role of standardization in ethical AI deployment [4]. Similarly, cybersecurity concerns related to IoT ecosystems necessitate robust risk mitigation strategies to prevent large-scale vulnerabilities [5]. This chapter examines ethical governance, security, regulatory frameworks, and sustainable deployment strategies that support responsible AI and IoT implementation.

To provide a structured approach to these challenges, this chapter first explores the Foundations and Ethical Frameworks of AI and IoT, highlighting their role in sustainability, justice, and ethical digital transformation. It then delves into Ethical Governance and Implementation, discussing the regulatory and governance structures that guide responsible AI and IoT deployment. The Security Architecture and Privacy section examines the vulnerabilities of IoT networks, AI model security risks, and privacy-preserving technologies that can safeguard sensitive data. The chapter further discusses Resource Justice and Equitable Access, focusing on AI-driven equitable resource distribution, IoT-based monitoring systems, and the role of digital technologies in global sustainability efforts. This is followed by an analysis of Regulatory Frameworks and Compliance, detailing policies, accountability mechanisms, and governance models shaping AI–IoT adoption. Finally, the Implementation Roadmap provides phased strategies, governance maturity models, and future research directions, offering a structured pathway for ensuring AI and IoT contribute to a just, secure, and sustainable future. Figure 12.1 presents an overview of the key themes discussed in this chapter.

12.2 Foundations and Ethical Frameworks

12.2.1 AI and IoT Technological Convergence and Sustainability

12.2.1.1 Synergistic Effect of IoT–AI Integration

The integration of IoT and AI is driving a paradigm shift beyond their individual capabilities. IoT serves as the sensory layer, continuously collecting real-time data

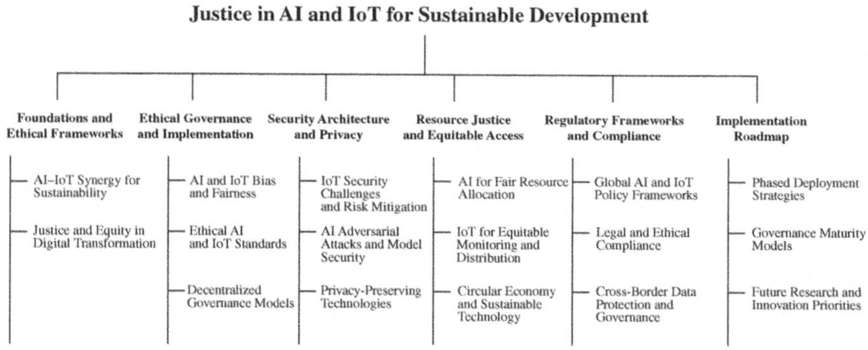

Figure 12.1 Key categories in justice for AI and IoT.

Figure 12.2 AI–IoT synergy flowchart.

from various environments, while AI acts as the analytical layer, processing and extracting meaningful insights from these data streams. This synergistic relationship enables the development of intelligent, adaptive systems capable of optimizing processes, improving efficiency, and driving sustainability across multiple domains [6].

For example, predictive maintenance in industrial settings demonstrates this synergy effectively. IoT sensors deployed in manufacturing equipment monitor vibration, temperature, and operational parameters, continuously transmitting data to AI systems. The AI models analyze these data streams to identify anomalies and predict potential failures, enabling proactive maintenance actions that reduce downtime, optimize resource usage, and extend equipment lifespan [7]. Similarly, in smart cities, IoT-powered traffic-monitoring systems collect real-time data from sensors, while AI algorithms optimize traffic flow, minimize congestion, and reduce carbon emissions by dynamically adjusting signal timings and suggesting alternative routes to drivers [8].

As AI systems evolve, their integration with IoT will enable self-learning systems that improve sustainability outcomes over time. With continuous data collection and refinement of learning models, future AI–IoT solutions will be more adaptive, further enhancing resource utilization, energy efficiency, and environmental conservation [9]. Figure 12.2 illustrates the synergy between AI and IoT, depicting the flow from IoT-driven data collection to AI-processing and real-world applications. The process begins with IoT devices collecting data, which is then processed at the edge or cloud before undergoing AI/ML analysis. The resulting insights enable intelligent applications such as smart cities and predictive maintenance.

12.2.1.2 Evolution of Sustainable Development Through Digital Technologies

The role of digital technologies in sustainable development has evolved through distinct phases, from basic monitoring to comprehensive AI-driven optimization. Initially, digital transformation was primarily focused on monitoring environmental impact and tracking resource consumption. However, advancements in AI and IoT have enabled real-time analysis, automated optimization, and predictive modeling, significantly enhancing sustainability initiatives across multiple sectors [10]. In the energy sector, smart grids powered by IoT sensors and AI analytics have revolutionized power distribution, allowing dynamic balancing of electricity loads, seamless integration of renewable energy sources, and real-time demand–response adjustments. These intelligent energy systems optimize energy consumption, improve efficiency, reduce greenhouse gas emissions, and enhance grid resilience, fostering a more sustainable energy infrastructure [11].

Similarly, AI-driven water management solutions leverage IoT-enabled sensors for real-time leak detection, quality monitoring, and demand-responsive distribution. By utilizing machine learning models, cities can analyze water consumption patterns, optimize supply chains, and minimize resource waste, contributing to long-term water sustainability [12]. Beyond infrastructure, digital transformation has also reshaped agriculture, healthcare, and education, facilitating equitable access to essential resources. Precision agriculture, for example, integrates IoT sensors, satellite imaging, and AI-driven predictive models to optimize irrigation, reduce pesticide usage, and improve crop yields. These technologies are particularly beneficial to smallholder farmers in developing nations, enabling data-driven decision-making and sustainable farming practices [13].

In the healthcare sector, telemedicine platforms powered by AI diagnostics and IoT-enabled remote patient monitoring are bridging the healthcare accessibility gap in remote and underserved regions. AI algorithms analyze real-time patient data collected from wearable IoT devices, enabling early disease detection, proactive intervention, and improved patient outcomes. These advancements reduce the burden on healthcare facilities while ensuring efficient resource allocation [14].

Despite these benefits, the widespread adoption of AI–IoT technologies in sustainable development necessitates a strong ethical and governance framework to ensure fair, inclusive, and responsible technological deployment. It is critical to mitigate digital inequalities, safeguard data privacy, and ensure environmental and social justice in the pursuit of sustainability [15].

12.2.1.3 Impact Metrics and Measurement Frameworks

Measuring the impact of IoT–AI integration on sustainability requires comprehensive evaluation frameworks that capture environmental, social, and economic dimensions. These frameworks provide quantifiable insights into the effectiveness of digital sustainability initiatives and help organizations refine their strategies for maximum impact [16].

Environmental Metrics

Key indicators of environmental sustainability include:

- Carbon footprint reduction achieved through AI-optimized energy grids.
- Resource efficiency improvements via IoT-enabled waste and water management.
- Energy consumption per capita in smart buildings, factories, and transportation networks.

For instance, a smart factory leveraging IoT–AI systems can track real-time energy consumption, emissions, and waste generation. AI models analyze this data to recommend operational adjustments, reducing carbon emissions and unnecessary energy consumption.

Social Metrics

From a social justice standpoint, impact assessment must include:

- Equitable access to digital infrastructure (e.g. Internet access, telemedicine services).
- Community engagement in sustainability initiatives (e.g. participatory urban planning).
- Reduction in healthcare disparities through AI–IoT-powered medical interventions.

Economic Metrics

Economic sustainability can be measured through:

- Cost savings from predictive maintenance and AI-driven operational efficiencies.
- Job creation in AI–IoT-driven sectors such as renewable energy and automation.
- Return on Sustainability Investment (ROSI), a metric that considers both financial and social returns [17].

By developing standardized sustainability metrics, organizations and policymakers can ensure data-driven decision-making, fostering a balanced approach to digital transformation and sustainability.

To comprehensively measure and communicate the impact of IoT–AI integration on sustainability, organizations should:

1. Align sustainability goals with established global frameworks such as the UN SDGs.
2. Identify key metrics and indicators spanning environmental, social, and economic dimensions.
3. Establish baseline measurements and set data-driven targets for ongoing improvement.
4. Continuously collect and analyze data, using AI-driven insights for optimization.
5. Engage stakeholders transparently, ensuring inclusive participation in measurement and decision-making processes.
6. Leverage impact assessments to refine strategies and drive sustainable innovation.

By embracing integrated measurement frameworks, organizations can demonstrate the tangible benefits of their IoT–AI investments, build stakeholder confidence, and accelerate global progress toward a sustainable future. Figure 12.3 shows a comparative analysis of sustainability impact metrics across key industries, highlighting the role of AI–IoT solutions in reducing carbon footprints, improving efficiency, and optimizing resource utilization [18].

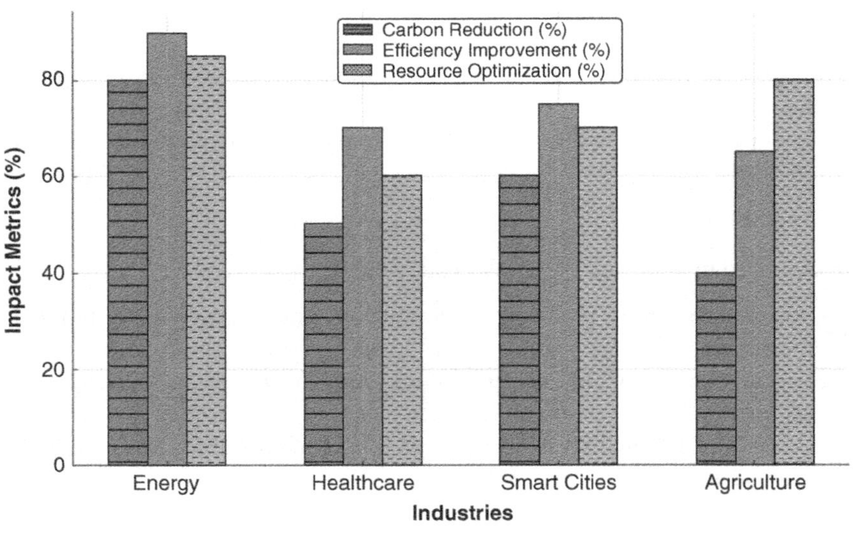

Figure 12.3 Comparative analysis of sustainability impact metrics across key industries.

12.2.1.4 Case Study: Smart City Sustainability Initiatives

Barcelona's transformation into a leading smart city highlights how AI–IoT integration can drive urban sustainability. The city has implemented a wide range of digital solutions to enhance energy efficiency, waste management, water conservation, and urban mobility. These initiatives have contributed to measurable improvements in sustainability, economic efficiency, and citizen engagement [19].

One of Barcelona's most notable smart city interventions is its AI-powered intelligent lighting system. By deploying IoT sensors across the city, the system dynamically adjusts street lighting based on pedestrian movement and traffic flow. This has led to a 30% reduction in energy consumption, lower municipal electricity costs, and decreased light pollution, contributing to improved urban living conditions [20].

In the area of waste management, Barcelona has introduced IoT-enabled smart waste bins equipped with sensors that monitor fill levels in real time. This AI-driven system optimizes waste collection routes, significantly reducing fuel consumption and lowering carbon emissions from garbage trucks. Additionally, automated waste sorting has increased recycling efficiency and minimized landfill dependency, further enhancing the city's environmental impact.

Water conservation has also been a key focus of Barcelona's smart city initiatives. The city has implemented IoT-based water meters and AI-driven irrigation systems to optimize water usage. These smart systems detect leaks, monitor real-time consumption patterns, and adjust irrigation in public parks based on soil moisture levels. As a result, Barcelona has achieved a 30% reduction in water wastage, substantial cost savings in municipal water management, and improved sustainability of urban green spaces [21].

Despite the success of these initiatives, Barcelona has faced several challenges in implementing its smart city vision. One of the main concerns has been addressing the digital divide, particularly among older citizens who may have limited access to digital services. To mitigate this, the city has launched public awareness campaigns and digital literacy programs to promote citizen engagement. Additionally, concerns over privacy and data security have required the implementation of strict governance policies to ensure transparency in AI–IoT data collection and usage.

The deployment of these smart city solutions has been underpinned by a robust digital infrastructure, including a vast network of IoT sensors, high-speed communication networks, and cloud-based platforms for real-time data analysis. Advanced AI algorithms continuously process and interpret this data, enabling proactive decision-making and optimizing city operations. Barcelona's approach to sustainability has also fostered a collaborative environment where citizens can participate in urban planning efforts, provide feedback, and contribute to shaping the city's future.

Barcelona's smart city transformation serves as a model for cities worldwide, demonstrating how AI–IoT integration can lead to enhanced sustainability, economic growth, and improved quality of life. By investing in digital infrastructure, citizen participation, and transparent governance, cities can replicate Barcelona's success and accelerate their transition toward smarter, more resilient urban environments.

12.2.2 Justice Framework in the Digital Age

In the rapidly evolving landscape of the digital age, the concept of justice has taken on new dimensions and complexities. As the convergence of IoT and AI technologies transform various aspects of our lives, it is crucial to examine the implications of these developments and establish a framework that ensures equity, access, and ethical governance in the digital realm [22].

12.2.2.1 Defining Technological Justice Dimensions

To understand the justice framework in the digital age, we must first define the key dimensions of technological justice. These dimensions encompass the principles and considerations that should guide the development, deployment, and governance of IoT and AI technologies to ensure that they promote fairness, inclusion, and the well-being of all individuals and communities.

Equity Considerations: At the core of technological justice lies the principle of equity. In the context of IoT and AI, equity refers to the fair distribution of the benefits and risks associated with these technologies. It means ensuring that the opportunities and advantages offered by digital innovations are accessible to all, regardless of socioeconomic status, geographic location, or other demographic factors [23].

Access Paradigms: In the digital age, access to technology has become a fundamental prerequisite for full participation in society, from education and employment to healthcare and social services. However, significant disparities exist both within and between countries. To establish a just access paradigm, universal access to digital technologies must be prioritized through infrastructure development, affordable internet connectivity, and digital literacy programs [24].

Ethical Governance Structures: As IoT and AI technologies become increasingly pervasive, the need for ethical governance structures becomes paramount. Ethical governance refers to frameworks, policies, and practices that ensure the responsible development, deployment, and use of these technologies in ways that align with societal values and promote the well-being of all [25].

12.2.2.2 Data Colonialism Challenges

As IoT and AI technologies advance, they are increasingly reliant on the collection and analysis of vast amounts of data. While this data-driven approach has the potential to unlock significant social and economic benefits, it also raises concerns about data colonialism, where powerful actors extract and commodify data from marginalized communities without equitable benefit-sharing [26]. Addressing data colonialism requires new models of data governance that prioritize data sovereignty and ensure communities maintain control over their digital resources.

12.2.2.3 Bias Amplification Risks

One of the most significant justice challenges in the digital age is the risk of bias amplification in IoT and AI systems. Bias amplification occurs when systemic inequities are embedded into technological solutions, reinforcing societal discrimination in areas such as hiring, lending, healthcare, and law enforcement [8]. To mitigate this, rigorous bias detection frameworks, transparent AI model design, and diverse representation in the tech sector are essential.

12.2.2.4 Case Study: Global Digital Divide Impacts

The global digital divide exemplifies the justice challenges of the digital age. The divide refers to the gap between those who have access to digital technologies and those who do not, both within and between countries. This disparity limits opportunities for education, employment, and social mobility, disproportionately affecting marginalized populations [9]. Addressing this issue requires infrastructure investment, inclusive technology policies, and initiatives such as public–private partnerships to expand internet access and promote digital literacy worldwide.

An additional challenge is the unequal representation of developing nations in the global technology economy. Many digital innovations are designed primarily for developed markets, leaving lower income regions with limited access to tailored solutions that address their specific needs. To bridge this gap, collaborative efforts must focus on affordable digital infrastructure, context-specific technological solutions, and knowledge-sharing initiatives that empower local communities to participate in the digital transformation actively. Figure 12.4 illustrates the Comparative Analysis of Digital Equity Initiatives, highlighting key digital equity metrics across developed and developing nations. The data showcases disparities in Internet Access, Digital Literacy, AI Adoption, Infrastructure Investment, and Policy Reforms, reflecting the digital divide and the need for strategic interventions to promote equitable digital growth worldwide. The figure underscores how developed nations exhibit higher percentages in digital literacy, infrastructure, and AI adoption, while developing nations face challenges in accessibility and policy implementation [27].

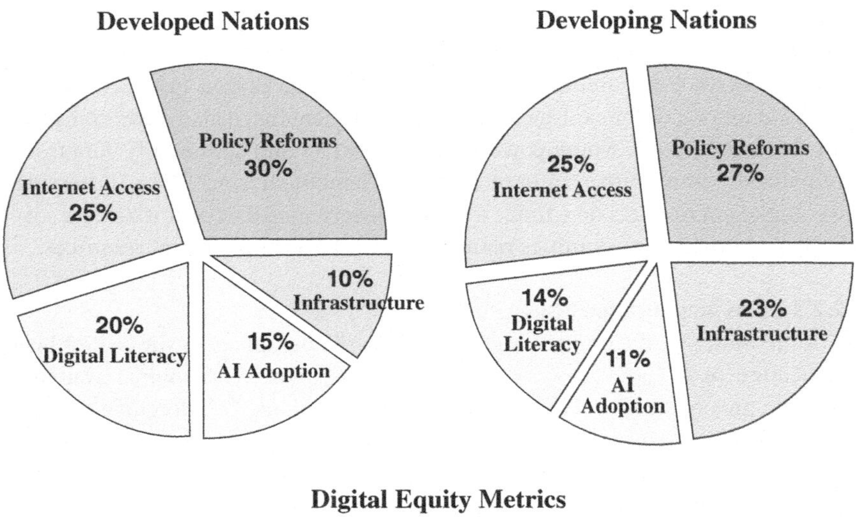

Figure 12.4 Comparative analysis of digital equity initiatives.

12.3 Ethical Governance and Implementation

12.3.1 AI-specific Ethical Frameworks

As AI systems become increasingly integrated into daily life, ethical frameworks guiding AI development and deployment are critical. These frameworks must address challenges such as algorithmic bias, privacy, transparency, fairness, security, and accountability. AI governance requires balancing technological advancement with ethical safeguards to ensure that AI systems serve society without reinforcing societal inequalities or infringing on fundamental rights.

One significant concern in AI ethics is algorithmic bias, where AI systems inadvertently reinforce societal discrimination. A notable example is AI-powered recruiting tools, which have demonstrated biases against underrepresented groups due to historical training data reflecting existing inequalities. In some cases, resumes linked to specific institutions or demographics were unfairly downgraded, underscoring the need for robust ethical guidelines to identify and mitigate such biases.

To counteract algorithmic bias, researchers have developed various approaches, including explainable AI (XAI) and federated learning. XAI enhances transparency by making AI decision-making processes interpretable, enabling stakeholders to identify and correct biases. Federated learning, on the other hand, decentralizes AI training, allowing models to learn from diverse data sources without centralizing sensitive information. This approach enhances data privacy while reducing the risk of bias amplification.

However, AI governance faces global challenges due to regulatory fragmentation. While the OECD AI Principles provide a foundational guideline for AI governance, different regions have adopted varying regulatory frameworks. For example:

- The European Union's AI Act follows a risk-based approach, imposing strict regulations on high-risk AI applications.
- The United States' AI Bill of Rights emphasizes fairness, transparency, and human oversight but lacks enforceable legislation.
- China's AI regulatory framework prioritizes real-time government auditing and strict state oversight of AI-driven decision-making.

These regulatory divergences pose challenges for international AI governance, requiring harmonization across jurisdictions to ensure responsible AI adoption without regulatory loopholes.

Implementing ethical AI frameworks requires embedding ethics into every stage of AI development, from data collection and model training to deployment and monitoring. Organizations should establish clear ethical guidelines, conduct regular audits, and foster a workplace culture emphasizing transparency and accountability. Beyond industry efforts, public awareness and regulatory frameworks play a vital role in ensuring ethical AI use. Governments must enforce responsible AI regulations, while public education initiatives can increase societal understanding of AI's risks and benefits. Ethical AI development requires interdisciplinary collaboration and continuous refinement of governance strategies.

Despite these efforts, a key gap remains in global regulatory coordination. While frameworks like the OECD AI Principles encourage the best practices, enforcement mechanisms vary widely, leading to inconsistencies in AI governance. The lack of global consensus on AI accountability, particularly in high-stakes applications such as autonomous weapons or predictive policing, underscores the urgency for internationally binding agreements. Addressing this gap requires global coordination to ensure that AI ethics principles translate into enforceable regulations across jurisdictions.

Decentralized governance models, which are well discussed in IoT, are also gaining traction in AI. Blockchain-based transparency tools can be integrated into AI systems to track decision-making processes, ensuring accountability in high-risk AI applications. Similarly, Decentralized Autonomous Organizations (DAOs) are emerging as governance structures where AI regulation can be community driven. Blockchain-powered AI auditing frameworks provide traceable, immutable records of AI decision-making, increasing trust and transparency in AI governance. Figure 12.5 illustrates the governance landscape of AI and IoT ethics, depicting interactions between key stakeholders, including governments,

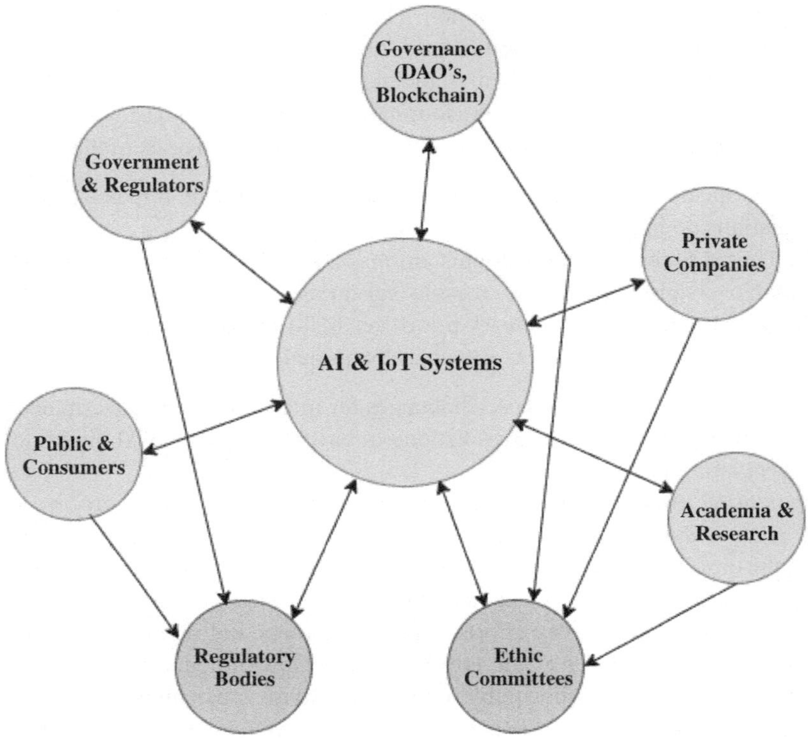

Figure 12.5 AI and IoT governance models and stakeholder interactions.

regulatory bodies, private companies, AI and IoT developers, decentralized governance structures (such as DAOs and blockchain networks), and public stakeholders.

A case study in AI-assisted healthcare highlights the importance of ethical AI governance. A healthcare organization deploying an AI-based diagnostic tool must:

1. Define ethical guidelines ensuring fairness and privacy.
2. Assess training data for bias before model development.
3. Implement explainable AI to enhance transparency.
4. Adopt federated learning to safeguard patient data.
5. Monitor and audit AI performance for unintended disparities.
6. Engage stakeholders, including patients and medical professionals, to address concerns.

By adopting such measures, AI systems can serve society equitably while balancing innovation with ethical considerations.

12.3.2 Ethical Considerations in IoT

The rapid growth of IoT technologies has necessitated new infrastructure and systems to support billions of connected devices. However, as IoT integrates into daily life and critical industries, ethical concerns surrounding sustainability, responsible sourcing, e-waste management, privacy, governance, and equity must be addressed. Ensuring that IoT solutions are implemented responsibly requires comprehensive ethical frameworks that consider both technological and societal impacts.

One of the significant ethical concerns is hardware sustainability. The surge in research, development, and production of IoT devices has led to new engineering and manufacturing lifecycles, which depend on various raw materials. A critical issue in this context is the use of conflict minerals, including tin, tantalum, tungsten, and gold (collectively known as 3TG). These minerals are essential for producing consumer electronics, automotive components, and industrial IoT solutions. However, they are often sourced from conflict-affected regions, particularly the Democratic Republic of Congo (DRC), where mining operations frequently involve forced labor, environmental destruction, and human rights violations. The mining process also exposes workers to serious health risks such as respiratory diseases, skin burns, and toxic metal poisoning. Without stringent ethical controls, increased IoT adoption indirectly fuels these exploitative supply chains. To mitigate such risks, regulatory frameworks like the European Union's Conflict Minerals Regulation require companies to conduct due diligence in their mineral sourcing practices. Noncompliance results in sanctions, which vary by country. For instance, German firms face penalties up to €50 000, while other EU states may impose import bans or blacklist noncompliant businesses. Despite these regulations, the global demand for these minerals remains high, underscoring the need for stricter ethical oversight in supply chains [28, 29].

Another growing concern is electronic waste (e-waste) management. IoT devices have relatively short lifespans, leading to an increase in e-waste, which is projected to rise from 53.6 million metric tons in 2019 to 74.7 million metric tons by 2030. The improper disposal of these devices leads to environmental degradation and poses health risks, particularly in developing nations that lack formal waste management systems. In such countries, informal recycling processes dominate, with workers – including an estimated 12.9 million women – being exposed to hazardous substances such as lead and cadmium without proper protective measures. These practices not only threaten worker safety but also contribute to long-term environmental pollution. The World Health Organization (WHO) has emphasized the urgent need for structured e-waste management policies, increased government oversight, and the establishment of safe recycling infrastructure. Potential solutions include corporate take-back programs, extended

Table 12.1 Comparative E-waste governance frameworks across countries.

Country/region	E-waste regulations	Compliance mechanisms	Incentives for E-waste disposal	Challenges
European Union (EU)	WEEE Directive (Waste Electrical and Electronic Equipment) mandates producer responsibility	Strict producer responsibility, mandatory collection targets, penalties for noncompliance	Financial incentives for proper recycling, eco-design regulations	Cross-border e-waste movement, illegal dumping, compliance disparities between EU nations
United States (US)	State-level regulations (e.g. California e-waste recycling law), no federal law	Varies by state, some states have disposal bans, limited enforcement	Retailer take-back programs, tax incentives for e-waste collection in some states	Lack of national law, inconsistent state-level policies, corporate reluctance
Japan	Home Appliance Recycling Law mandates manufacturers to recycle specific appliances	Government-mandated recycling fees, strict manufacturer take-back requirements	Recycling subsidies for consumers, strict eco-friendly product guidelines	Consumer participation in e-waste recycling, enforcement gaps in rural areas
Developing Nations	Limited regulations, often informal sector-led recycling with health risks	Weak enforcement, reliance on informal recyclers, significant environmental hazards	Minimal to no incentives, reliance on local NGOs, and international funding for waste management	High informal sector involvement, unsafe recycling practices, lack of public awareness

producer responsibility (EPR) laws, and incentive-based e-waste collection models that encourage proper disposal and recycling. Table 12.1 provides a comparative analysis of global e-waste governance, showing variations in regulations, compliance mechanisms, and incentives for responsible IoT device disposal across different regions.

Furthermore, IoT's role in labor displacement and job automation is an emerging challenge. The increasing adoption of smart IoT systems in industrial automation, logistics, and customer service has led to workforce reductions in several sectors. Ethical IoT governance must include policies that balance automation with employment protections, such as upskilling programs and social safety nets.

To address ethical challenges in IoT, decentralized governance models are emerging as an alternative to traditional regulatory approaches [30]. These models distribute decision-making power across multiple stakeholders, promoting transparency, accountability, and security in IoT ecosystems. Key mechanisms include:

- **Blockchain for Ethical Supply Chains**: Blockchain technology allows real-time tracking of raw materials used in IoT device manufacturing, ensuring compliance with ethical sourcing standards. Immutable transaction records help prevent fraudulent claims and unethical practices.
- **Token-based Incentive Systems**: Digital tokens can be used to encourage responsible recycling and sustainable IoT production. Companies and consumers can earn rewards for returning old devices to certified disposal centers.
- **Decentralized Autonomous Organizations (DAOs)**: DAOs facilitate community-driven governance, allowing IoT manufacturers, regulators, and environmental organizations to collaborate on sustainability initiatives without centralized oversight.
- **Edge Computing for Energy Efficiency**: Unlike cloud-based models, which rely on large data centers, edge computing enables localized data processing, reducing energy consumption and lowering the carbon footprint of IoT networks.

Smart cities serve as a critical use case where ethical governance models must be actively integrated into IoT infrastructure. While IoT enhances urban planning, traffic optimization, and public services, it also raises concerns about privacy, surveillance, and data security. Decision-making transparency is essential in preventing cities from becoming panoptic surveillance environments where mass data collection erodes personal freedoms. Ethical smart city implementations must address:

1. **Transparency and Accountability**: Ensuring transparency in data collection, device operation, and decision-making is crucial to building trust in IoT systems. Governments and businesses must clearly define how IoT devices collect, store, and process user data. Mechanisms should be in place to audit and verify IoT operations, ensuring compliance with ethical guidelines.
2. **Privacy and Data Protection**: IoT devices continuously collect vast amounts of personal, behavioral, and biometric data. Without proper safeguards, this data can be misused for surveillance, profiling, or commercial exploitation. Stringent privacy policies, encryption standards, and user consent mechanisms must be implemented to protect individual rights.

3. **Equity and Social Justice**: IoT solutions should be designed to benefit all communities, particularly underrepresented and economically disadvantaged populations. Unequal access to smart city technologies, healthcare IoT, and digital infrastructure can widen societal disparities. Policies must ensure that IoT benefits are distributed fairly, preventing technological exclusion.
4. **Sustainability and Environmental Responsibility**: The energy consumption and carbon footprint of IoT networks must be minimized. Implementing eco-friendly materials, energy-efficient processors, and circular economy models can reduce the environmental impact of IoT. Governments and corporations should adopt strict regulations on e-waste management and promote sustainable manufacturing practices.
5. **Avoiding Algorithmic Bias**: Many IoT systems use AI-driven analytics to make decisions in healthcare, security, and urban planning. However, if these AI models are trained on biased data, they can reinforce discriminatory patterns, leading to unfair outcomes. Ethical AI governance, bias audits, and diverse data representation are essential to prevent bias in IoT applications.
6. **Ethical AI and Automation**: IoT-enabled automation, such as autonomous vehicles and smart surveillance, raises ethical concerns regarding job displacement, human oversight, and liability in decision-making. Ethical frameworks must define responsibility, accountability, and human intervention in automated IoT systems.

A comparative analysis of e-waste management practices highlights global disparities in regulation and enforcement. In low-income countries such as Tanzania and Nepal, informal e-waste recycling leads to widespread toxic exposure due to inadequate safety regulations [31]. In contrast, high-income nations like the United States, Germany, and Australia have established strict e-waste management frameworks, leveraging advanced recycling technologies and corporate accountability measures. Countries such as Sweden have implemented urban mining initiatives to recover valuable metals from discarded electronics, reducing reliance on newly mined resources. In the United States, California's Electronic Waste Recycling Act funds e-waste disposal through consumer fees on electronics purchases, while New York enforces mandatory collection programs for electronic waste. These case studies underscore the importance of regulatory consistency and international collaboration in mitigating the negative impacts of e-waste.

In ensuring the ethical deployment of IoT, governments, industry leaders, and consumers must work together to implement sustainable policies and technological safeguards. While decentralized governance models and emerging innovations like blockchain and federated learning offer promising solutions, continuous oversight, public engagement, and regulatory enforcement are essential for maintaining ethical standards in IoT development.

12.3.3 Cross-cutting Governance Tools

The increasing complexity of AI and IoT governance requires robust frameworks that promote interoperability, ethical responsibility, and structured decision-making. Several governance tools, including the OECD AI Principles, EU Ethics Guidelines for Interoperability, and smart grid governance models, provide regulatory clarity and encourage transparency, accountability, and global cooperation in technology governance.

The OECD AI Principles, first adopted in 2019 and revised in May 2024, set forth key guidelines to ensure that AI systems are developed responsibly, with a focus on inclusivity, fairness, transparency, and accountability. These principles align AI policies across different nations while ensuring that democratic values and human rights are upheld. Countries such as Brazil and Japan have incorporated these principles into their legal frameworks, demonstrating their relevance beyond OECD member states. By promoting cross-border collaboration, these principles mitigate governance inconsistencies and establish common ethical and technical standards for AI deployment worldwide.

Another significant governance tool is the EU Ethics Guidelines for Interoperability, which provides a structured approach to assessing cross-border digital public services within the European Union. These guidelines emphasize transparent and accountable interoperability assessments, ensuring that digital interactions between Member States are seamless and equitable. Interoperability tests, mandated by the Interoperable Europe Act, address legal, organizational, semantic, and technical challenges by harmonizing digital governance across diverse jurisdictions. Stakeholder participation is a core aspect of these guidelines, ensuring that assessments incorporate input from government agencies, citizens, and service implementers. Best practices include forming multidisciplinary teams to evaluate the ethical implications of interoperability and fostering a culture of continuous improvement by documenting assessments and facilitating peer learning.

A case study on smart grid governance models highlights the practical application of governance frameworks in the energy sector. The Beethoven–Noord smart grid project, conducted using the ILTIAD (Institutional, Legal, and Technological Institutional Analysis and Development) framework, offers valuable insights into decision-making and regulatory challenges in local energy projects. The study employed qualitative research methodologies, including participant observation, semi-structured interviews, and document analysis, to assess how stakeholders navigate institutional and legal barriers in smart grid deployment. The research was structured around three key consistency challenges:

- **Legal Institution Consistency**: Ensuring that smart grid policies align with existing legal frameworks.

- **Action Situation Consistency**: Analyzing the decision-making processes among stakeholders.
- **Level Consistency**: Evaluating governance structures across different layers of administration.

The study's findings underscored critical behavioral and institutional factors that shape decision-making in smart grid projects. By applying a Comparative Case Analysis approach, researchers identified the importance of contract ownership, governance structures, and participatory decision-making in optimizing smart grid deployment. The ILTIAD framework proved effective in highlighting regulatory inconsistencies and potential governance pitfalls that could hinder the adoption of smart grid solutions.

These governance tools – OECD AI Principles, EU Ethics Guidelines for Interoperability, and smart grid governance models – demonstrate the necessity of cross-sectoral regulatory alignment. By fostering transparency, accountability, and multistakeholder collaboration, these frameworks contribute to the ethical and sustainable development of AI, IoT, and smart infrastructure globally.

12.4 Security Architecture and Privacy

12.4.1 IoT Security Vulnerabilities

The rapid expansion of IoT networks across industries, including healthcare, manufacturing, and smart home automation, has introduced significant security challenges. The growing number of interconnected devices presents multiple attack surfaces that threat actors can exploit. As IoT systems handle sensitive personal and business data, securing these networks is imperative to prevent unauthorized access, data breaches, and large-scale cyberattacks [32].

12.4.1.1 Network Security Challenges

12.4.1.1.1 Common Vulnerabilities

A primary security concern in IoT networks is the presence of widespread vulnerabilities across devices. Many IoT devices are shipped with weak default credentials such as "admin" or "password," which users often fail to change. These default settings make it easier for attackers to gain unauthorized access. Additionally, the heterogeneous nature of IoT ecosystems – ranging from industrial sensors to consumer smart cameras – makes it difficult to implement uniform security standards [33].

12.4.1.1.2 Data Privacy Concerns

IoT devices continuously collect, process, and transmit sensitive data, making them attractive targets for cybercriminals. Privacy breaches can expose personal, financial, or medical records, leading to regulatory noncompliance and reputational damage. For example, in the healthcare sector, unauthorized access to IoT medical devices could lead to severe consequences, including identity theft and fraudulent insurance claims. Regulations such as the General Data Protection Regulation (GDPR) and the California Consumer Privacy Act (CCPA) require organizations to secure personal data and maintain user consent mechanisms [34].

12.4.1.1.3 Attack Vectors

IoT systems face a variety of attack vectors:

- **Unauthorized Access**: Attackers exploit weak credentials and outdated firmware to gain control over IoT devices.
- **Distributed Denial of Service (DDoS) Attacks**: Compromised IoT devices can be recruited into botnets, such as the infamous Mirai botnet, which was used to launch large-scale DDoS attacks, disrupting major online services.
- **Man-in-the-middle (MitM) Attacks**: Hackers intercept communications between IoT devices to eavesdrop, manipulate data, or inject malicious commands.

12.4.1.1.4 Network Security Measures

Given the critical role of IoT devices in modern infrastructure, robust network security practices are necessary:

- **Network Segmentation**: Isolating IoT devices from business-critical networks to contain potential breaches.
- **Encryption**: Securing communication channels using Transport Layer Security (TLS) to prevent eavesdropping.
- **Device Authentication**: Enforcing multifactor authentication (MFA) and role-based access control (RBAC) to restrict unauthorized access.

12.4.1.1.5 Regulatory Compliance and Standards

To mitigate security risks, global regulatory bodies are enacting IoT security standards. For example, California's SB 327 law mandates that all IoT devices must have "reasonable security features" to prevent unauthorized access. Similarly, the European Telecommunications Standards Institute (ETSI) EN 303 645 outlines the best security practices for IoT manufacturers, including requiring unique passwords and regular security updates [35].

12.4.1.2 Real-life Examples of IoT Attacks

IoT security breaches have demonstrated the urgent need for stronger security protocols. Some notable incidents include:

- **Smart Security Cameras (Xiaomi Mijia Breach)**: A vulnerability in Xiaomi's Mijia smart cameras allowed users to view feeds from random homes, raising concerns about consumer privacy.

- **Ransomware on IoT Devices**: Cybercriminals have exploited weakly secured IoT devices as entry points for ransomware attacks, demanding payments to restore access.
- **MitM Attacks on Smart Homes**: Hackers have intercepted unencrypted communications between smart home devices, compromising user privacy and system integrity.

These incidents underscore the critical importance of implementing security-by-design principles in IoT ecosystems.

12.4.1.3 Sensor Vulnerability Management

IoT sensors serve as primary data collection points and are often the weakest links in security. Securing these devices requires continuous monitoring and updating. Key security measures include:

Strong Authentication and Access Control

- Unique credentials should be used for each sensor, and default passwords must be changed.
- MFA and RBAC should be implemented.

Secure Data Transmission

- End-to-end encryption (TLS/SSL) should be enforced to protect sensor data.
- Virtual Private Networks (VPNs) or private networks should be used for sensitive data transmission.

Firmware Updates and Patch Management

- IoT manufacturers should push automatic security updates to patch vulnerabilities.
- Users should be notified of firmware updates and encouraged to apply them promptly.

Physical Security Controls

- Tamper-resistant enclosures should be used for sensors deployed in public or vulnerable areas.
- Digital signatures and authentication mechanisms should be employed to detect rogue sensors.

AI-based Threat Detection

- Machine learning algorithms can identify unusual sensor behaviors and detect potential cyber threats.
- Automated response systems can be deployed to mitigate attacks in real time.

By integrating these security measures, organizations can enhance the resilience of IoT sensor networks against cyber threats.

12.4.1.4 Cascading Failure Prevention

IoT networks are highly interconnected, making them susceptible to cascading failures. If one device is compromised, it can trigger a chain reaction affecting the entire system. Preventive measures include:

- **Redundancy Mechanisms**: Ensuring that critical systems have backup devices to prevent single points of failure.
- **Network Segmentation**: Isolating different sections of an IoT network to prevent a breach from spreading.
- **AI-based Anomaly Detection**: Using machine learning to detect early signs of failure and proactively mitigate disruptions.
- **Automated Recovery Mechanisms**: Implementing self-healing network protocols that can autonomously reroute traffic and restore functionality after an attack.

These measures help maintain operational continuity and minimize downtime in IoT environments.

12.4.1.5 Case Study: Industrial IoT Security Breaches – The Stuxnet Attack (2010)

One of the most well-documented cyberattacks on IoT infrastructure was the Stuxnet attack, which targeted Iran's nuclear enrichment program in 2010. Stuxnet was a state-sponsored cyberweapon believed to have been developed by the United States and Israel to disrupt Iran's uranium enrichment processes. Unlike traditional cyberattacks that target data theft or financial gain, Stuxnet was designed to cause physical destruction by manipulating industrial control systems (ICS). It specifically targeted Siemens Supervisory Control and Data Acquisition (SCADA) systems, which were used to control nuclear centrifuges at Iran's Natanz facility.

The malware spread via infected USB drives and exploited multiple zero-day vulnerabilities in Microsoft Windows, enabling it to infiltrate air-gapped networks that were otherwise isolated from the internet. Once inside a system, Stuxnet identified Siemens PLCs (Programmable Logic Controllers) and altered the rotational speed of nuclear centrifuges in a way that would degrade them over time while displaying normal readings to operators. This deception allowed the attack to go undetected until substantial physical damage had already been done.

It is estimated that Stuxnet destroyed nearly 1 000 centrifuges, significantly setting back Iran's nuclear program. However, the attack did not remain confined to its intended target. Stuxnet accidentally spread beyond the Natanz facility, infecting computers worldwide and drawing global attention to the vulnerabilities of industrial IoT infrastructure. The incident had profound implications, highlighting the risks of cyber warfare and the importance of securing critical infrastructure against cyber threats. In response, nations worldwide strengthened their cybersecurity policies, and organizations involved in industrial operations adopted advanced security measures to prevent similar attacks. Stuxnet remains a landmark case in cybersecurity history, demonstrating how cyberattacks on IoT-connected industrial systems can lead to real-world physical consequences.

12.4.2 AI Security Considerations

The increasing integration of AI across industries has brought significant advancements but also introduced new security challenges. AI systems process vast amounts of data, automate decision-making, and power critical infrastructure, making them an attractive target for cyber threats. Unlike traditional IT security, AI security extends beyond protecting data and systems – it also involves securing AI models, mitigating adversarial attacks, and ensuring governance mechanisms are in place to prevent unintended consequences [36].

As AI continues to evolve, addressing its security risks becomes crucial, particularly in high-risk sectors such as healthcare, finance, autonomous systems, and national security. This section explores the emerging threats, security frameworks, and risk mitigation strategies required to ensure the safe deployment of AI technologies.

12.4.2.1 The CIG Framework: Confidentiality, Integrity, and Governance

AI security risks can be categorized into three fundamental areas, forming the Confidentiality, Integrity, and Governance (CIG) Framework. Each aspect ensures that AI systems function securely and align with ethical and regulatory standards.

- **Confidentiality Risks**: AI models process-sensitive personal and enterprise data, making them vulnerable to data leaks, model inversion attacks, and membership inference attacks. A breach in an AI-based financial risk assessment system could expose proprietary trading algorithms, while AI-powered healthcare analytics tools might inadvertently reveal patient identities.
- **Integrity Risks**: AI-generated outputs can be manipulated through data poisoning, adversarial inputs, and malicious training data, leading to biased predictions, security vulnerabilities, and loss of trust. For instance, autonomous vehicles relying on AI-driven perception models can be deceived by adversarial perturbations, leading to the misidentification of road signs or obstacles.

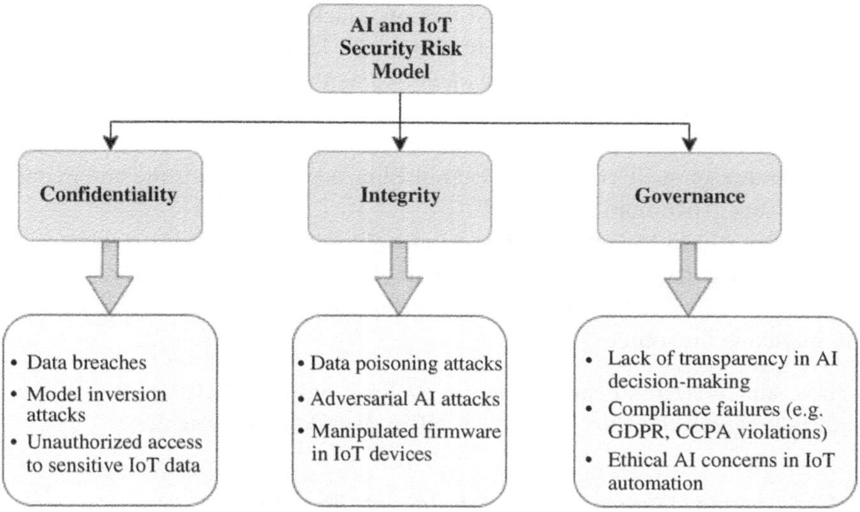

Figure 12.6 AI and IoT security risk model.

- **Governance Risks**: The opaque nature of AI decision-making raises concerns about accountability and regulatory compliance. AI models used in hiring, credit scoring, or law enforcement must be transparent and subject to rigorous audits to ensure fairness, explainability, and accountability [37].

Figure 12.6 illustrates the major AI and IoT security risks, categorizing threats under CIG. A well-structured governance model ensures multi-stakeholder oversight, continuous audits, and adherence to established AI security standards such as the OECD AI Principles, NIST AI Risk Management Framework, and the EU AI Act.

12.4.2.2 Emerging Threat Vectors in AI Security

AI security risks extend beyond traditional cyber threats, as attackers develop novel techniques to exploit AI vulnerabilities. Several emerging attack vectors highlight the unique security concerns in AI systems:

- **Adversarial AI Attacks**: Attackers manipulate input data to deceive AI models. For example, subtle pixel modifications in images can cause facial recognition systems to misidentify individuals, allowing unauthorized access to secure locations.
- **Data Poisoning Attacks**: Attackers inject malicious data during the AI training process, leading to manipulated outcomes. This is especially dangerous in AI-based fraud detection, healthcare analytics, and cybersecurity, where tainted training data can render models ineffective.

- **Model Inversion and Data Extraction**: Hackers can reverse-engineer AI models to reconstruct original training data, leading to privacy violations. AI-powered voice assistants and medical diagnostic systems are particularly vulnerable to such attacks.
- **AI-enabled Cyberattacks**: Malicious actors use AI-driven tools to automate cyberattacks, such as AI-enhanced phishing, deepfake fraud, and automated malware generation.
- **Supply Chain Vulnerabilities**: AI models depend on third-party datasets, pretrained models, and cloud infrastructure, making them vulnerable to supply chain attacks, such as backdoor implants in AI models or data manipulation at the source.

These attack vectors demonstrate the evolving nature of AI security threats, requiring organizations to continuously adapt their defense strategies.

12.4.2.3 Secure AI Development and Deployment Strategies

To mitigate AI security risks, organizations must adopt secure development practices throughout the AI lifecycle.

A multilayered security approach ensures that AI models remain robust against adversarial threats, data breaches, and manipulation.

- **Threat Modeling for AI Systems**: AI applications require custom threat modeling based on their domain of operation. For example, AI-driven biometric authentication must consider risks such as spoofing attacks and adversarial perturbations.
- **Secure Data Handling and Model Training**: Protecting training data is critical to ensuring AI integrity. Techniques such as federated learning, differential privacy, and homomorphic encryption can enhance security while maintaining AI performance.
- **Robust AI Governance Models**: AI models must undergo bias audits, fairness testing, and regulatory compliance checks to prevent security failures. Organizations should implement audit logs and explainability tools to track AI decision-making processes.
- **AI Access Control and Authentication**: AI models must have RBAC and MFA to restrict unauthorized modifications.
- **AI Model Resilience Testing**: AI security must include adversarial testing, red teaming, and stress testing to evaluate model robustness against cyber threats.

By embedding security principles throughout AI development, organizations can mitigate risks and ensure trustworthy, transparent, and resilient AI systems.

12.4.2.4 AI Security for Sustainable Development and Critical Applications

AI is increasingly used in climate monitoring, renewable energy, and smart city management. However, securing AI in these domains requires a balance between robust security measures and resource efficiency.

- **Energy-efficient AI Security Models**: AI security mechanisms must be designed for low-power consumption in off-grid and IoT-based sustainability applications.
- **Cybersecurity for Smart Infrastructure**: AI-driven smart grids, water management, and waste management systems must integrate zero-trust architectures, intrusion detection systems, and secure data transmission protocols.
- **AI-enabled Disaster Prediction Security**: AI models for climate change forecasting, flood prediction, and resource optimization must be protected against manipulation, misinformation, and unauthorized access.

Ensuring strong AI security measures in sustainability applications is vital to prevent cyber threats that could disrupt essential environmental protection efforts.

12.4.2.5 Case Study: AI Security Breach in Environmental Monitoring

A real-world case highlights the importance of securing AI models in environmental applications. A cloud-hosted AI system for tracking deforestation and illegal logging was compromised by adversarial input manipulation, causing AI models to misclassify illegal activities as natural deforestation. This led to delayed interventions, allowing deforestation to continue unchecked.

Following the attack, multiple security measures were implemented, including:

- Encryption of satellite imagery to prevent tampering.
- AI adversarial training to enhance robustness against manipulated inputs.
- MFA for system access control.
- Automated anomaly detection to flag suspicious classification patterns.

These enhancements improved system accuracy, resilience, and protection against adversarial attacks, ensuring reliable environmental monitoring. This case underscores the necessity of integrating security best practices into AI-driven sustainability applications to prevent cyber threats and ensure long-term effectiveness.

12.4.2.6 AI Security Policy and Regulatory Compliance

The development and deployment of AI security frameworks must align with global standards and regulations. Governments and industry leaders are establishing AI security policies to safeguard AI applications from cyber threats. Some of the key regulatory frameworks include:

- **The EU AI Act**: Establishes strict security guidelines for high-risk AI applications.
- **The NIST AI Risk Management Framework**: Provides security best practices for AI model validation, robustness testing, and bias mitigation.
- **The OECD AI Principles**: Focus on accountability, fairness, and transparency in AI governance.
- **US AI Executive Order (2023)**: Mandates federal agencies to implement secure AI development practices.

Adopting international AI security standards ensures that AI systems remain trustworthy, compliant, and resilient against emerging cyber threats. Table 12.2 presents a comparative table highlighting key regulatory frameworks governing AI and IoT security across different regions.

By embedding robust security frameworks, continuous monitoring, and adversarial resilience strategies, AI developers and policymakers can safeguard AI systems against cyber risks while ensuring ethical, fair, and sustainable deployment.

12.4.3 Privacy Considerations

The rapid expansion of AI and IoT technologies has significantly transformed data-driven decision-making across various industries. However, these advancements bring critical privacy concerns that must be addressed to ensure ethical deployment. AI and IoT systems continuously collect, process, and analyze vast amounts of data, often containing sensitive personal information. The challenge lies in striking a balance between leveraging this data for innovation while safeguarding user privacy. Without robust privacy measures, these technologies can lead to invasive surveillance, algorithmic discrimination, and unauthorized data exploitation.

One of the primary privacy risks in AI and IoT ecosystems is unauthorized data collection and usage. Many AI systems operate on large datasets sourced from user interactions, IoT sensors, and connected devices, often without explicit user consent. IoT devices, such as smart home assistants and wearable health monitors, continuously gather behavioral and biometric data, increasing exposure to breaches and misuse. AI-powered recommendation engines and behavioral profiling algorithms further complicate this landscape by personalizing services at the cost of individual privacy. Organizations must implement stringent data minimization strategies, ensuring only necessary information is collected and processed. Additionally, differential privacy and federated learning approaches can help mitigate risks by allowing AI models to learn from decentralized data without directly accessing user information.

A growing concern is the emergence of AI-driven surveillance and predictive analytics. Governments and private companies increasingly rely on AI

Table 12.2 Comparative regulatory frameworks for AI and IoT security.

Regulation	Region	Key focus areas	Enforcement mechanisms	Impact on AI and IoT security
GDPR (General Data Protection Regulation)	European Union	Data protection, user consent, right to erasure, privacy-by-design	Heavy fines for noncompliance, data breach notifications	Strengthens AI and IoT data privacy, enforces accountability in handling personal data
CCPA (California Consumer Privacy Act)	United States (California)	Consumer data rights, opt-out provisions, transparency requirements	Legal penalties, consumer lawsuits	Enhances AI-driven user data privacy, imposes restrictions on IoT device data collection
NIST AI Risk Management Framework	United States	AI system security, risk mitigation, bias detection, transparency	Voluntary compliance framework	Provides guidelines for AI safety and IoT security best practices
EU AI Act	European Union	Risk-based classification of AI systems, bans on high-risk AI applications, fairness and accountability	Strict regulations for high-risk AI, penalties for noncompliance	Directly impacts AI-driven IoT governance, ensures ethical AI development
China's AI Governance Framework	China	Real-time government oversight, data security laws, social responsibility of AI	Strict state monitoring and enforcement	Strongest AI surveillance laws, strict IoT security enforcement
ISO/IEC 27001 (Information Security Standard)	Global	Cybersecurity, risk assessment, secure AI and IoT deployments	Internationally recognized certification standard	Enhances AI and IoT security posture for global enterprises

for monitoring public spaces, facial recognition, and social behavior analysis, leading to potential mass surveillance and algorithmic profiling. AI models can infer sensitive details such as political affiliations, health conditions, and financial stability through indirect data analysis, often without user awareness. This phenomenon, known as shadow profiling, raises ethical and legal challenges, as individuals may be categorized or targeted based on inferred attributes rather than explicit consent. Additionally, deepfake technologies present privacy risks by generating hyper-realistic synthetic content, blurring the lines between real and manipulated identities. Without proper regulations and transparency mechanisms, AI can be exploited for deceptive activities, including misinformation campaigns and identity theft.

Regulatory compliance and privacy-preserving technologies are essential in mitigating AI and IoT privacy risks. Global data protection laws such as the GDPR and the CCPA mandate stricter data governance, emphasizing user rights, consent mechanisms, and accountability. However, the effectiveness of these regulations varies across jurisdictions, creating enforcement challenges for multinational AI-driven enterprises. To address this, organizations must adopt privacy-enhancing technologies (PETs), such as homomorphic encryption, which enables computations on encrypted data without exposing its content, and zero-knowledge proofs, allowing data verification without revealing underlying information. By integrating these advanced techniques with strong governance frameworks, AI and IoT systems can foster trust, security, and ethical data stewardship, ensuring that privacy remains a core pillar of technological advancement. This comprehensive approach to privacy not only protects individual rights but also enhances the long-term viability of AI and IoT systems in a world increasingly reliant on digital intelligence.

12.5 Resource Justice and Equitable Access

12.5.1 Resource Justice and Equitable Access in AI Systems

The accelerating global adoption of AI presents a paradoxical challenge for sustainable development. While AI technologies offer unprecedented capabilities to address pressing environmental and social challenges, the resources required to develop, deploy, and benefit from these systems remain profoundly unequally distributed. This section presents the multifaceted dimensions of resource justice and equitable access in AI development for sustainability, addressing both the uneven distribution of technological infrastructure and the algorithmic biases that can perpetuate existing inequalities [38, 39].

12.5.1.1 The Resource Justice Imperative in AI Development

12.5.1.1.1 Understanding the AI Equity Gap

The promise of AI to transform societies and advance sustainable development faces a fundamental challenge: the deeply unequal distribution of resources necessary to participate in the AI revolution. Recent data from early 2025 reveals the stark reality of this disparity. According to the United Nations, only 2% of the world's data centers are located in Africa, despite the continent being home to approximately 17% of the global population. Furthermore, only 5% of African AI innovators have access to sufficient computing resources for AI research and development [40]. These structural inequalities prevent many regions from fully engaging in the AI economy and benefiting from technological progress.

Economic disparities further widen the AI gap. The US AI sector received approximately $67.2 billion in investments in 2023, while leading technology hubs in Africa, such as Kenya and Nigeria, received only $15 million and $2.9 million, respectively. This imbalance perpetuates technological dominance by wealthier nations, reinforcing patterns of economic and digital dependence [41].

12.5.1.2 Resource Justice as a Foundational Principle

Resource justice in AI development transcends mere technical considerations to encompass the fair distribution of both tangible and intangible resources across the AI lifecycle. This includes equitable access to digital infrastructure, computing power, educational opportunities, and decision-making authority in shaping AI governance. Unlike traditional conceptualizations of digital equity that focus primarily on end-user access to technology, resource justice addresses the deeper structural inequities that determine who can participate in creating AI systems and defining their purpose [42].

The environmental dimensions of resource justice are particularly relevant when considering AI's rapidly growing ecological footprint. Training large AI models requires enormous computational resources and energy consumption, raising questions about the environmental justice implications of AI development. Communities already vulnerable to climate change may bear disproportionate burdens from increased energy demand and e-waste generation while receiving minimal benefits from the resulting technologies [43].

To counteract these disparities, community-driven AI models that leverage open-source infrastructure and collaborative funding can democratize AI innovation. Decentralized AI commons, where multiple stakeholders contribute resources, data, and computing power, can support underrepresented regions in AI research and applications [44].

Fundamentally, resource justice in AI demands recognition that technological resources are not merely neutral tools but embody specific values, priorities, and power relations. Achieving equitable outcomes requires confronting these embedded dimensions and reimagining AI development as a process that actively redistributes technological power rather than concentrating it.

12.5.1.3 Global Disparities in AI Resources and Capabilities

12.5.1.3.1 Infrastructure and Connectivity Challenges

The physical infrastructure supporting AI development – data centers, high-speed connectivity, computing clusters – remains heavily concentrated in wealthy nations. This geographical concentration creates structural advantages for entities in these regions while systematically keeping communities in the Global South at a disadvantage. The implications extend beyond economic opportunity to influence whose needs AI systems are designed to address and whose perspectives shape their development [45].

Connectivity barriers remain particularly acute in rural and remote areas globally. Even where basic internet access exists, the bandwidth and reliability required for meaningful participation in AI development or utilization of AI-powered services may be insufficient. These infrastructure limitations reinforce existing patterns of marginalization, as communities already underserved by traditional development efforts find themselves further excluded from technological advancement [46].

Energy security presents another critical dimension of infrastructure inequality. AI development requires stable, abundant energy resources, yet approximately 759 million people worldwide lack reliable electricity access. This creates a paradoxical situation where communities most in need of sustainable development solutions may be least able to participate in creating AI systems to address these challenges due to fundamental energy constraints.

12.5.1.3.2 Knowledge and Capacity Disparities

Beyond physical infrastructure, profound inequalities exist in access to AI education, skills development, and technical expertise. Educational systems in many regions lack sufficient resources to provide quality instruction in computer science, data science, and related fields essential for AI development. This creates a self-reinforcing cycle where regions with established technological advantages continue to develop specialized human capital while others fall further behind.

AI literacy programs must include culturally relevant training materials, translations in local languages, and adaptation of AI applications to meet diverse linguistic needs. The dominance of English-based AI development disproportionately excludes non–English-speaking communities, limiting the global accessibility of AI-powered tools.

12.5.1.4 Strategies for Promoting Equitable Access to AI Technologies

Building Inclusive Digital Infrastructure

- Expanding broadband access in underserved regions through a combination of fiber optics, wireless networks, and satellite connectivity.
- Developing community-based digital hubs to facilitate access to computing resources, training, and technical support.
- Implementing affordable technology initiatives, such as subsidized hardware and reduced-cost data plans, to lower financial barriers to AI access.

Democratizing AI Knowledge and Skills

- Providing open-source AI educational resources that are accessible to all, regardless of geographic or financial limitations.
- Supporting multilingual AI education initiatives to address language barriers and enable broader participation.
- Establishing partnerships between universities and industries to enhance AI research and workforce development in underrepresented regions.

Financial Resources and Investment Mechanisms

- Encouraging impact investment in AI projects that prioritize equitable access and social goods.
- Implementing technology transfer initiatives to support AI education and development in emerging economies.
- Establishing AI commons initiatives to provide shared resources, such as datasets and computational infrastructure, to resource-constrained communities.

12.5.1.5 Algorithmic Justice and Fair Distribution of AI Benefits

12.5.1.5.1 Addressing Bias in AI Systems for Sustainability

Algorithmic bias presents significant challenges for equitable AI development, particularly in sustainability applications where data gaps and historical prejudices may distort outcomes. Examples abound climate risk platforms underrepresent informal urban settlements in sub-Saharan Africa, skewing adaptation funding toward wealthier districts. Smart city deployments have excluded majority-Black districts due to algorithmic prioritization of tax revenue potential over public health needs. In addition to diverse oversight committees, AI fairness frameworks should incorporate transparent auditing mechanisms, bias detection tools, and impact assessments to measure real-world outcomes. This ensures AI applications remain accountable and do not disproportionately harm marginalized communities.

12.5.1.5.2 Environmental Justice Dimensions of AI

The current trajectory of AI infrastructure relies on high-energy consumption data centers, disproportionately impacting low-income communities. Efforts to develop energy-efficient AI models, such as federated learning and edge AI, can mitigate this problem. Future AI sustainability efforts should prioritize eco-friendly data centers, alternative computing methods like quantum computing for energy efficiency, and stronger AI sustainability policies. AI-driven solutions should not only optimize energy use but also ensure equitable access to renewable energy sources for underserved regions.

12.5.2 IoT for Equitable Resource Distribution and Monitoring

The IoT has emerged as a transformative technology for ensuring equitable resource distribution and efficient monitoring across various sectors. By leveraging interconnected devices, sensors, and real-time data analysis, IoT enhances decision-making, optimizes resource allocation, and reduces inefficiencies. With the number of connected IoT devices projected to reach 41.4 billion by 2025, their role in sustainable resource management continues to grow.

12.5.2.1 Applications of IoT in Resource Distribution

12.5.2.1.1 Smart Agriculture

IoT-enabled sensors facilitate precision farming by monitoring soil conditions, optimizing irrigation, and regulating fertilization. These technologies improve crop yields, minimize resource wastage, and enhance food security in regions affected by climate change.

12.5.2.1.2 Environmental Monitoring

IoT-based systems track air, water, and soil quality in real time, allowing for timely interventions to combat pollution and mitigate environmental degradation. By providing continuous data streams, IoT supports climate resilience efforts and regulatory compliance.

12.5.2.1.3 Smart Grids and Decentralized Energy Systems

IoT-driven smart grids optimize electricity distribution by balancing supply and demand, reducing power losses, and integrating renewable energy sources. Decentralized energy networks, enabled by self-healing mechanisms, improve reliability and sustainability while enhancing energy access for underserved communities.

12.5.2.1.4 Smart Metering

IoT-enabled smart meters allow for real-time energy consumption monitoring, improving billing accuracy and forecasting demand patterns. These systems empower consumers with insights into their energy usage while enabling utility providers to optimize distribution and reduce inefficiencies.

12.5.2.1.5 Water Management

IoT sensors detect water leaks, regulate distribution, and improve conservation efforts, particularly in drought-prone areas. These technologies ensure efficient water allocation and equitable access, supporting long-term water security.

12.5.2.1.6 Waste Reduction in Manufacturing

IoT-powered monitoring systems track production lines in real time, allowing manufacturers to reduce defects, minimize waste, and optimize resource utilization. These capabilities promote sustainable industrial practices while lowering operational costs.

12.5.2.2 Challenges and Limitations of IoT for Equitable Resource Distribution

12.5.2.2.1 Data Privacy and Security Risks

IoT devices collect vast amounts of sensitive data, raising concerns about unauthorized access, cyber threats, and data exploitation. In regions with limited digital literacy and weak regulatory frameworks, individuals may be unaware of the risks associated with data collection. Robust security measures and governance frameworks are necessary to protect user data and ensure ethical use of IoT technologies.

12.5.2.2.2 Social and Economic Disparities

While IoT improves resource management, access to these technologies remains unequal. Wealthier regions and large corporations often benefit the most from IoT-driven efficiencies, while low-income populations, small farmers, and underdeveloped regions face barriers to adoption. Without targeted interventions, IoT may exacerbate existing economic inequalities rather than bridging them.

12.5.2.2.3 Infrastructure and Deployment Costs

IoT implementation requires substantial financial investment in hardware, connectivity, and integration with existing systems. High initial costs, software licensing fees, and maintenance expenses present challenges, particularly in developing economies where infrastructure gaps persist.

12.5.2.2.4 Ethical and Governance Issues

IoT governance must address concerns related to data ownership, decision-making algorithms, and regulatory compliance. The rapid evolution of technology often outpaces existing policies, requiring continuous updates to ensure fair and transparent resource management.

12.5.2.2.5 Integration and Scalability Challenges
Managing large-scale IoT networks requires flexibility and interoperability to prevent vendor lock-in and ensure future adaptability. Organizations that fail to implement modular and scalable systems risk losing long-term value as IoT technology evolves.

12.5.2.3 Future Trends in IoT for Equitable Resource Distribution

12.5.2.3.1 Predictive Modeling for Climate Resilience
IoT-driven predictive analytics analyze historical and real-time environmental data to forecast climate-related risks. These models support early warning systems and enable proactive responses to disasters, water shortages, and extreme weather events.

12.5.2.3.2 Automation and Smart Systems
The increasing automation of IoT networks improves efficiency but also raises concerns about job displacement and economic shifts. The focus must be on integrating human-centered automation to enhance workforce skills and promote socio-economic stability.

12.5.2.3.3 Sustainability and Green Technologies
IoT plays a crucial role in sustainability initiatives by enabling real-time emissions tracking, energy-efficient monitoring, and waste reduction strategies. Green IoT solutions, such as low-power sensors and carbon-neutral data centers, are being explored to reduce environmental impact.

12.5.2.3.4 Collaborative Resource Management
Governments, industries, and communities are adopting blockchain and decentralized IoT for transparent and collaborative resource governance. These frameworks improve trust, accountability, and equitable resource distribution in supply chains, energy networks, and water management systems.

12.5.2.3.5 AI and IoT Integration
The convergence of IoT and AI enhances real-time decision-making, automation, and self-optimizing systems. AI-driven IoT solutions analyze complex datasets, optimize resource allocation, and improve overall efficiency, ensuring sustainable and equitable development.

12.5.2.3.6 Expansion of Smart Cities and Urban Hubs
With increasing urbanization, IoT-powered smart cities are integrating intelligent transportation systems, smart grids, and adaptive infrastructure to improve urban resource distribution. These technologies promote energy efficiency, reduce congestion, and enhance quality of life while ensuring fair access to urban services.

12.5.3 Global Resource Management

The rapid expansion of AI and IoT technologies raises critical concerns regarding the equitable distribution and sustainable management of global resources. As these technologies become more embedded in economic, social, and environmental systems, ensuring fair access to energy, infrastructure, and supply chains is essential. This section explores the key challenges and strategies for resource justice in AI and IoT deployment, focusing on energy consumption, infrastructure disparities, and sustainable resource management.

12.5.3.1 Escalating Energy Demands in AI and IoT

The energy footprint of AI and IoT systems has become a pressing sustainability issue. AI computing alone is projected to consume 0.5% of global electricity by 2027, equivalent to the energy consumption of an entire country. The training of advanced AI models requires weeks or months of continuous computation across thousands of processors, leading to significant energy investments. While individual AI interactions appear minimal – such as generating an AI-generated image consuming roughly the energy needed to charge a smartphone – billions of interactions daily result in a massive cumulative energy demand.

IoT infrastructure also adds to global energy consumption. Networks of sensors, data centers, and communication systems operate continuously, consuming substantial power for applications such as environmental monitoring, industrial automation, and smart city management. The challenge is ensuring that these systems are powered sustainably, particularly in regions with limited access to clean and affordable energy.

12.5.3.2 Resource Distribution and Digital Colonialism

Digital resource colonialism refers to the unequal distribution of AI and IoT infrastructure, where powerful transnational corporations dominate software, hardware, network connectivity, and data ownership. This concentration of control reinforces technological dependence, limiting local innovation in less-developed regions.

AI and IoT development are primarily centered in high-income nations, with resource-constrained countries relying on imported technologies, often with little say in their design or application. This raises concerns about who benefits from AI-driven advancements and how economic and technological dependencies shape global inequalities.

Data ownership is another key issue in global resource management. As AI models are trained on diverse global datasets, questions arise regarding the fair compensation and benefit-sharing for communities that contribute their data. Without inclusive data governance frameworks, resource-rich but technology-poor nations risk becoming mere data suppliers rather than equal stakeholders in AI innovation.

12.5.3.3 Infrastructure Disparities in AI and IoT Development

The physical infrastructure needed for AI and IoT – including data centers, high-speed connectivity, and energy grids – remains heavily concentrated in high-income countries. This unequal infrastructure distribution creates significant barriers for low-income and developing regions to access or implement AI and IoT solutions effectively.

Reliable electricity access is a key limiting factor. More than 759 million people worldwide lack stable electricity, making it nearly impossible to build local AI research hubs, IoT-enabled systems, or digital industries. While smart grids and decentralized energy solutions offer potential solutions, they require substantial investment that many underdeveloped regions cannot afford.

Smart grid technology demonstrates both the opportunities and challenges of AI-driven resource management. By integrating AI and IoT, smart grids optimize energy distribution, improve efficiency, and support renewable energy integration. However, deploying these grids requires financial resources, policy support, and infrastructure that many developing countries do not yet possess, further widening the global technology divide.

12.5.3.4 Strategies for Equitable Resource Management

12.5.3.4.1 Optimizing Resource Efficiency

One of the most effective ways to mitigate AI and IoT resource demands is through energy-efficient computing and smarter infrastructure management. AI-driven predictive analytics help optimize energy use, reduce waste, and improve resource allocation. For example, smart water meters and AI-based energy management analyze consumption patterns to optimize distribution, reducing both operational costs and environmental impact.

Developing low-energy AI models and edge computing solutions can significantly cut down on energy consumption. Unlike traditional cloud-based systems that rely on centralized data centers, edge AI processes data locally, reducing both power usage and latency. These innovations help ensure AI and IoT applications remain accessible to regions with limited infrastructure.

12.5.3.4.2 Circular Economy Approaches

A circular economy approach in AI and IoT focuses on designing technologies that prioritize durability, repairability, and recyclability. This reduces material waste while extending the lifecycle of technological products. IoT-enabled circular systems improve tracking of product lifecycles, waste management, and resource reuse to enhance sustainability.

Examples of circular economy strategies in AI and IoT include:

- Modular IoT devices that allow for component upgrades instead of full replacements.
- Product-as-a-service models, where companies retain ownership of devices and ensure maintenance.
- AI-driven material recovery systems that use machine learning to optimize e-waste recycling.

By integrating circular design principles, AI and IoT technologies can minimize environmental footprints while expanding access to sustainable innovation.

12.5.3.4.3 Responsible Supply Chain Management

Ensuring ethical sourcing of raw materials for AI and IoT components is crucial for sustainable development. The mining of rare minerals (such as cobalt, lithium, and rare earth elements) needed for semiconductors, batteries, and electronic components has led to environmental degradation and human rights violations in various parts of the world.

AI and blockchain technologies provide potential solutions for supply chain transparency. Real-time monitoring and smart contracts ensure ethical labor practices, sustainable sourcing, and fair-trade compliance throughout the technology supply chain. These solutions prevent exploitative practices and support more equitable resource allocation in the global market.

12.5.3.4.4 Global Cooperation and Governance Frameworks

Effective resource justice in AI and IoT requires international cooperation, including policy harmonization, financial support for underdeveloped regions, and knowledge-sharing frameworks. Public–private partnerships, South–South collaborations, and AI governance alliances help ensure that resource-constrained communities have opportunities to participate in AI and IoT innovation.

Key strategies include:

- Technology transfer initiatives that provide AI education, research funding, and digital infrastructure in emerging economies.
- Equitable investment models that prioritize social impact AI and IoT projects over purely profit-driven ventures.
- International standards for sustainable AI and IoT development, ensuring compliance with environmental, labor, and data governance principles.

By addressing global resource inequities, these strategies create pathways for inclusive AI and IoT deployment, ensuring that technological advancements benefit all communities, not just those with existing infrastructure advantages.

12.6 Regulatory Frameworks and Compliance

The expansion of AI and the IoT has created complex regulatory challenges requiring legal, ethical, and governance frameworks to ensure fair, transparent, and accountable deployment. The evolution of global AI and IoT regulations seeks to address issues related to data protection, resource justice, algorithmic accountability, and environmental sustainability. This section examines key legal frameworks, accountability systems, and international governance efforts aimed at ensuring ethical AI and IoT development.

12.6.1 Legal Landscape Evolution

12.6.1.1 EU Algorithmic Accountability Act

The EU Algorithmic Accountability Act (AAA) represents one of the most significant regulatory measures to date, aiming to govern AI deployment in high-risk applications such as automated decision-making, resource allocation, and sustainability planning. Key provisions include:

- **Third-party Impact Assessments**: Mandates independent audits for AI systems influencing public policy, financial services, and environmental regulations.
- **Right to Explanation**: Ensures that individuals affected by AI decisions receive clear, transparent, and actionable justifications for automated decisions.
- **High-risk AI Moratoriums**: Temporarily bans predictive policing, biometric surveillance, and AI-driven redlining in sectors prone to bias and discrimination.

By setting compliance benchmarks, the EU AAA has influenced global AI policy, with nations including Canada, India, and Australia exploring similar regulatory structures to safeguard algorithmic fairness.

12.6.1.2 Rights of Nature Enforcement

In parallel with AI regulations, several jurisdictions have begun enforcing Rights of Nature (RoN) laws that establish legal personhood for ecosystems affected by AI-driven resource extraction, land use, and energy consumption. This approach has gained traction in Ecuador, New Zealand, and Colombia, where courts have ruled against corporations deploying AI-driven mining and logging systems without environmental accountability.

Governments are now integrating AI-specific environmental justice frameworks, ensuring that AI-driven industrial processes comply with sustainability metrics before being deployed in forests, water basins, and indigenous lands.

12.6.1.3 ISO/IEC Standardization

The International Organization for Standardization (ISO) and the International Electrotechnical Commission (IEC) have developed technical standards for AI and IoT governance. Key initiatives include:

- **ISO/IEC 42001**: Establishes AI governance frameworks, focusing on bias mitigation, transparency, and explainability.
- **ISO 30141**: Defines the IoT Reference Architecture, ensuring secure and scalable IoT implementations.
- **ISO/IEC 27563**: Introduces privacy guidelines for AI-enabled smart environments, governing data collection, anonymization, and ethical data use.

12.6.1.4 Case Study: Cross-border Data Protection

The GDPR and the AI Act have set the foundation for cross-border AI governance. However, data sovereignty conflicts continue to emerge, especially in cloud-based AI and IoT deployments.

A notable case in 2024 involved a major smart city project in Nairobi, where EU-based cloud services were found storing African citizens' biometric data without consent. The dispute led to Kenya's AI Data Protection Law, which mandates that all AI-driven resource allocation systems store local data within national boundaries. This case highlights the growing importance of data localization, ethical AI deployment, and cross-jurisdictional regulatory harmonization.

12.6.2 Accountability Systems

12.6.2.1 AI Model Cards

AI model cards function as comprehensive documentation tools for AI systems, detailing training data sources, performance metrics, bias assessments, and intended uses. Originally developed by Google, model cards have now become a global standard for algorithmic transparency.

Regulatory bodies such as the US National Institute of Standards and Technology (NIST) and the EU AI Office now mandate model cards for public-sector AI procurement, ensuring that automated decision-making systems disclose potential risks before deployment.

12.6.2.2 IoT Lifecycle Certification

Ensuring IoT device security, energy efficiency, and data privacy requires rigorous certification protocols. In response, the IEEE P2668 Sustainable IoT Framework has introduced mandatory lifecycle assessments, which:

- Certify IoT sustainability, ensuring energy-efficient operations.
- Mandate ethical sourcing of IoT components, reducing reliance on conflict minerals.

- Evaluate long-term recyclability, encouraging circular economy practices in IoT manufacturing.

By 2026, over 45% of smart city deployments will be required to pass IoT life-cycle certification before government approval.

12.6.2.3 Impact Assessment Frameworks

To measure AI and IoT effects on resource distribution, economic fairness, and sustainability, impact assessment frameworks (IAFs) have become a regulatory requirement. These frameworks evaluate:

- Bias risk assessments for AI-driven resource allocation.
- IoT energy footprints, ensuring compliance with sustainability goals.
- Public transparency requirements, allowing affected communities to review AI decisions.

12.6.2.4 Case Study: Urban IoT Data Trusts

Cities including Barcelona, Amsterdam, and Seoul have pioneered Urban IoT Data Trusts, a new model for community-driven data governance. These trusts:

- Enable citizens to control how IoT data is used in public infrastructure planning.
- Enforce ethical data monetization, preventing private corporations from exploiting smart city data.
- Ensure equitable AI-driven resource management, preventing algorithmic redlining.

By 2027, urban data trusts are expected to become a global best practice for AI and IoT accountability in public governance.

12.6.3 Policy and Governance Frameworks for Resource Justice

12.6.3.1 International Cooperation and Multilateral Initiatives

Addressing AI resource disparities requires multilateral governance efforts beyond national policies. Key summits shaping global AI governance include:

- **AI Action Summit (Paris, 2025)**: Focuses on ethical AI procurement and sustainability protocols.
- **Global AI Summit on Africa (Rwanda, 2025)**: Establishes AI development roadmaps for emerging economies.
- **Hamburg Sustainability Conference**: Works toward an international AI sustainability treaty.
- **AI for Good Global Summit (Geneva, 2025)**: Prioritizes AI solutions for climate adaptation and resource justice.

These initiatives aim to ensure Global South leadership in AI and IoT governance, promoting equitable technology access and inclusive regulatory frameworks.

12.6.3.2 Regulatory Frameworks and Standards

Emerging regulatory strategies focus on justice-based AI governance, such as:

- IEEE P2668 Sustainable IoT Framework, mandating justice impact assessments for municipal tech projects.
- Public procurement policies that require ethically sourced AI models.
- Global licensing initiatives that redistribute AI patent revenues to marginalized communities.

12.6.3.3 Alternative Economic Models and Innovative Frameworks

Several alternative governance approaches aim to redistribute AI and IoT benefits fairly:

- **Data Cooperatives**: Community-led governance models ensuring fair compensation for data contributors.
- **Reparative Licensing**: Requires corporations to allocate a percentage of AI patent revenues to historically marginalized communities.
- **Green Workforce Transition Programs**: Ensures that workers displaced by automation receive training and access to new jobs.

These frameworks support a more just, inclusive, and transparent AI ecosystem, reinforcing ethical AI governance for all stakeholders. Table 12.3 summarizes key regulatory frameworks, their objectives, and example applications in AI and the IoT.

Table 12.3 Regulatory mechanisms in AI and IoT.

Regulatory mechanism	Purpose	Example application
EU Algorithmic Accountability Act	AI risk mitigation and transparency	High-risk AI impact assessments
Rights of Nature Laws	Environmental protection from AI	AI-driven deforestation monitoring
ISO/IEC Standards	AI and IoT governance standardization	ISO 42001 AI fairness assessment
AI Model Cards	Algorithmic transparency	NIST AI procurement guidelines
IoT Lifecycle Certification	Ethical IoT deployment	Sustainable smart city projects
Urban IoT Data Trusts	Citizen-controlled AI data	Barcelona's Smart City Governance

12.7 Implementation Roadmap

12.7.1 Phased Deployment Strategy

A structured deployment strategy is critical to ensuring the success of AI and IoT-driven projects. The phased approach enables organizations to mitigate risks, incorporate stakeholder feedback, and align technological advancements with regulatory frameworks.

12.7.1.1 Ten-year Implementation Framework

12.7.1.1.1 Objectives and Desired Outcomes

A well-defined roadmap is essential for guiding AI and IoT initiatives toward meaningful impact. Clearly identifying objectives, whether improving efficiency, reducing costs, or promoting sustainability – helps prevent scope creep and ensures alignment with long-term goals.

12.7.1.1.2 Comprehensive Planning and Stakeholder Involvement

Effective implementation requires robust planning and active stakeholder engagement. Collaboration between policymakers, industry leaders, and communities enhances decision-making, minimizes resistance, and ensures smooth deployment. Transparent communication channels should be established to keep stakeholders informed throughout the process.

12.7.1.1.3 Standardization and Interoperability

Ensuring compliance with widely accepted standards and protocols is vital for interoperability among IoT systems. Protocols such as MQTT and CoAP for data transmission and standardized data formats like JSON and XML enhance compatibility across platforms. Investment in monitoring tools ensures system stability and predictive maintenance.

12.7.1.1.4 Agile Methodologies and Iterative Development

Adopting Agile methodologies enhances responsiveness to changing requirements and user feedback. Frameworks like Scrum allow iterative deployment, ensuring that AI and IoT solutions remain adaptive and user-centric. This iterative approach minimizes risks and promotes continuous refinement.

12.7.1.1.5 Phased Expansion and Continuous Improvement

A gradual rollout strategy enables organizations to test AI and IoT solutions on a smaller scale before full deployment. Pilot projects help identify challenges, refine processes, and optimize system performance. Key performance indicators (KPIs) related to operational efficiency, energy consumption, and environmental impact should be continuously monitored.

12.7.1.2 Justice-aligned Technology Metrics

12.7.1.2.1 Defining Justice Metrics in Technology

Justice-aligned technology metrics ensure that AI and IoT applications uphold fairness, accountability, and transparency. These metrics focus on mitigating algorithmic bias, protecting user privacy, and promoting equitable access to technological benefits.

12.7.1.2.2 AI and IoT in Criminal Justice Systems

The adoption of AI in criminal justice, such as predictive policing and facial recognition, raises ethical concerns. Justice-aligned metrics guide the responsible use of these technologies, ensuring minimal intrusion, fairness, and adherence to human rights principles.

12.7.1.2.3 Ethical Considerations and Governance

Algorithmic decision-making must be scrutinized for biases that could disproportionately impact marginalized groups. Ethical frameworks should be developed in collaboration with policymakers, technologists, and civil society organizations to ensure fairness, explainability, and accountability in AI deployment.

12.7.1.3 Governance Maturity Index

12.7.1.3.1 Framework for IoT Governance

The Governance Maturity Index (GMI) provides a structured model for evaluating an organization's ability to manage IoT implementations. As IoT adoption grows, structured governance is critical for addressing security risks, data privacy concerns, and operational efficiency.

12.7.1.3.2 Key Maturity Levels

The GMI categorizes governance capabilities into different levels:

- **Basic**: Organizations operate IoT systems with minimal oversight.
- **Intermediate**: Governance structures are in place but require optimization.
- **Advanced**: Mature governance frameworks ensure transparency, compliance, and risk mitigation.

12.7.1.3.3 Assessment and Continuous Monitoring

Regular audits and assessments help organizations refine their IoT governance strategies. Data security policies, regulatory compliance, and ethical considerations should be integrated into the decision-making process.

12.7.1.4 Case Study: UN AI for Good Initiatives

12.7.1.4.1 The UN AI Actions Report

The United Nations' AI for Good initiative highlights the role of AI in addressing global challenges, aligning technological advancements with the Sustainable Development Goals (SDGs). The report documents over 400 AI projects across all 17 SDGs, focusing on poverty reduction, climate action, and equitable access to technology.

12.7.1.4.2 Global Impact and Cross-sector Collaboration

The success of AI-driven solutions depends on collaboration between governments, private enterprises, and research institutions. The UN emphasizes ethical AI deployment through multi-stakeholder partnerships, ensuring that AI solutions benefit both high-income and low-income regions equitably.

12.7.2 Future Directions

12.7.2.1 Emerging Challenges

As AI and IoT adoption grows, new challenges such as data sovereignty, ethical AI decision-making, and cybersecurity risks must be addressed. Regulatory frameworks should evolve to protect individual rights and ensure fair access to technological advancements.

12.7.2.2 Research Priorities

Future research should focus on energy-efficient AI models, decentralized IoT architectures, and sustainable smart infrastructure. AI ethics and bias mitigation techniques should also be prioritized to ensure fairness and transparency.

12.7.2.3 Stakeholder Collaboration Models

Public–private partnerships, academic research consortia, and government-led initiatives are crucial for scaling responsible AI and IoT deployments. Stakeholder collaboration should ensure that technology benefits are equitably distributed across different socio-economic groups.

12.7.2.4 Policy Recommendations

Governments must implement robust AI governance policies that enforce:

- Transparency in algorithmic decision-making.
- Privacy protection for IoT-driven data collection.
- Sustainability benchmarks for AI-powered infrastructure.
- Global cooperation in technology transfer and innovation sharing.

By integrating these policy frameworks, ethical governance models, and collaborative approaches, the global community can foster an AI and IoT ecosystem that is equitable, sustainable, and responsible.

References

1 European Commission. (2021). *Proposal for a Regulation Laying Down Harmonised Rules on Artificial Intelligence (Artificial Intelligence Act)*. European Commission.

2 UNESCO. (2021). *Recommendation on the Ethics of Artificial Intelligence*. UNESCO.

3 Weber, R.H. (2010). Internet of Things – new security and privacy challenges. *Computer Law & Security Review* 26 (1): 23–30.

4 Binns, R. (2017). "Fairness in Machine Learning: Lessons from Political Philosophy." *Decision-Making in Computational Design & Technology eJournal* 81: 1–11.

5 Gubbi, J., Buyya, R., Marusic, S., and Palaniswami, M. (2013). Internet of things: a vision, architectural elements, and future directions. *Future Generation Computer Systems* 29 (7): 1645–1660.

6 Floridi, L. (2020). *Ethics of Artificial Intelligence*. Oxford University Press.

7 IEEE. (2019). *Ethically Aligned Design*. IEEE Standards Association.

8 Marr, B. (2019). *Artificial Intelligence in Practice*. Wiley.

9 Barcelona City Council. (2021). *Barcelona Smart City Strategy*.

10 Caragliu, A., Del Bo, C., and Nijkamp, P. (2011). Smart cities in Europe. *Journal of Urban Technology* 18 (2): 65–82.

11 Neirotti, P., De Marco, A., Cagliano, A.C., et al. (2014). Current trends in smart city initiatives: some stylised facts. *Cities* 38: 25–36.

12 Kourtit, K., Nijkamp, P., and Arribas, D. (2020). Smart cities in perspective – a comparative European study by means of self-organizing map. *European Journal of Social Science Research* 25: 229–246.

13 European Commission. (2021). *AI Act Proposal*. European Commission.

14 ISO/IEC 27001. (2021). *Cybersecurity Standards*.

15 United Nations. (2022). *Digital Sustainability Report*.

16 GSMA. (2021). *Mobile for Development*.

17 World Economic Forum. (2020). *AI for Sustainable Cities*.

18 Sachs, J., Schmidt-Traub, G., Kroll, C. et al. (2023). *The Sustainable Development Goals and AI–IoT Impact Measurement*. Sustainable Development Report.

19 Dameri, R.P. (2017). *Smart City Implementation: Creating Economic and Public Value in Innovative Urban Systems*. Springer.

20 Bifulco, F., Tregua, M., Amitrano, C., and D'Auria, F. (2018). Smart mobility and public transport: opportunities and challenges in smart city initiatives. *Cities* 78: 66–75.

21 Batty, M. (2020). The computable city. *International Journal of Urban Sciences* 24 (2): 155–169.

22 Eubanks, V. (2018). *Automating Inequality: How High-Tech Tools Profile, Police, and Punish the Poor*. St. Martin's Press.

23 Benjamin, R. (2019). *Race After Technology: Abolitionist Tools for the New Jim Code*. Polity.

24 Floridi, L. (2018). The ethics of artificial intelligence: principles, challenges, and opportunities. *Philosophical Transactions of the Royal Society A* 376 (2133).

25 Couldry N., and Mejias, U.A. (2019). *The Costs of Connection: How Data Is Colonizing Human Life and Appropriating It for Capitalism*. Stanford University Press.

26 O'Neil, C. (2016). *Weapons of Math Destruction: How Big Data Increases Inequality and Threatens Democracy*. Broadway Books.

27 OECD. (2023). *Measuring the Digital Transformation: A Roadmap for the Future*. OECD iLibrary.

28 Circularise. (2024). The EU conflict minerals regulation for responsible sourcing. *Circularise Blog*. https://www.circularise.com/blogs/the-eu-conflict-minerals-regulation-for-responsible-sourcing (accessed Mar 10 2025).

29 Mining Digital. (2023). Conflict minerals: navigating ethical sourcing challenges. https://miningdigital.com/sustainability/conflict-minerals-navigating-ethical-sourcing-challenges (accessed Mar 10 2025).

30 Cryptsy. (2023). Governance in decentralized systems. https://cryptsy.com/governance-in-decentralized-systems (accessed Mar 10 2025).

31 Nguyen, K. (2020). A case study comparison of waste management strategy in industrialized and developing nations for better health outcomes. *SIR Journal of Research*. https://www.sirjournal.org/research/2020/11/17/a-case-study-comparison-of-waste-management-strategy-in-industrialized-and-developing-nations-for-better-health-outcomes (accessed Mar 10 2025).

32 Google. (2023). Cybersecurity advancements – secure AI framework (SAIF). *Google Safety & Security*. https://safety.google/cybersecurity-advancements/saif/ (accessed Mar 10 2025).

33 Gupta, P. (2024). *Sustainable Information Security in the Age of Green Computing*. IGI Global. https://www.igi-global.com/book/sustainable-information-security-age-green/349903 (accessed Mar 10 2025).

34 Internet Crime Complaint Center (IC3). (2023). Cybersecurity and AI threats report. *IC3 Reports*. https://www.ic3.gov/CSA/2023/231128.pdf (accessed Mar 10 2025).

35 Software Engineering Institute (SEI). (2023). Weaknesses and vulnerabilities in modern AI: why security and safety are so challenging. *SEI Blog*. https://insights.sei.cmu.edu/blog/weaknesses-and-vulnerabilities-in-modern-ai-why-security-and-safety-are-so-challenging/ (accessed Mar 10 2025).

36 Microsoft. (2021). Best practices for AI security risk management. *Microsoft Security Blog*. https://www.microsoft.com/en-us/security/blog/2021/12/09/best-practices-for-ai-security-risk-management/ (accessed Mar 10 2025).

37 United Nations General Assembly (UNGA). (2024). Steering the use of safe, secure, and trustworthy AI for sustainable development. *IISD SDG Knowledge Hub*. https://sdg.iisd.org/news/

unga-steers-use-of-safe-secure-trustworthy-ai-for-sustainable-development/ (accessed Mar 10 2025).

38 United Nations. (2024). *AI and Sustainable Development Goals.*

39 World Economic Forum. (2025). *Bridging the AI Divide: Global Infrastructure Report.*

40 Crawford, K. (2021). *Atlas of AI: Power, Politics, and the Planetary Costs of Artificial Intelligence.* Yale University Press.

41 Smolenski, P. (2024). Economic barriers in AI innovation: a global perspective. *Journal of AI Ethics* 18 (2): 45–62.

42 Greenfield, A. (2017). *Radical Technologies: The Design of Everyday Life.* Verso.

43 Johnson, S. (2025). Environmental impacts of AI: a sustainability perspective. *IEEE Transactions on Green Computing* 12 (3): 89–101.

44 Zuboff, M. (2019). *The Age of Surveillance Capitalism.* PublicAffairs.

45 Smith, C. (2025). Digital infrastructure gaps and AI access. *International Journal of Technology and Society* 22 (1): 34–49.

46 Brown, J. (2025). IoT and energy justice: policy implications. *Energy Policy Review* 45 (4): 99–112.

Part IV

Future Directions and Innovations

13

Advances, Challenges, and Future Directions in IoT Architectures for Smart Environments: A Comprehensive Review

Manuel J.C.S. Reis[1,], Carlos Serôdio[2,3], and Nishu Gupta[4]*

[1] *Engineering Department/IEETA, University of Trás-os-Montes e Alto Douro, Vila Real, Portugal*
[2] *Department of Engineering, School of Sciences and Technology, University of the Trás-os-Montes e Alto Douro, Vila Real, Portugal*
[3] *Center ALGORITMI, Universidade do Minho, Campus de Azurém, Guimarães, Portugal*
[4] *VTT Technical Research Centre of Finland Ltd., Oulu, Finland*

*Corresponding author: mcabral@utad.pt

13.1 Introduction

The Internet of Things (IoT) is revolutionizing smart environments by enabling interconnected devices to communicate autonomously and perform tasks. IoT architecture in smart environments integrates sensors, networks, and platforms for efficient operations across sectors like manufacturing [1], transportation [2], agriculture [3], healthcare [4], industry [5], and smart cities [6]. However, challenges such as scalability, security, interoperability, and privacy persist due to the complex and fragmented IoT ecosystem. Addressing these issues is crucial for developing robust and efficient IoT systems. We have identified several open problems, including energy efficiency, data privacy, real-time data processing, edge computing, context awareness, and standardization, which continue to drive ongoing research.

This comprehensive review discusses the eight most pressing issues: scaling IoT architectures to accommodate device growth, ensuring security in the context of diverse and often unsecured devices, addressing privacy concerns due

AI and ML Techniques in IoT-based Communication: A Path to Sustainable Development Goals, First Edition. Edited by Sumita Mishra, Nishu Gupta and Polat Goktas.

to sensitive data, and achieving interoperability across heterogeneous systems. Furthermore, challenges like optimizing energy consumption, real-time analytics, integrating edge computing, and creating adaptive systems are explored alongside the need for global IoT standards. Solving these problems is critical to future-proofing IoT systems and ensuring seamless and efficient smart environments.

We will examine the core technical and operational challenges of integrating IoT devices in smart environments. These challenges include the need to support scalability and interoperability, the importance of energy efficiency, the critical role of data privacy and security, the demand for real-time data processing, the benefits and complexities of edge and fog computing, the requirement for context awareness and adaptability, the limitations due to lack of standardization, and the difficulties in network management and optimization.

As the number of IoT devices continues to grow, architectures must scale to accommodate them while maintaining cross-platform interoperability. Integrating different IoT protocols, such as MQTT and CoAP, across varied platforms and vendor ecosystems is difficult, given the heterogeneity of devices and communication standards [7, 8]. Energy efficiency is also crucial, especially for battery-operated devices in smart environments like cities, where extending the device's lifetime through efficient communication protocols, processing techniques, and low-power sensors is essential [9, 10].

Data privacy and security concerns are heightened in IoT systems due to the vast amounts of sensitive data they gather. Protecting this data, whether stored locally or in the cloud, is challenging because IoT systems are decentralized and often resource-constrained, complicating the deployment of robust security measures [11–13]. Furthermore, real-time data processing is crucial for traffic management and health monitoring applications. However, achieving low latency and handling high data volumes remains challenging, especially at the edge. Current systems often struggle with latency and the inability to process large data streams at the edge [14–17].

Integrating edge and fog computing helps reduce latency and bandwidth usage, yet implementing these layers in IoT architectures, particularly where resources are limited, poses additional difficulties. Ensuring optimal placement of computation and storage across the edge, fog, and cloud layers is an unsolved problem [18, 19]. Smart environments require IoT systems to be context-aware and adaptable to environmental changes, dynamically responding to shifts in context or user behavior to improve user experiences. However, developing predictive and adaptive architectures remains a research priority. Enabling systems to predict and adapt based on contextual information (e.g. user location, preferences) without user intervention is critical to enhancing user experience [20–22].

Standardization is another obstacle, as many IoT systems operate on proprietary technologies, limiting interoperability and scalability. The fragmentation in the IoT ecosystem across vendors, communication protocols, and platforms limits scalability and interoperability [23–25]. Managing and optimizing IoT networks with numerous interconnected devices is similarly complex, as maintaining optimal performance, low latency, and energy efficiency in environments like smart cities requires effective network management [26, 27].

To address these issues, machine learning techniques for scalability, blockchain-based solutions for security, and interoperability frameworks have been suggested. Future research is expected to focus on artificial intelligence (AI)-driven automation, resilient security frameworks, and open-source platforms to unlock the full potential of IoT in smart environments.

The aforementioned open problems in IoT architecture for smart environments represent ongoing challenges for researchers and engineers. As IoT continues to evolve and smart environments become more complex, addressing these issues will be crucial for realizing the full potential of IoT systems.

The following sections will discuss these problems, possible solutions, and future research directions. The final section of this study presents the main conclusions.

13.2 Scalability and Interoperability

Scalability in IoT refers to the system's ability to handle a growing number of devices, services, and data without degradation in performance. Interoperability, on the other hand, refers to the ability of diverse systems and devices to work together seamlessly, regardless of their underlying architecture or communication protocols.

With the growing number of connected devices in intelligent environments (e.g. smart cities or industrial IoT), ensuring systems can scale effectively is a significant challenge. Interoperability is also critical, as various devices often use different communication standards (e.g. ZigBee, Bluetooth, Wi-Fi) and must work together seamlessly [28–32].

Concerning challenges of scalability, as the number of IoT devices continues to grow (with estimates of tens of billions of devices in the near future), managing large-scale networks becomes complex. Each additional device generates data, requires connectivity, and places a strain on processing resources. Traditional centralized architectures, where all data is sent to the cloud, struggle to cope with this increased load due to bandwidth, latency, and computational limitations.

Interoperability challenges are typically related to the fact that different manufacturers with proprietary technologies and standards often design IoT devices. Integrating these into a unified system is difficult due to variations in communication protocols (e.g. MQTT, CoAP, ZigBee, Bluetooth), data formats, and security mechanisms. Achieving interoperability requires standardization, middleware solutions, and protocol translation mechanisms. Nonetheless, this diversity is present at each phase, from configuring devices to collecting and processing data, highlighting the need for a scalable, interoperable, and economical solution, like Service-Oriented Architectures (SOA) or other additional middleware [33–35]. IoT applications that employ SOA are gaining popularity, as they inherently support seamless interoperability and composability. With the emergence of SOA, microservices architectures have seen great development, since these, in contrast to monolithic applications, are services deployed autonomously, offering a variety of benefits. However, migrating from a traditional concept to microservices is a challenging task [36].

Other solutions, based on syntactic and semantic interoperability, like the Open Messaging Interface (O-MI) and Open Data Format (O-DF), One Data Model (OneDM) Semantic, could be employed to obtain features like data balancing, interoperability, and data semantics in addition to scalability [37, 38].

Potential solutions to these problems may include responses on Decentralized Architectures, Middleware Solutions, and Standardized Protocols and APIs. Edge and fog computing (discussed below) help manage scalability by distributing processing closer to the data sources. Middleware platforms, like FIWARE or oneM2M, are proposed to provide interoperability across different IoT systems by abstracting underlying communication protocols and devices. Efforts to create interoperable APIs and standardized communication protocols (e.g. MQTT, CoAP, and DDS) are vital. Open-source IoT ecosystems like EdgeX Foundry can act as a glue between different devices and platforms.

In this context, the ability for different devices and data to work together across platforms facilitates interaction among IoT systems focused on specific and diverse areas. Once cross-platform interoperability is established, it guarantees cross-domain compatibility, allowing diverse platforms from different sectors to collaborate and create integrated IoT applications [39].

A different approach to supporting interoperability by IoT systems is proposed by [40], which proposes the use of a Model-Driven Architecture (MDA), standardized by the Object Management Group (OMG), in which a high level of abstraction to deal with heterogeneous devices, applying concepts such as extensibility and reusability, both to deal with Device-to-Device (D2D) and Device-to-Server (D2S) protocols, such as CoAP, HTTP, MQTT, AMQP, and XMPP.

Interoperability problems can also be solved at a data level, acting at the data level defining ontologies, which facilitate tasks such as acquisition, representation, dissemination, and transfer of data between nodes, and ensures cross-platform and cross-device compatibility. Based on these principles, [41], proposes a distributed multilayered data platform, with modular architecture, on an OSGi layer, based on OM2M & RESTful to improve interoperability, making it dynamic and distributed, also improving issues such as system latency.

Integrating cloud and edge computing technologies greatly supports the expansion of IoT systems. Cloud platforms provide flexible storage and computational power, while edge computing enables localized data processing, minimizing delays and reducing bandwidth consumption. Combining these approaches ensures effective data handling and analysis, paving the way for scalable and high-performing IoT infrastructures.

Recent advances in AI, machine learning, and the deployment of 5G networks will play a distinct role in driving the growth of IoT. These innovations will enhance the intelligence and efficiency of IoT solutions, enabling improved connectivity and data processing capabilities. As these developments unfold, scalability will remain a key factor for the success of IoT initiatives.

However, frequently, interoperability and scalability work at odds. In other words, the greater the interoperability characteristics of an IoT system, the lower its scalability and overall performance. This idea is defended by [38], who bases his study on scalability and support for heterogeneity. Your work translates into a trade-off study considering interfaces, data format (Open Connectivity Foundation (OCF), Open Mobile Alliance (OMA), Smart Applications Reference (SAREF) and Sensor Network Ontology (SSN) LWM2M data models over CoAP, JSON or UltraLight over HTTP/MQTT, OPC-UA, Sigfox LoRaWAN), communication protocols (MQTT brokers, OPC-UA servers, Sigfox Back-end, Modbus slaves, or CAN nodes), and supporting platforms (FIWARE/NGI, ThingsBoard, and Konker).

Future research directions will indeed include the development of scalable algorithms for managing large-scale IoT networks and global standardization efforts to ensure seamless interoperability.

Table 13.1 highlights key contributions to IoT research, focusing on scalability, interoperability, and technological advancements. It summarizes diverse approaches, including standards-based frameworks, blockchain integrations, distributed architectures, and microservices. The outcomes demonstrate significant advancements such as improved data sharing, enhanced device communication, and scalable platforms. However, the studies also reveal persistent challenges, including balancing security with scalability, addressing heterogeneous device integration, and overcoming computational overheads. These insights underscore the complexity of IoT system design and the need for further research to address open issues in achieving seamless, secure, and scalable IoT ecosystems.

Table 13.1 Summary of IoT research contributions: scalability, interoperability, and associated challenges.

Paper	Scalability and interoperability	Technology solution	Outcomes	Challenges
24	Standards-based scalable IoT data sharing	Cloud integration and monetization	Improved data sharing and scalability	Ensuring standards compliance
25	Cybersecurity and IoT certification	Proposed template for IoT certification	Framework for IoT certification processes	Complex IoT certification landscape
26	Challenges in IoT security certifications	Security certification strategies	Addressing security gaps	Lack of unified security approaches
27	IoT survey: scalability and challenges	Architectures and frameworks	Enhanced understanding of IoT challenges	Addressing scalability issues
28	Conceptual IoT definition	IoT definition and components	Broad definition of IoT capabilities	Ambiguity in IoT definitions
29	Blockchain for IoT scalability and privacy	Blockchain and deep learning	Enhanced IoT threat intelligence	Balancing security and scalability
30	Enhanced MQTT protocol for IoT scalability	Distributed machine learning	Improved IoT security protocols	Computational overhead concerns
31	Framework for service discovery	Distributed IoT frameworks	Efficient service discovery	Managing distributed IoT components
32	Trade-offs in open IoT platforms	Platform optimization strategies	Scalable and interoperable platforms	Balancing interoperability and scalability
33	Guidelines for managing IoT scalability	NIST recommendations	Privacy and security risk management	Implementing comprehensive guidelines
34	Microservices for IoT device interoperability	Microservice-based architecture	Enhanced device communication	Complex microservice deployments
35	Technologies for critical infrastructure IoT	Middleware solutions	Improved monitoring solutions	Ensuring real-time monitoring
36	Platform for heterogeneous IoT devices	Scalable IoT platform	Support for diverse IoT devices	Handling device heterogeneity

Paper	Scalability and interoperability	Technology solution	Outcomes	Challenges
37	Microservices using knowledge graphs	Knowledge graph extraction	Efficient system migration	Complexity of knowledge extraction
38	Standards for industrial IoT data exchange	Data exchange protocols	Reliable industrial IoT communication	Standard adoption in industries
39	Software architecture for IoT interoperability	Next-gen gateway solutions	Interoperability across devices	Integrating diverse device protocols
40	Taxonomies for IoT interoperability	Interoperability frameworks	Identification of open challenges	Addressing multiple interoperability gaps
41	Model-driven IoT device interoperability	Model-driven architecture	Enhanced device interactions	Implementing model-driven standards
42	Distributed IoT architecture	Distributed IoT systems	IoT advancements	Scalability in IoT systems

13.3 Energy Efficiency

Energy efficiency in IoT refers to optimizing energy consumption in IoT devices and systems to prolong sensor battery life, reduce operational costs, and maintain system functionality.

Many IoT devices operate on constrained power sources, particularly in intelligent environments. The need to transmit, process, and store large amounts of data while maintaining low energy consumption presents a significant challenge. Techniques such as energy-efficient protocols, sleep scheduling, and edge computing are being explored to address this [42–46].

Consequently, energy efficiency, management, and control are now top priorities in the IoT field. The focus is deploying solutions and algorithms that enhance battery efficiency and lower energy use in IoT sensors and devices, ultimately extending their operational life [47, 48].

Another constraint addressed is related to the optimization of energy-efficient routing protocols or techniques [49], where most recent works apply SDN with AI systems (cloud, fog, or edge), optimize network scalability, lower-level routing, and load balancing using Genetic Algorithm (GA), Particle Swarm Optimization (PSO), and Artificial Bee Colony (ABC).

The first challenge faced by IoT is powering the devices. Many IoT devices, such as sensors and actuators, are deployed in environments where frequent battery replacement or recharging is impractical. For example, smart agriculture sensors might be placed in hard-to-reach fields, and in industrial IoT systems, downtime for maintenance can be costly. Contrasting with this is the fact that these devices are being exposed to more and more energy-intensive operations. Continuous data transmission, especially in real-time systems, consumes significant energy. The need for constant connectivity and high data throughput can rapidly drain battery life. While energy-saving mechanisms like sleep cycles and duty cycling reduce power consumption, they can introduce latency, affecting system responsiveness and performance. As such, energy vs. performance trade-offs is a real problem.

Potential solutions involve energy-efficient communication protocols and energy-harvesting methods. Protocols such as Low Power Wide Area Networks (LPWAN), Narrowband IoT (NB-IoT), SigFox, and 6LoWPAN are designed to reduce power consumption in IoT systems. Energy harvesting technologies (e.g. solar, wind, kinetic energy, RF, etc.) allow devices to replenish power from their surroundings, reducing dependence on batteries. Developing energy-efficient processors and sensors, such as ARM Cortex-M processors or Intel's Loihi neuromorphic chips, can help reduce energy usage without compromising performance. Additionally, adaptive power management systems that optimize when and how devices consume energy, such as Dynamic Voltage and Frequency Scaling (DVFS), can help improve battery life without affecting system responsiveness.

Future research directions include developing ultra-low-power processors, sensors, communication modules, and intelligent algorithms for dynamic power management and task scheduling that balance energy consumption and performance.

Table 13.2 synthesizes recent research on energy-efficient IoT systems, highlighting diverse technological solutions and their associated advantages and challenges. The studies cover strategies such as multi-hop routing, mathematical programming, cross-layer frameworks, and AI-driven approaches. The outcomes emphasize significant advancements in optimizing energy consumption, enhancing scalability, and improving resource utilization. Advantages include improved system adaptability, reliability, and energy savings across various IoT environments. However, the studies also reveal ongoing challenges, such as integrating new energy-efficient frameworks with existing systems, balancing security with energy optimization, and managing computational overhead in advanced routing algorithms. These insights illustrate the need for continued innovation to achieve sustainable and efficient IoT architectures.

Table 13.2 Summary of research on energy efficiency in IoT systems: technologies, outcomes, and challenges.

Paper	Energy efficiency	Technologies solutions	Outcomes	Advantages	Disadvantages	Challenges
9	Energy-efficient mobile distributed sensing platform	On-demand distributed sensing platform with location and activity awareness	Reduced energy consumption during sensing operations	Improved sensing efficiency and resource utilization	Complexity in platform deployment	Ensuring consistent energy efficiency at scale
10	Energy considerations in smart IoT systems	IoT frameworks tailored to smart campus applications	Optimized energy use across smart campus applications	Enhanced system scalability and adaptability	Challenges in integrating diverse IoT devices	Balancing energy optimization with performance demands
26	Energy efficiency in IoT architectures	Various energy-efficient IoT components	Improved IoT scalability and energy performance	Broader application of IoT energy principles	Scalability issues in older frameworks	Adapting architectures to modern IoT demands
27	Conceptual focus on energy optimization	IoT energy strategies and definitions	Clear focus on IoT energy-related goals	Systematic energy focus in IoT definitions	Limited technical details on energy optimization	Maintaining relevance with evolving technologies
43	Overview of energy-efficient industrial IoT	Routing and resource optimization algorithms	Efficient IoT systems for industrial applications	Optimized energy use in industrial IoT setups	Challenges in adapting to diverse industrial needs	Ensuring flexibility in industrial environments
44	Secure cross-layer IoT energy-efficient framework	Integrated cross-layer and secure framework	Enhanced security and energy savings	Cross-layer design improves both security and energy use	Implementation complexity in cross-layer models	Balancing security with energy efficiency

(Continued)

Table 13.2 (Continued)

Paper	Energy efficiency	Technologies solutions	Outcomes	Advantages	Disadvantages	Challenges
45	Mathematical programming for WSN design	Mathematical models for sensor network design	Improved network design for energy efficiency	Scalable and efficient WSN design	Potential computational overhead	Addressing heterogeneity in WSNs
46	Multi-hop routing for efficient resource allocation	Resource-efficient multi-hop routing protocols	Effective resource allocation for IoT networks	Better use of resources in constrained environments	Latency concerns in multi-hop networks	Optimizing routing under resource constraints
47	Energy-efficient routing algorithms in IoT	Routing techniques and algorithm surveys	Enhanced energy performance in routing	Comprehensive understanding of IoT routing	Complexity in deploying algorithms	Deploying efficient algorithms at scale
48	Optimizing energy in sensor networks	Mobile sinks for sensor optimization	Increased energy reliability in sensor networks	Enhanced reliability and sensor efficiency	Device dependency on mobile sinks	Handling scalability with sensor movement
49	Energy-efficient routing in IoT and fog computing	Fog-enabled routing in IoT applications	Improved energy use in IoT-fog environments	Energy savings with fog computing integration	Integration complexity with fog layers	Adapting routing for large fog-enabled networks
50	Optimized routing with SDN-AI in large-scale IoT	AI-driven SDN for large-scale industrial IoT	Efficient routing in large-scale IoT setups	Improved energy optimization at scale	High initial cost of implementation	Integrating AI with SDN for IoT challenges

13.4 Data Privacy and Security

Data privacy refers to ensuring that users' personal or sensitive data is collected, processed, and stored securely and protected against unauthorized access. Security in IoT systems involves securing data transmission and device authentication and protecting against threats such as data breaches or cyberattacks.

IoT systems collect massive amounts of sensitive data in real time, posing privacy and security risks. Ensuring end-to-end encryption, authentication, and secure storage is essential, but resource constraints on IoT devices make implementing robust security challenging [50–58].

The first concern is related to privacy. IoT devices often collect sensitive information, such as location data, health metrics, or personal habits (e.g. smart home cameras or wearables). In smart cities, IoT data could include traffic monitoring, which could reveal individuals' movements. Protecting these data from unauthorized access is critical to prevent privacy violations. Security challenges in resource-constrained devices are another important issue. Many IoT devices have limited computational power and memory, making it difficult to implement complex encryption algorithms or intrusion detection mechanisms, such as end-to-end encryption, secure boot, or intrusion detection. As a result, these devices are vulnerable to attacks such as distributed denial-of-service (DoS/DDoS), man-in-the-middle attacks, data breaches, information gathering, malware attacks, or even physical tampering. The infamous Mirai Botnet attack exploited unsecured IoT devices to launch one of the largest DDoS attacks in history.

Potential solutions include Lightweight Encryption Protocols, Blockchain for IoT Security, IoT Security Frameworks, and Privacy-Preserving Techniques. Encryption protocols, such as Elliptic Curve Cryptography (ECC) and Lightweight AES, provide robust security without overloading resource-constrained devices. Blockchain and distributed ledger technologies (DLTs) are being explored as decentralized mechanisms for securing IoT networks, ensuring data integrity, secure device identification, and trust without a central authority. Solutions like IoT Security Foundation (IoTSF) and Trusted IoT Alliance are working to establish best practices for security for IoT device manufacturers and service providers. Techniques such as federated learning, which trains AI models on local devices while keeping data decentralized, can be applied to IoT environments to protect user privacy while still enabling data analytics.

Future research directions will undoubtedly include the study of secure and efficient protocols for device authentication and data encryption that work in low-power environments and privacy-preserving machine-learning techniques to analyze IoT data without compromising user privacy (e.g. federated learning).

Table 13.3 presents an overview of research contributions addressing data privacy and security in IoT systems, focusing on diverse technological solutions

Table 13.3 Summary of research on data privacy and security in IoT systems: technologies, outcomes, and challenges.

Paper	Data privacy and security	Technologies solution	Outcomes	Constraints	Challenges
11	Transport security in IoT	IP-based IoT transport security mechanisms	Enhanced IoT data transport protection	High implementation complexity	Ensuring universal applicability
12	Security, privacy, and trust in IoT	IoT trust and privacy frameworks	Improved trust and data security in IoT	Scalability of trust frameworks	Balancing privacy and performance
13	5G vulnerabilities and protections	5G-specific security protocols	Greater understanding of 5G security	Limited adoption of 5G standards	Securing dynamic 5G environments
51	Comprehensive IoT security overview	AI and IoT integration for security	Detailed security and privacy insights	Resource-intensive security measures	Maintaining security in diverse IoT apps
52	Blockchain-based IoT security	Blockchain for IoT threat mitigation	Increased IoT network resilience	Blockchain scalability issues	Adapting blockchain to IoT constraints
53	IoT privacy and countermeasures	Privacy mechanisms for IoT	Mitigated privacy risks in IoT	Complexity in IoT integration	Protecting sensitive data
54	Privacy-preserving trends in IoT	Privacy-preserving IoT architectures	Identification of trends and solutions	Cost and implementation challenges	Overcoming privacy-preservation limitations
55	Advanced security measures in IoT	Best practices for IoT security	Advanced security solutions for IoT	Adoption barriers in IoT ecosystems	Scalability of advanced IoT security
56	Security in Industry 4.0	Industry 4.0-specific security protocols	Enhanced Industry 4.0 security measures	Integration challenges with Industry 4.0	Meeting Industry 4.0 security needs
57	Cybersecurity dataset for IoT and IIoT	Federated learning for IoT security	Realistic data for IoT threat modeling	Limited real-world dataset availability	Real-world applicability of datasets
58	Industrial IoT security requirements	Fog computing for IoT security	Better industrial IoT security measures	Fog deployment costs and complexity	Managing fog and edge security
59	Blockchain for IoT security	Blockchain applications in IoT	Improved security in decentralized IoT	Blockchain energy consumption concerns	Addressing blockchain vulnerabilities

such as blockchain, federated learning, and privacy-preserving architectures. The studies highlight advancements like enhanced threat mitigation, improved data transport security, and increased resilience against vulnerabilities in IoT networks. Key outcomes include the development of trust frameworks, detailed security insights, and industry-specific protocols. Despite these achievements, the research reveals significant challenges, such as scalability of security measures, blockchain energy consumption, and the integration of fog and edge security solutions. These findings emphasize the need for ongoing research to create scalable, efficient, and robust privacy-preserving IoT systems.

13.5 Real-time Data Processing and Analytics

Real-time data processing refers to the capability of processing data streams and providing analysis, feedback, and decisions almost instantly as IoT devices generate the data.

IoT data's sheer volume and velocity in intelligent environments require frameworks and architectures that can handle real-time processing without causing significant delays. Latency issues in real-time analytics can lead to critical failures in applications like traffic control or emergency response systems [53, 59–61].

One of the most critical problems in IoT applications is Latency-Sensitive. Many IoT applications (e.g. autonomous vehicles, industrial automation, and smart grids) require real-time responses. Data transmission, processing, or decision-making delays can lead to catastrophic outcomes such as traffic accidents or machine failures. Another critical problem is Data Explosion. IoT systems generate massive amounts of data, often requiring real-time or near-real-time processing. The data must be filtered, processed, and acted upon in milliseconds. Still, centralized cloud architectures can introduce delays due to bandwidth limitations and the distance from data sources, which is problematic for critical applications like autonomous vehicles, smart grids, and real-time health monitoring. Centralized processing in cloud systems can introduce latency and bandwidth constraints, as all data must be sent to a remote data center for processing.

The potential solutions may involve Edge and Fog Computing, In-Memory Computing, and Stream Processing Frameworks. Processing data closer to the source, edge, and fog computing reduces latency and enables faster decision-making. For example, edge devices in a smart city might process traffic camera data locally to adjust traffic lights in real time without sending all data to the cloud. In-Memory Computing accelerates real-time analytics by storing and processing data in memory (rather than disks). Systems like Apache Kafka or Apache Flink allow real-time data streaming, processing, and analysis, helping IoT architectures handle high-velocity data without bottlenecks.

Future research directions will include developing distributed, scalable algorithms for real-time IoT data analytics and machine learning models that adapt to dynamic IoT data streams.

Table 13.4 summarizes recent research on real-time data processing and analytics in IoT systems, focusing on technologies such as privacy-preserving techniques, integrated platforms, and cloud-based frameworks. The outcomes showcase advancements in improving privacy during data analytics, enhancing data-driven decision-making, and optimizing IoT operations through big data tools. However, significant constraints remain, including the high computational cost of privacy-preserving methods, scalability issues in integrated platforms, and latency concerns in cloud-dependent frameworks. These challenges underscore the need for innovative solutions that balance performance, scalability, and privacy in real-time IoT analytics.

Table 13.4 Summary of research on real-time data processing and analytics in IoT systems: technologies, outcomes, and challenges.

Paper	RT data processing and analytics	Technologies solutions	Outcomes	Constraints	Challenges
54	Trends in privacy-preserving IoT data processing	Privacy-preserving techniques for IoT analytics	Improved privacy in IoT data analytics	High computational cost for privacy preservation	Balancing privacy and analytics performance
60	IoT-enabled evolution and analytics,	Integrated IoT platforms for data analytics,	Enhanced data-driven decision-making	Scalability limitations in integrated platforms	Integration across diverse IoT systems
61	Data processing with IoT and cloud	Cloud-based IoT frameworks	Optimized operations	Cloud dependency and latency issues	Ensuring real-time processing at scale
62	Big data management and real-time analytics	Big data frameworks for real-time analytics	Advanced tools for managing and analyzing data	Complexity in big data system implementations	Handling the volume and variety of big data

13.6 Edge Computing and Fog Computing Integration

Edge computing brings data processing closer to IoT devices to reduce latency and bandwidth usage. Fog computing extends cloud computing by providing services such as storage, computing, and networking between the edge and the cloud.

Integrating edge and fog computing into IoT architectures helps reduce latency and optimize resource consumption, but the design of these systems must balance the computational load between the cloud, fog, and edge nodes. This is incredibly challenging in resource-constrained environments [62–65].

Deciding what tasks should be processed at the edge, fog, or cloud layer is non-trivial. Balancing factors like computational load, latency, and energy consumption requires dynamic and intelligent algorithms. Processing data at the edge (i.e. near the devices generating data) helps to minimize latency, improve responsiveness, and reduce bandwidth consumption by sending only critical information to the cloud. For example, in healthcare, edge devices can monitor patient data locally and only send critical alerts to healthcare providers. Fog computing extends edge computing by adding a processing layer between the edge and the cloud. Fog nodes might process intermediate results and handle tasks that are too computationally expensive for edge devices but too time-sensitive for the cloud.

One of the most critical challenges IoT devices face is related to Distributed Processing Complexity. Efficient management of resources across edge, fog, and cloud layers is complex. Edge devices often have limited computational power and storage, while fog nodes serve as intermediaries and can be geographically distributed. Managing tasks across cloud, fog, and edge layers can be complex, especially when optimizing the placement of tasks and managing resource allocation dynamically. Additionally, security must be assured at multiple layers. Data must be protected as it traverses from edge to fog to cloud, increasing the complexity of implementing consistent security policies.

The potential solutions include task offloading, edge AI, and hierarchical architectures. In task offloading, adaptive algorithms that intelligently decide where to offload tasks (based on network conditions, processing requirements, and energy availability) are essential for optimizing edge-fog-cloud architectures. For example, machine learning models can be dynamically offloaded to the cloud during high-demand periods but run on edge devices during low-demand periods. Embedding AI capabilities in edge devices allows for localized, real-time decision-making. For example, AI-based object detection in smart cameras can operate entirely on the edge without sending video feeds to the cloud. Using hierarchical architectures for managing resources ensures that computational tasks are distributed optimally across different system layers (edge, fog, and cloud). For example, fog nodes can process medium-complexity tasks, while the cloud handles intensive analytics.

Another important feature of IoT networks related with interoperability and scalability is the capacity of elasticity, which means the capability to adjust and adapt resources flexibly based on demand. This functionality can be achieved through the inclusion of cloud technology, which depending on the strategy used (edge, fog, cloud) allows IoT ecosystems to efficiently manage resource allocation to accommodate constantly changing workloads, maintaining high performance and minimizing costs.

Future research directions will include optimizing fog-to-cloud and edge-to-cloud communication models for efficiency and security and developing fault-tolerant architectures for real-time task offloading and distributed processing.

Table 13.5 provides an overview of research on the integration of edge and fog computing in IoT systems, focusing on technologies such as energy-aware routing algorithms, secure cross-layer frameworks, and middleware for seamless IoT integration. The studies highlight outcomes like improved energy efficiency, optimized routing, and enhanced real-time processing capabilities. Despite these advancements, several constraints remain, including the complexity of implementing cross-layer models, high computational demands in heterogeneous networks, and challenges in scaling edge solutions across diverse IoT environments. Addressing issues like latency in multi-hop routing and ensuring compatibility with legacy systems will be critical for maximizing the potential of edge and fog computing in IoT ecosystems.

13.7 Context Awareness and Adaptability

Context awareness refers to the ability of IoT systems to sense and respond to the environment or user-specific parameters (e.g. location, preferences). Adaptability refers to the system's capacity to change behavior dynamically based on context.

In smart environments, IoT systems must understand and adapt to changing environmental factors and user needs in real time. This requires advanced algorithms for machine learning, context sensing, and dynamic decision-making [66–68].

Other essential features are automatic orchestration and collaboration. Since specific components include proprietary data, software, computing responsibilities, and hardware resources, it is necessary to integrate all these elements and facilitate coordinated communication between them [69].

Context Awareness is a major aspect. IoT systems in smart environments must sense and interpret contextual data – such as user location, device status, and environmental conditions – and use it to make decisions dynamically. For example, a smart thermostat might adjust heating based on occupancy or external weather conditions. Another critical facet is adaptability. IoT systems should adapt to changing environments or user preferences without requiring manual input. For example, smart lighting systems adapt to human presence or ambient light.

Table 13.5 Summary of research on edge and fog computing integration in IoT: technologies, outcomes, and challenges.

Paper	Edge and fog computing integration	Technologies solutions	Outcomes	Constraints	Challenges
43	Energy-efficient edge and fog solutions for IIoT	Resource optimization algorithms in edge/fog IIoT	Improved energy efficiency in industrial IoT	Limited scalability in some IIoT settings	Balancing energy efficiency and performance
44	Cross-layer framework for secure fog and edge IoT	Secure cross-layered fog/edge framework	Enhanced security and energy efficiency	Complexity in implementing cross-layer models	Ensuring scalability and security
45	Mathematical models for edge-enabled WSNs	Programming approaches for heterogeneous networks	Scalable and efficient WSN design	High computational demands in edge systems	Handling heterogeneity in IoT networks
46	Survey on multi-hop routing for fog/edge IoT	Protocols for resource-efficient multi-hop routing	Optimized routing and resource allocation	Latency concerns in multi-hop routing setups	Managing latency in fog-enabled networks
47	Energy-efficient IoT routing in fog environments	Algorithms for energy-aware routing in fog setups	Better energy utilization in IoT environments	Complexity in adapting to fog environments	Deploying algorithms in large IoT networks
63	Comprehensive architecture survey of fog computing	Architectural models for scalable fog computing	Comprehensive understanding of fog architectures	Integration challenges in legacy systems	Addressing interoperability in fog systems
64	Middleware and stack design for fog/edge IoT	Interface technologies for fog/edge computing	Streamlined IoT integration and middleware design	Cost-intensive middleware implementation	Ensuring middleware compatibility
65	Edge computing for Internet of Everything	Edge-driven IoT integration frameworks	Broader IoT ecosystem integration	Resource allocation in edge-heavy setups	Scaling edge solutions across IoT devices
66	Edge computing architectures for IoT applications	IoT-specific edge architecture models	Enhanced real-time processing capabilities	Design complexity for diverse IoT needs	Designing flexible architectures for IoT

The significant challenges include Accurate Context Sensing, Dynamic Adaptation, and Resource Constraints. Context-aware IoT systems must constantly gather and interpret data about their environment (e.g. user location, temperature, device status), which requires sophisticated sensors and processing algorithms. A key challenge is capturing context accurately and in real time without overwhelming the system. IoT systems must adapt to changing environments without human intervention. For example, a smart home system should adjust lighting or heating based on occupancy and external weather conditions. However, making intelligent decisions based on this context requires advanced AI and machine learning models. However, continuous context sensing and real-time adaptation consume significant resources (e.g. energy and processing power). Managing these constraints while providing intelligent, real-time responses is difficult, especially for resource-limited devices.

The potential solutions will undoubtedly include developments in Context-Aware Middleware, Adaptive Machine Learning Models, and Efficient Context Sensing. Middleware solutions that provide context abstraction and management (e.g. CA-RISMA and SOCAM) help IoT devices and applications understand and act on contextual data without requiring the devices themselves to handle complex processing. AI models that can adapt dynamically to changing contexts, such as reinforcement learning, can help IoT systems make better decisions based on their current environment. These models can also learn from user preferences over time. Additionally, using efficient context-sensing techniques (e.g. event-driven sensing) reduces the amount of data that needs to be continuously processed. For example, a smart system can switch between low-power and high-power modes based on detected events.

Future research directions will devise new techniques for real-time context sensing and dynamic decision-making in resource-constrained IoT environments and integrate AI and context-awareness to enable predictive services in smart environments.

Table 13.6 summarizes recent advancements in context-aware and adaptable IoT systems, focusing on technologies such as dynamic service migration, cognitive QoS management, and automated context-handling frameworks. The studies demonstrate outcomes like improved service continuity, sustainable communication, and enhanced decision-making in IoT applications. Advantages include better service reliability, energy efficiency, and reduced manual intervention. However, constraints such as latency in service migration, integration complexity, and scalability challenges remain significant barriers. Addressing these issues requires innovative approaches to balance computational costs, ensure adaptability at scale, and maintain performance under varying conditions in diverse IoT environments.

Table 13.6 Summary of research on context awareness and adaptability in IoT: technologies, outcomes, and challenges.

Paper	Context awareness and adaptability	Technologies solutions	Outcomes	Advantages	Constraints	Challenges
21	Dynamic service migration in wearable IoT	Edge-based dynamic service migration techniques	Improved service continuity and adaptability	Better service reliability	Latency in service migration processes	Ensuring real-time migration without disruption
22	Cognitive QoS management for networking	Cognitive algorithms for QoS-aware transmission	Enhanced video transmission quality	Optimized QoS for real-time video streaming	High computational requirements	Maintaining QoS under varying conditions
67	Surveys on context-aware intelligent IoT services	Context-aware service frameworks for IoT	Broad review of IoT context-aware services	Comprehensive understanding of context-aware IoT	Diverse implementation challenges	Addressing heterogeneity in IoT environments
68	Context-aware messaging for sustainable IoT	Messaging protocols for adaptive IoT communication	Sustainable communication in IoT networks	Energy efficiency and improved communication	Integration complexity in IoT networks	Ensuring sustainability of messaging frameworks
69	Context-aware computing in IoT environments	Computing models for context-aware IoT	Improved decision-making in IoT applications	Enhanced user-centric IoT experiences	Scalability in large-scale IoT setups	Balancing computational costs and performance
70	Automation framework for context-aware IoT systems	Automated context handling frameworks	Increased automation and adaptability in IoT	Reduction in manual intervention for IoT	Limited automation standards in IoT	Implementing context-aware frameworks at scale

13.8 Standardization

Standardization in IoT refers to establishing standard protocols, formats, and interfaces that allow diverse IoT devices and systems to communicate and work together without proprietary restrictions.

The lack of widely adopted standards in IoT creates fragmentation, making integrating different systems and technologies in smart environments difficult. International efforts, such as developing standard communication protocols and certification frameworks, are ongoing, but challenges remain [70–73].

A lack of universally adopted standards in IoT results in fragmented ecosystems where different devices and platforms cannot easily communicate with each other. This creates "vendor lock-in," making it hard to scale IoT deployments. Without standard communication, data exchange, and security protocols, it is hard to build large-scale, interoperable IoT systems, particularly across different industries and domains (e.g. smart homes, smart cities, industrial IoT).

As such, one of the most important challenges is the existence of multiple standards. Organizations such as the IEEE, IETF, and ISO are working on IoT standards, but competing standards from different alliances (e.g. Thread, ZigBee, Z-Wave) create confusion and slow down adoption.

The potential solutions will certainly rely on developing common standards and frameworks such as oneM2M, Open Connectivity Foundation (OPC-UA), or OMA-LWM2M that provide consistent communication and data-sharing protocols. Platforms like FIWARE, EdgeX Foundry, and OpenIoT promote open, standardized IoT ecosystems. Developers can build inherently compatible applications with multiple devices and services by adopting these platforms. Government-backed initiatives, such as the European IoT Large-Scale Pilots (IoT-LSP) or NIST Cybersecurity Framework, aim to promote common standards across industries and borders.

Future research directions shall develop efforts towards global standardization, with collaborations between government bodies, industry, and research communities and protocols that seamlessly integrate various IoT ecosystems while allowing future expansion.

Table 13.7 provides an overview of research addressing standardization challenges in IoT, covering areas such as cloud integration mechanisms, IoT certification frameworks, and interoperability standards for industrial IoT and smart cities. The outcomes demonstrate advancements like improved integration, scalable data-sharing practices, and enhanced IoT security through standard protocols. Key advantages include structured certification processes, better interoperability, and clarity in IoT system definitions. However, the research also highlights constraints such as the complexity of integrating cloud and IoT systems, incompatibilities in existing protocols, and scalability challenges in urban IoT. Achieving seamless standardization requires innovative solutions that harmonize global frameworks, address gaps in implementation, and ensure adaptability to the evolving IoT landscape.

Table 13.7 Summary of research on standardization in IoT: technologies, outcomes, and challenges.

Paper	Standardization	Technologies solutions	Outcomes	Constraints	Challenges
7	Integration of cloud and IoT standards	Cloud and IoT integration mechanisms	Improved integration of cloud and IoT systems	Complexity in integrating cloud and IoT	Achieving seamless standard integration
8	Security protocol standardization for IoT	Existing IoT security protocols	Enhanced IoT security through standard protocols	Incompatibilities in existing protocols	Ensuring protocol compatibility in IoT
23	Standards-based scalable IoT data sharing	Standardized data-sharing ecosystems	Efficient and scalable data-sharing practices	Scalability limitations in data-sharing	Balancing scalability and standardization
24	IoT certification frameworks	Template-based IoT certification schemes	Structured IoT certification landscapes	Limited coverage of certification schemes	Harmonizing EU standards with global needs
25	Challenges in IoT security certifications	Approaches for security certifications	Identification of security gaps and solutions	Fragmentation in certification practices	Developing unified certification approaches
26	Standard frameworks for IoT definitions	Definitional frameworks for IoT systems	Improved understanding of IoT structures	Ambiguities in IoT standard definitions	Maintaining relevance with evolving IoT tech
27	Broad IoT standardization definitions	Conceptual IoT models for standards	Greater clarity in IoT standard definitions	Lack of universal IoT standards	Addressing interoperability across devices
71	Interoperability standards for IIoT	Taxonomies and frameworks for IIoT	Enhanced interoperability in industrial IoT	Adapting standards to diverse IIoT needs	Creating inclusive IIoT standard frameworks
72	IETF standardization efforts in IoT	Protocols developed by IETF for IoT	Increased adoption of IoT standards	Gaps in implementation of IETF standards	Overcoming barriers in IoT standard adoption
73	IoT standard practices	IoT practices tailored to smart cities	Better IoT practices for setups	Resource constraints in smart city IoT	Aligning standards with IoT demands
74	IoT applications in smart city standards	Smart city IoT and service frameworks	Comprehensive view of IoT city services	Scalability challenges in urban IoT	Implementing standards in large urban areas

13.9 Network Management and Optimization

Network management in IoT involves controlling, coordinating, and optimizing the communication infrastructure that connects IoT devices. Optimization refers to techniques that reduce latency, minimize energy usage, and ensure reliability.

IoT networks in smart environments, such as smart cities, often consist of a massive number of devices transmitting data over wireless networks. Managing and optimizing these networks to ensure performance and reliability is a significant challenge, particularly as the number of devices grows [74–77].

Smart environments often involve massive, dynamic networks of IoT devices communicating over various wireless technologies. Managing these networks – optimizing for latency, throughput, and energy – requires intelligent coordination. IoT networks must guarantee specific QoS parameters (e.g. reliability, delay, and bandwidth) to ensure the smooth operation of time-sensitive applications (e.g. real-time monitoring in healthcare or smart grids).

IoT environments involve different network technologies (e.g. cellular, Wi-Fi, Bluetooth) operating together. Managing these heterogeneous systems while ensuring consistent performance remains a challenge. Additionally, in smart cities, for example, traffic from emergency services should be prioritized over regular traffic, but implementing such QoS guarantees is challenging. IoT networks are often dynamic, with devices constantly entering or leaving the network (e.g. mobile devices in smart cities). Maintaining a robust, resilient, and self-healing network in such dynamic environments is challenging.

The potential solutions include Software-Defined Networking (SDN), Network Function Virtualization (NFV), and Machine Learning-Based Optimization. SDN can dynamically manage IoT traffic, optimize routing, and prioritize critical data. NFV allows virtualized network functions to be created, reducing reliance on physical network infrastructure and enabling better resource allocation and scaling. It can help deploy services on demand based on real-time traffic patterns. AI-based algorithms can optimize network resources by learning patterns in network usage and traffic demands.

Within the realm of network security, SDN and NFV are increasingly influential. They serve as essential components for the on-demand delivery of security solutions, in line with the Security-as-a-Service (SECaaS) model [78].

Future research directions will certainly persist in developing intelligent, self-optimizing networks that autonomously manage IoT devices and dynamically allocate resources and in researching new communication protocols designed for large-scale, heterogeneous IoT environments.

Table 13.8 summarizes the research on network management and optimization in IoT, highlighting key technologies such as IoT certification frameworks, model-driven architectures, deep learning for QoS enhancement, and SDN/NFV

Table 13.8 Summary of research on network management and optimization in IoT: technologies, outcomes, and challenges.

Paper	Network management and optimization	Technologies solutions	Outcomes	Constraints	Challenges
25	Security-focused network optimization in IoT	IoT certification and security frameworks	Improved security and network reliability	Fragmentation in security certification approaches	Developing unified security certifications
26	Foundational concepts for IoT network optimization	IoT network protocols and standardizations	Enhanced understanding of IoT network dynamics	Lack of uniformity in IoT standards	Maintaining relevance with evolving IoT tech
40	Interoperability-driven network architecture for IIoT	Model-driven architectures for seamless IoT networks	Greater interoperability in industrial IoT systems	Complexity in implementing model-driven systems	Ensuring interoperability across varied devices
41	Distributed network architectures	Scalable and interoperable IoT architectures	Efficient management of urban IoT networks	Scalability issues in large scale IoT deployments	Balancing scalability and efficiency in networks
1	Automotive IoT network management strategies	Integration of smart materials and IoT networks	Streamlined IoT networks in automotive applications	High costs of automotive IoT implementation	Integrating IoT seamlessly into vehicles
75	Quality of Service (QoS) improvements in IoT	QoS protocols for real-time IoT applications	Enhanced QoS in IoT applications	Latency concerns in real-time QoS management	Addressing latency and reliability in QoS protocols

(Continued)

Table 13.8 (Continued)

Paper	Network management and optimization	Technologies solutions	Outcomes	Constraints	Challenges
76	Deep learning applications for IoT QoS enhancement	Deep learning models for IoT QoS optimization	Optimized QoS using advanced machine learning	Computational overhead in applying deep learning	Efficient deployment of deep learning models
77	Comprehensive network optimization in IoT	Optimization algorithms for IoT networks	More efficient IoT network designs	Challenges in handling diverse IoT data streams	Adapting optimization techniques to real-world use
78	Reviews on IoT network optimization techniques	Network resource management techniques	Practical insights into IoT resource management	Resource allocation inefficiencies in IoT networks	Ensuring consistent network performance in IoT
79	SDN and NFV mechanisms for secure IoT networks	SDN and NFV frameworks for IoT security	Strengthened IoT network security via SDN/NFV	Implementation challenges in SDN and NFV	Overcoming adoption barriers for SDN/NFV

mechanisms for secure networks. The outcomes emphasize improved network reliability, enhanced understanding of IoT network dynamics, and optimized QoS for real-time applications. These advancements offer greater interoperability, efficient resource management, and improved security in IoT systems. However, challenges persist, such as fragmentation in security certification approaches, scalability issues in large IoT deployments, and the complexity of integrating deep learning models in real-time systems. Overcoming these constraints is essential for ensuring the consistent performance and seamless operation of IoT networks at scale.

13.10 Conclusion

The IoT is reshaping smart environments by allowing connected devices to autonomously communicate and perform various tasks. IoT architecture in these environments integrates sensors, networks, and platforms to optimize operations in industries such as healthcare, smart cities, and manufacturing. However, due to its fragmented nature, IoT faces major hurdles, including scalability, security, privacy, and interoperability. Tackling these issues is essential for creating efficient and secure IoT systems and driving ongoing research to develop innovations and standards.

This review highlights challenges such as managing expanding networks, securing heterogeneous devices, ensuring interoperability, and safeguarding sensitive data. Potential solutions include using machine learning for scalability, blockchain for security, and global frameworks for standardization. Future research focuses on advancing AI-driven automation, building stronger security frameworks, and improving open-source platforms to foster interoperability and privacy.

Addressing these challenges is crucial as IoT systems continue to grow in complexity and become embedded in critical real-world applications. The evolving nature of IoT systems requires continued research, new technologies, and standardized solutions to realize their full potential in smart environments. These ongoing efforts reflect the intricate task of building scalable, secure, and adaptable IoT architectures.

Author Contributions

All authors equally contributed to this research work.

Funding

This research received no external funding.

Conflicts of Interest

The authors declare no conflicts of interest.

References

1 Ammar, M., Haleem, A., Javaid, M., et al. (2022). Significant applications of smart materials and Internet of Things (IoT) in the automotive industry. *Materials Today: Proceedings* 68: 1542–1549. https://doi.org/10.1016/j.matpr.2022.07.180.

2 Ushakov, D., Dudukalov, E., Kozlova, E., and Shatila, K. (2022). The internet of things impact on smart public transportation. *Transportation Research Procedia* 63: 2392–2400. https://doi.org/10.1016/j.trpro.2022.06.275.

3 Sinha, B.B., and Dhanalakshmi, R. (2022). Recent advancements and challenges of Internet of Things in smart agriculture: a survey. *Future Generation Computer Systems* 126: 169–184. https://doi.org/10.1016/j.future.2021.08.006.

4 Kashani, M.H., Madanipour, M., Nikravan, M., et al. (2021). A systematic review of IoT in healthcare: applications, techniques, and trends. *Journal of Network and Computer Applications* 192: 103164. https://doi.org/10.1016/j.jnca.2021.103164.

5 Kalsoom, T., Ahmed, S., Rafi-ul-Shan, P.M., et al. (2021). Impact of IoT on manufacturing industry 4.0: a new triangular systematic review. *Sustainability* 13: 12506. https://doi.org/10.3390/su132212506.

6 Ghazal, T.M., Hasan, M.K., Alshurideh, M.T., et al. (2021). IoT for smart cities: machine learning approaches in smart healthcare—a review. *Future Internet* 13: 218. https://doi.org/10.3390/fi13080218.

7 Botta, A., de Donato, W., Persico, V., and Pescapé, A. (2016). Integration of cloud computing and internet of things: a survey. *Future Generation Computer Systems* 56: 684–700. https://doi.org/10.1016/j.future.2015.09.021.

8 Granjal, J., Monteiro, E., and Sa Silva, J. (2015). Security for the internet of things: a survey of existing protocols and open research issues. *IEEE Communications Surveys and Tutorials* 17: 1294–1312. https://doi.org/10.1109/COMST.2015.2388550.

9 Perera, C., Talagala, D.S., Liu, C.H., and Estrella, J.C. (2015). Energy-efficient location and activity-aware on-demand mobile distributed sensing platform for sensing as a service in IoT clouds. *IEEE Transactions on Computational Social Systems* 2: 171–181. https://doi.org/10.1109/TCSS.2016.2515844.

10 Domínguez-Bolaño, T., Barral, V., Escudero, C.J., and García-Naya, J.A. (2024). An IoT system for a smart campus: challenges and solutions illustrated over several real-world use cases. *Internet of Things* 25: 101099. https://doi.org/10.1016/j.iot.2024.101099.

11 Garcia-Macias, J.A. (2023). Transport in the IP-based internet of things: status report. *Procedia Computer Science* 224: 18–25. https://doi.org/10.1016/j.procs.2023.09.006.

12 Sicari, S., Rizzardi, A., Grieco, L.A., and Coen-Porisini, A. (2015). Security, privacy and trust in internet of things: the road ahead. *Computer Networks* 76; 146–164. https://doi.org/10.1016/j.comnet.2014.11.008.

13 Sousa, A., and Reis, M.J.C.S. (2024). 5G Security features, vulnerabilities, threats, and data protection in IoT and mobile devices: a systematic review. *Evolutionary Studies in Imaginative Culture* 8.2: 414–427. https://doi.org/10.70082/esiculture.vi.1054.

14 Oladimeji, D., Gupta, K., Kose, N.A., et al. (2023). Smart transportation: an overview of technologies and applications. *Sensors* 23: 3880. https://doi.org/10.3390/s23083880.

15 Hussain, A., Ali, T., Althobiani, F., et al. (2021). Security framework for IoT based real-time health applications. *Electronics* 10: 719. https://doi.org/10.3390/electronics10060719.

16 Yeri, V., and Shubhangi, D.C. (2020). IoT based real time health monitoring. In: *Proceedings of the 2020 Second International Conference on Inventive Research in Computing Applications (ICIRCA)* (pp. 980–984). Coimbatore India: IEEE.

17 Gubbi, J., Buyya, R., Marusic, S., and Palaniswami, M. (2013). Internet of Things (IoT): a vision, architectural elements, and future directions. *Future Generation Computer Systems* 29: 1645–1660. https://doi.org/10.1016/j.future.2013.01.010.

18 Chiang, M., and Zhang, T. (2016). Fog and IoT: an overview of research opportunities. *IEEE Internet of Things Journal* 3: 854–864. https://doi.org/10.1109/JIOT.2016.2584538.

19 Yi, S., Li, C., and Li, Q. (2015). A survey of fog computing: concepts, applications and issues. In: *Proceedings of the Proceedings of the 2015 Workshop on Mobile Big Data* (pp. 37–42). Hangzhou China: ACM.

20 Baker, S.B., Xiang, W., and Atkinson, I. (2017). Internet of things for smart healthcare: technologies, challenges, and opportunities. *IEEE Access* 5: 26521–26544. https://doi.org/10.1109/ACCESS.2017.2775180.

21 Raza, I., Hussain, S.A., and Latif, S. (2024). EdgeMob: a context-aware dynamic service migration for wearable IoT in healthcare. In: *Proceedings of the 2024 International Conference on Computing, Internet of Things and Microwave Systems (ICCIMS)* (pp. 1–6). Gatineau, QC, Canada: IEEE.

22 Cheng, B., Wang, M., Lin, X., and Chen, J. (2021). Context-aware cognitive QoS management for networking video transmission. *IEEE/ACM Transactions on Networking* 29: 1422–1434. https://doi.org/10.1109/TNET.2021.3066262.

23 Figueredo, K., Seed, D., and Wang, C. (2022). A scalable, standards-based approach for IoT data sharing and ecosystem monetization. *IEEE Internet of Things Journal* 9: 5645–5652. https://doi.org/10.1109/JIOT.2020.3023035.

24 Khurshid, A., Alsaaidi, R., Aslam, M., and Raza, S. (2022). EU cybersecurity act and IoT certification: landscape, perspective and a proposed template scheme. *IEEE Access* 10: 129932–129948. https://doi.org/10.1109/ACCESS.2022.3225973.

25 Cirne, A., Sousa, P.R., Resende, J.S., and Antunes, L. (2022). IoT security certifications: challenges and potential approaches. *Computers & Security* 116: 102669. https://doi.org/10.1016/j.cose.2022.102669.

26 Atzori, L., Iera, A., and Morabito, G. (2010). The internet of things: a survey. *Computer Networks* 54: 2787–2805. https://doi.org/10.1016/j.comnet.2010.05.010.

27 Minerva, R., Biru, A., and Rotondi, D. (2015). Towards a definition of the Internet of Things (IoT). *IEEE Internet Initiative* 1: 1–86.

28 Kumar, P., Kumar, R., Gupta, G.P., et al. (2022). P2TIF: a blockchain and deep learning framework for privacy-preserved threat intelligence in industrial IoT. *IEEE Transactions on Industrial Informatics* 18: 6358–6367. https://doi.org/10.1109/TII.2022.3142030.

29 Alotaibi, N.S., Sayed Ahmed, H.I., Kamel, S.O.M., and ElKabbany, G.F. (2024). Secure enhancement for MQTT protocol using distributed machine learning framework. *Sensors* 24: 1638. https://doi.org/10.3390/s24051638.

30 Sivrikaya, F., Ben-Sassi, N., Dang, X.-T., et al. (2019). Internet of smart city objects: a distributed framework for service discovery and composition. *IEEE Access* 7: 14434–14454. https://doi.org/10.1109/ACCESS.2019.2893340.

31 Ottolini, D., Zyrianoff, I., and Kamienski, C. (2022). Interoperability and scalability trade-offs in open IoT platforms. In: *Proceedings of the 2022 IEEE 19th Annual Consumer Communications & Networking Conference (CCNC)* (pp. 1–6). Las Vegas, NV, USA: IEEE.

32 Boeckl, K., Fagan, M., Fisher, W., et al. (2019). *Considerations for Managing Internet of Things (IoT) Cybersecurity and Privacy Risks*. Gaithersburg, MD: National Institute of Standards and Technology, NIST IR 8228.

33 Rath, C.K., Mandal, A.K., and Sarkar, A. (2023). Microservice based scalable IoT architecture for device interoperability. *Computer Standards & Interfaces* 84: 103697. https://doi.org/10.1016/j.csi.2022.103697.

34 Alonso, L., Barbarán, J., Chen, J., et al. (2018). Middleware and communication technologies for structural health monitoring of critical infrastructures: a survey. *Computer Standards & Interfaces* 56: 83–100. https://doi.org/10.1016/j.csi.2017.09.007.

35 Javed, A., Malhi, A., Kinnunen, T., and Framling, K. (2020). Scalable IoT platform for heterogeneous devices in smart environments. *IEEE Access* 8: 211973–211985. https://doi.org/10.1109/ACCESS.2020.3039368.

36 Li, Z., Shang, C., Wu, J., and Li, Y. (2022). Microservice extraction based on knowledge graph from monolithic applications. *Information and Software Technology* 150: 106992. https://doi.org/10.1016/j.infsof.2022.106992.

37 Madhikermi, M., Yousefnezhad, N., and Framling, K. (2018). Data exchange standard for industrial internet of things. In: *Proceedings of the 2018 3rd International Conference on System Reliability and Safety (ICSRS)* (pp. 53–61). Barcelona, Spain: IEEE.

38 Resende, C., Moreira, W., and Almeida, L. (2022). NextGenGW: a software-based architecture targeting IoT interoperability. In: *Proceedings of the 2022 IEEE 27th International Conference on Emerging Technologies and Factory Automation (ETFA)* (pp. 1–4). Stuttgart, Germany: IEEE.

39 Noura, M., Atiquzzaman, M., and Gaedke, M. (2019). Interoperability in internet of things: taxonomies and open challenges. *Mobile Networks and Applications* 24: 796–809. https://doi.org/10.1007/s11036-018-1089-9.

40 Amjad, A., Azam, F., and Anwar, M.W. (2022). Device interoperability for industrial IoT using model-driven architecture. In: *Proceedings of the 2022 24th International Multitopic Conference (INMIC)* (pp. 1–6). Islamabad, Pakistan: IEEE.

41 Pranavasri, V., Francis, L., Mogadali, U., et al. (2023). Scalable and interoperable distributed architecture for IoT in smart cities. In: *Proceedings of the 2023 IEEE 9th World Forum on Internet of Things (WF-IoT)* (pp. 1–6). Aveiro, Portugal: IEEE.

42 Mao, W., Zhao, Z., Chang, Z., et al. (2021). Energy-efficient industrial internet of things: overview and open issues. *IEEE Transactions on Industrial Informatics* 17: 7225–7237. https://doi.org/10.1109/TII.2021.3067026.

43 Mustafa, R., Sarkar, N.I., Mohaaghegh, M., and Pervez, S. (2024). A secure and energy-efficient cross-layer framework for internet of things. In: *Proceedings of the 2024 International Conference on Information Networking (ICOIN)* (pp. 661–666). Ho Chi Minh City Vietnam: IEEE.

44 Taparci, E., Olcay, K., Akmandor, M.O., et al. (2024). A mathematical programming approach for IoT-enabled, energy-efficient heterogeneous wireless sensor network design and implementation. *Sensors* 24: 1457. https://doi.org/10.3390/s24051457.

45 Awais, M., Mateen, A., Khan, Z.A., and Javaid, N. (2018). A survey on multi-hop routing protocols for efficient resource allocation in IoTs. In: *Proceedings of the 2018 Fifth HCT Information Technology Trends (ITT)* (pp. 239–244). Dubai, United Arab Emirates: IEEE.

46 Gamal, M. (2023). Energy efficiency routing algorithms in IoT: a survey. In: *Proceedings of the 8th International Conference on Advanced Intelligent Systems and Informatics 2022* (Vol. 152, pp. 668–679) (eds. A.E. Hassanien, V. Snášel, M. Tang et al.). *Lecture Notes on Data Engineering and Communications Technologies*. Cham: Springer International Publishing.

47 Ali, S., Nadeem, M., Ahmed, S., and Tahir, M. (2024). IoT-based framework for optimizing energy efficiency and reliability in acoustic sensor networks using mobile sinks. *Scientific Reports* 14: 24122. https://doi.org/10.1038/s41598-024-74664-2.

48 Kaur, L., and Kaur, R. (2021). A survey on energy efficient routing techniques in WSNs focusing IoT applications and enhancing fog computing paradigm. *Global Transitions Proceedings* 2: 520–529. https://doi.org/10.1016/j.gltp.2021.08.001.

49 Udayaprasad, P.K., Shreyas, J., Srinidhi, N.N., et al. (2024). Energy efficient optimized routing technique with distributed SDN-AI to large scale I-IoT networks. *IEEE Access* 12: 2742–2759. https://doi.org/10.1109/ACCESS.2023.3346679.

50 Aouedi, O., Vu, T.-H., Sacco, A., et al. (2024). A survey on intelligent internet of things: applications, security, privacy, and future directions. *IEEE Communications Surveys and Tutorials* 1. https://doi.org/10.1109/COMST.2024.3430368.

51 Khan, M.A., and Salah, K. (2018). IoT security: review, blockchain solutions, and open challenges. *Future Generation Computer Systems* 82: 395–411. https://doi.org/10.1016/j.future.2017.11.022.

52 Elhoseny, M., Thilakarathne, N.N., Alghamdi, M.I., et al. (2021). Security and privacy issues in medical internet of things: overview, countermeasures, challenges and future directions. *Sustainability* 13: 11645. https://doi.org/10.3390/su132111645.

53 Safaei Yaraziz, M., Jalili, A., Gheisari, M., and Liu, Y. (2023). Recent trends towards privacy-preservation in internet of things, its challenges and future directions. *IET Circuits, Devices & Systems* 17: 53–61. https://doi.org/10.1049/cds2.12138.

54 USENIX Association (ed.). (2017). *Proceedings of the 26th USENIX Security Symposium: August 16-18, 2017, Vancouver, BC, Canada*. Berkeley, CA: USENIX Association.

55 Jamai, I., Ben Azzouz, L., and Saidane, L.A. (2020). Security issues in Industry 4.0. In: *Proceedings of the 2020 International Wireless Communications and Mobile Computing (IWCMC)* (pp. 481–488). Limassol Cyprus: IEEE.

56 Ferrag, M.A., Friha, O., Hamouda, D., et al. (2022). Edge-IIoTset: a new comprehensive realistic cyber security dataset of IoT and IIoT applications for centralized and federated learning. *IEEE Access* 10: 40281–40306. https://doi.org/10.1109/ACCESS.2022.3165809.

57 Tange, K., De Donno, M., Fafoutis, X., and Dragoni, N. (2020). A systematic survey of industrial internet of things security: requirements and fog computing opportunities. *IEEE Communications Surveys and Tutorials* 22: 2489–2520. https://doi.org/10.1109/COMST.2020.3011208.

58 Wang, Q., Zhu, X., Ni, Y., et al. (2020). Blockchain for the IoT and industrial IoT: a review. *Internet of Things* 10: 100081. https://doi.org/10.1016/j.iot.2019.100081.

59 Bauer, M., Sanchez, L., and Song, J. (2021). IoT-enabled smart cities: evolution and outlook. *Sensors* 21: 4511. https://doi.org/10.3390/s21134511.

60 Su, P., Chen, Y., and Lu, M. (2022). Smart city information processing under internet of things and cloud computing. *Journal of Supercomputing* 78: 3676–3695. https://doi.org/10.1007/s11227-021-03972-5.

61 Li, K.-C., Jiang, H., and Zomaya, A.Y. (eds.) (2017). *Big Data Management and Processing*, 1e. Chapman and Hall/CRC.

62 Habibi, P., Farhoudi, M., Kazemian, S., et al. (2020). Fog computing: a comprehensive architectural survey. *IEEE Access* 8: 69105–69133. https://doi.org/10.1109/ACCESS.2020.2983253.

63 Ali, O., Ishak, M.K., Bhatti, M.K.L., et al. (2022). A comprehensive review of internet of things: technology stack, middlewares, and fog/edge computing interface. *Sensors* 22: 995. https://doi.org/10.3390/s22030995.

64 Kong, X., Wu, Y., Wang, H., and Xia, F. (2022). Edge computing for internet of everything: a survey. *IEEE Internet of Things Journal* 9: 23472–23485. https://doi.org/10.1109/JIOT.2022.3200431.

65 Hamdan, S., Ayyash, M., and Almajali, S. (2020). Edge-computing architectures for internet of things applications: a survey. *Sensors* 20: 6441. https://doi.org/10.3390/s20226441.

66 Gil, D., Ferrández, A., Mora-Mora, H., and Peral, J. (2016). Internet of things: a review of surveys based on context aware intelligent services. *Sensors* 16: 1069. https://doi.org/10.3390/s16071069.

67 Almagrabi, A.O., and Al-Otaibi, Y.D. (2020). A survey of context-aware messaging-addressing for sustainable internet of things (IoT). *Sustainability* 12: 4105. https://doi.org/10.3390/su12104105.

68 Perera, C., Zaslavsky, A., Christen, P., and Georgakopoulos, D. (2014). Context aware computing for the internet of things: a survey. *IEEE Communications Surveys and Tutorials* 16: 414–454. https://doi.org/10.1109/SURV.2013.042313.00197.

69 Chegini, H., and Mahanti, A. (2019). A framework of automation on context-aware Internet of Things (IoT) systems. In: *Proceedings of the Proceedings of the 12th IEEE/ACM International Conference on Utility and Cloud Computing Companion* (pp. 157–162). Auckland New Zealand: ACM.

70 Hazra, A., Adhikari, M., Amgoth, T., and Srirama, S.N. (2023). A comprehensive survey on interoperability for IIoT: taxonomy, standards, and future directions. *ACM Computing Surveys* 55: 1–35. https://doi.org/10.1145/3485130.

71 Ishaq, I., Carels, D., Teklemariam, G., et al. (2013). IETF standardization in the field of the Internet of Things (IoT): a survey. *Journal of Sensor and Actuator Networks* 2: 235–287. https://doi.org/10.3390/jsan2020235.

72 Syed, A.S., Sierra-Sosa, D., Kumar, A., and Elmaghraby, A. (2021). IoT in smart cities: a survey of technologies, practices and challenges. *Smart Cities* 4: 429–475. https://doi.org/10.3390/smartcities4020024.

73 Rafiq, I., Mahmood, A., Razzaq, S., et al. (2023). IoT applications and challenges in smart cities and services. *The Journal of Engineering* 2023: e12262. https://doi.org/10.1049/tje2.12262.

74 Singh, M., and Baranwal, G. (2018). Quality of Service (QoS) in internet of things. In: *Proceedings of the 2018 3rd International Conference on Internet of Things: Smart Innovation and Usages (IoT-SIU)* (pp. 1–6). Bhimtal: IEEE.

75 Kimbugwe, N., Pei, T., and Kyebambe, M. (2021). Application of deep learning for quality of service enhancement in internet of things: a review. *Energies* 14: 6384. https://doi.org/10.3390/en14196384.

76 Srinidhi, N.N., Kumar, S.M.D., and Venugopal, K.R. (2019). Network optimizations in the internet of things: a review. *Engineering Science and Technology, an International Journal* 22: 1–21. https://doi.org/10.1016/j.jestch.2018.09.003.

77 Srivastava, A., and Kumar, A. (2022). A review of network optimization on the internet of things. In: *Innovations in Computer Science and Engineering* (Vol. 385, pp. 49–57) (eds. H.S. Saini, R. Sayal, A. Govardhan et al.). *Lecture Notes in Networks and Systems*. Singapore: Springer Singapore.

78 Farris, I., Taleb, T., Khettab, Y., and Song, J. (2019). A survey on emerging SDN and NFV security mechanisms for IoT systems. *IEEE Communications Surveys and Tutorials* 21: 812–837. https://doi.org/10.1109/COMST.2018.2862350.

14

Future Trends and Innovations in IoT-based Communication Through AI/ML for SDG

Shriram K. Vasudevan[1,*], *Sini Raj Pulari*[2], *and Subashri V.*[3]

[1] *Software Ecosystem Enablement, Office of the CTO (OCTO), Intel India Pvt. Ltd., Bengaluru, Karnataka, India*
[2] *BICT-Database, EDICT, Government Bahrain Technical University, The Kingdom of Bahrain, ISA Town*
[3] *Bank of America Continuum India Pvt. Ltd., Chennai, Tamil Nadu, India*

*Corresponding author: shriram.kris.vasudevan@intel.com

14.1 Introduction

The convergence of physical and digital realms through the Internet of Things (IoT) has revolutionized various industries, enabling unparalleled levels of automation, connectivity, and intelligence. The incorporation of artificial intelligence (AI) and machine learning (ML) has transformed IoT devices from simple data collectors into autonomous entities capable of pattern recognition, outcome prediction, and real-time process optimization. This evolution has led to improved operational efficiency, cost reduction, and enhanced decision-making across sectors such as healthcare, manufacturing, agriculture, and urban development [1]. AI-enhanced IoT communication facilitates seamless device interoperability, enabling adaptive control systems that respond dynamically to changing environmental and operational conditions, resulting in highly efficient and sustainable ecosystems.

The implementation of AI-driven IoT communication plays a vital role in advancing Sustainable Development Goals (SDGs) by promoting responsible resource management, reducing energy consumption, and enabling intelligent automation to minimize environmental impact. Smart grid systems utilize

AI and ML Techniques in IoT-based Communication: A Path to Sustainable Development Goals,
First Edition. Edited by Sumita Mishra, Nishu Gupta and Polat Goktas.
 Companion Website: http://www.wiley.com/go/mishra_aiml

AI-powered IoT sensors to track energy usage patterns, anticipate demand fluctuations, and optimize power distribution, contributing to SDG 7 (Affordable and Clean Energy) [2]. In agriculture, AI-integrated IoT solutions enhance precision farming by analyzing soil conditions, weather patterns, and crop health in real time, thereby minimizing water and fertilizer waste while maximizing yield, aligning with SDG 2 (Zero Hunger) and SDG 13 (Climate Action) [3]. Additionally, AI-driven predictive maintenance in industrial IoT (IIoT) prevents equipment failures by analyzing sensor-generated data to identify potential malfunctions, thus enhancing operational efficiency and reducing downtime, supporting SDG 9 (Industry, Innovation, and Infrastructure) [4].

In addition to its immediate impact on efficiency and sustainability, AI/ML-powered IoT communication is driving innovative advancements in next-generation network architectures and communication protocols. The integration of AI with 5G and 6G technologies is transforming wireless communication by dynamically optimizing bandwidth allocation, decreasing latency, and enhancing overall network performance [5]. AI-enabled edge computing allows for real-time data processing at the device level, reducing reliance on cloud infrastructure and improving security and privacy in IoT networks. Moreover, federated learning (FL) techniques ensure privacy-preserving AI model training across distributed IoT networks, mitigating data breaches and addressing ethical concerns related to centralized data storage [6]. As IoT systems continue to evolve, the synergy between AI, ML, and advanced communication technologies will redefine digital transformation across industries, paving the way for intelligent, resilient, and sustainable ecosystems that align with global sustainability objectives. This section examines these AI/ML-powered IoT advancements, their significant impact on communication networks, and the future trajectory of intelligent IoT systems.

14.2 Role of AI/ML in IoT-based Communication

IoT communication is being revolutionized by AI and ML through enhanced data processing, decision-making, and network efficiency. Several key advancements are transforming IoT-enabled communication systems:

14.2.1 Predictive Analytics and Anomaly Detection

IoT systems leverage AI-powered predictive analytics to forecast failures, enhance operational performance, and minimize unplanned downtime. By continuously monitoring sensor data, these systems detect patterns and anomalies that indicate potential failures before they occur. In industries where reliability is crucial, such as manufacturing and energy, predictive analytics plays a vital role in ensuring seamless operations. AI models trained on historical and real-time data analyze fluctuations in machine parameters like temperature, pressure, and vibration to identify early warning signs of degradation. This proactive approach prevents costly breakdowns and optimizes asset utilization, resulting in improved efficiency and extended equipment lifespan [3].

In the IIoT, predictive maintenance is a key application of AI-driven analytics. Traditional maintenance strategies, such as reactive or scheduled maintenance, often lead to either unexpected equipment failures or unnecessary servicing costs. Predictive maintenance, however, overcomes these inefficiencies by using ML algorithms to assess real-time sensor data, detect deviations from normal operational conditions, and estimate the remaining useful life of assets. When an anomaly is detected, automated alerts can be generated, allowing maintenance teams to address issues before they escalate. This capability is particularly beneficial in industries such as manufacturing, oil and gas, and power generation, where even a short downtime can lead to significant financial losses and safety hazards [3].

Beyond predictive maintenance, anomaly detection in IoT systems also enhances security, quality control, and process optimization. In cybersecurity, for instance, IoT devices monitor network traffic and detect irregular access patterns that may indicate cyber threats. In production environments, AI-powered anomaly detection identifies defects in materials or inconsistencies in manufacturing processes, reducing waste and improving product quality. Additionally, logistics and supply chain management benefit from predictive analytics by anticipating shipment delays or inventory shortages based on sensor data and external factors like weather or demand fluctuations. By integrating AI-driven predictive analytics into IoT ecosystems, businesses can achieve smarter decision-making, reduce operational risks, and enhance overall system resilience [3].

14.2.2 Edge AI and Edge Computing

Edge AI enables IoT devices to process data on-site rather than relying on centralized cloud computing, significantly reducing latency and improving real-time decision-making. Traditional cloud-based architectures require data to be transmitted to remote servers for processing, which can introduce delays and increase dependency on network connectivity. Edge AI, however, brings computational power closer to the source of data generation, allowing IoT devices to analyze and respond to information instantly. This local processing capability is essential in time-sensitive applications where even milliseconds of delay can impact performance, safety, and efficiency. By minimizing the need for constant cloud communication, Edge AI also enhances privacy and reduces bandwidth usage, making it a cost-effective solution for industries with high data-processing demands [4].

This approach is particularly advantageous in fields such as autonomous vehicles, where rapid decision-making is critical for navigation and collision avoidance. Self-driving cars rely on a combination of sensors, cameras, and AI models to interpret road conditions, detect obstacles, and make split-second driving decisions. Cloud-based processing introduces an unacceptable delay in such scenarios, whereas Edge AI enables immediate analysis within the vehicle itself. Similarly, in healthcare monitoring, wearable medical devices and remote patient-monitoring systems benefit from Edge AI by processing vital signs in real time. AI-powered wearables can detect abnormalities, such as irregular heartbeats or sudden drops in oxygen levels, and alert medical professionals instantly, improving patient outcomes and reducing the risk of complications [4].

Disaster management is another domain where Edge AI proves invaluable. In emergency situations such as wildfires, earthquakes, or floods, real-time data processing is crucial for coordinating rapid response efforts. Drones and IoT-enabled sensors deployed in disaster zones use Edge AI to analyze environmental data and detect potential hazards, providing first responders with timely and actionable insights. This capability enhances situational awareness and facilitates quicker decision-making, ultimately saving lives and resources. Furthermore, Edge AI plays a vital role in industrial automation, smart cities, and retail analytics, where localized data processing leads to faster insights and improved operational efficiency. By enabling IoT devices to act independently, Edge AI transforms the way modern systems operate, fostering innovation and resilience in a wide range of applications [4].

14.2.3 FL for Privacy-preserving IoT Communication

FL empowers IoT devices to train ML models locally, sharing only model updates rather than raw data, thereby enhancing security, privacy, and efficiency. Traditional AI models often require vast amounts of centralized data collection, which can pose risks related to data breaches, regulatory compliance, and network congestion. FL addresses these challenges by ensuring that data remains on individual devices while still contributing to a global model. This decentralized approach is particularly beneficial in environments where data sensitivity is a major concern, such as personal healthcare records, financial transactions, and industrial IoT applications. By reducing the dependency on cloud-based data aggregation, FL not only improves privacy but also minimizes latency and bandwidth consumption, making it an optimal choice for large-scale, distributed AI deployments [5].

One of the most impactful applications of FL is in healthcare, where patient data privacy is paramount. Wearable devices, remote patient-monitoring systems, and hospital networks can utilize FL to collaboratively improve disease detection models without exposing sensitive medical records. For instance, different hospitals can train AI models locally on patient data while sharing only the trained parameters with a central server, ensuring compliance with regulations such as HIPAA and GDPR. This method enables continuous learning from diverse datasets across institutions while safeguarding individual privacy. FL also enhances the accuracy of predictive healthcare models by leveraging localized patient data, enabling early diagnosis and personalized treatment recommendations. By eliminating the need to transfer vast amounts of medical data over networks, FL not only improves privacy but also accelerates real-time decision-making in clinical settings [5].

Smart city applications are also rapidly adopting FL to enhance security, traffic management, and urban planning. Connected infrastructure, such as traffic cameras, IoT-enabled streetlights, and environmental sensors, can collaboratively train AI models to optimize urban mobility, reduce congestion, and enhance public safety. Instead of transmitting raw video footage or sensor data to a central location, edge devices process the data locally and share only aggregated insights, reducing network load and maintaining citizen privacy. This approach is particularly valuable in intelligent transportation systems, where real-time adaptation to traffic patterns is essential. Additionally, FL is being explored for cybersecurity applications, enabling IoT devices to detect threats collaboratively without exposing sensitive data. As the adoption of FL grows, industries are leveraging its potential to build more secure, privacy-preserving, and efficient AI-driven ecosystems [5].

14.2.4 AI-optimized Communication Protocols

AI enhances IoT communication by dynamically selecting protocols and optimizing bandwidth allocation based on real-time network conditions. Traditional IoT communication relies on predefined protocols that may not efficiently adapt to fluctuating network environments, leading to congestion, latency, and inefficiencies. AI-driven solutions address these challenges by continuously monitoring network parameters such as signal strength, data traffic, and interference levels. ML algorithms analyze this data to adjust communication strategies dynamically, ensuring optimal connectivity and resource utilization. This intelligent adaptation is particularly valuable in large-scale IoT deployments, where diverse devices operate under varying network conditions. By leveraging AI, IoT systems can maintain stable and efficient communication without manual intervention, enhancing reliability and performance across applications such as smart grids, industrial automation, and connected vehicles [6].

In advanced network infrastructures like 5G and emerging 6G technologies, reinforcement learning (RL) plays a crucial role in optimizing adaptive communication protocols. Unlike traditional rule-based network management, RL enables IoT devices to learn and refine their communication strategies through continuous interaction with the network. By employing RL-based models, IoT systems can predict traffic congestion, dynamically switch between communication frequencies, and allocate bandwidth more efficiently. This capability is particularly useful in high-demand environments such as autonomous transportation networks, where seamless connectivity is critical for real-time vehicle-to-vehicle (V2V) and vehicle-to-infrastructure (V2I) communication. RL-based network optimization also enhances energy efficiency, as devices can intelligently adjust transmission power based on network demand, prolonging battery life in battery-powered IoT systems [6].

Beyond improving connectivity, AI-powered IoT communication enhances security, reliability, and scalability in next-generation networks. AI models can detect anomalies in network traffic, identifying potential cyber threats and mitigating risks before they escalate. In smart cities, AI-driven communication ensures that critical IoT applications – such as emergency response systems, public safety monitoring, and intelligent traffic control – operate seamlessly under varying network conditions. Additionally, AI enables efficient spectrum utilization by dynamically reallocating unused frequencies, reducing interference, and supporting massive IoT device connectivity. As 5G/6G networks continue to evolve, AI-driven communication technologies will play an integral role in shaping resilient, adaptive, and intelligent IoT ecosystems that meet the demands of modern digital infrastructure [6].

14.3 Emerging Trends in IoT-based Communication Through AI/ML

14.3.1 6G and AI-driven Network Optimization

The incorporation of AI into 6G networks seeks to deliver URLLC for IoT applications. Network slicing and intelligent spectrum allocation driven by AI enhance connectivity and energy efficiency [7].

14.3.2 Digital Twins and AI-powered IoT Simulation

By combining digital twin technology with AI, real-time simulation of IoT environments becomes possible, enabling industries to virtually test and refine smart systems prior to implementation [8].

14.3.3 Blockchain and AI for Secure IoT Communication

The integration of blockchain and AI bolsters IoT security through decentralized authentication and fraud detection. Trust and data integrity in blockchain-based IoT networks are enhanced by AI-powered anomaly detection [9].

14.3.4 Explainable AI for IoT Decision-making

Explainable AI (XAI) is becoming increasingly important to ensure transparency in AI-driven IoT decisions. It provides interpretability in automated decision-making, which is essential for regulatory compliance in sectors like healthcare and finance [10].

14.3.5 AI-powered Energy Optimization for Sustainable IoT

IoT energy management systems utilizing AI optimize energy usage by forecasting demand, automating load balancing, and incorporating renewable energy sources, contributing to SDG 7 (Affordable and Clean Energy) [11].

14.4 Challenges and Ethical Considerations

Although there have been significant advancements, the integration of AI/ML in IoT communication still faces numerous obstacles.

14.4.1 Threats to Data Security and Privacy

The extensive data generated by large-scale IoT networks raises concerns about privacy violations, cyber threats, and unauthorized data access. To address these issues, it is crucial to implement robust AI-driven encryption methods and differential privacy techniques [12].

14.4.2 Limitations in Computation and Infrastructure

The implementation of AI-powered IoT systems requires substantial computational resources, which presents challenges in resource-limited environments, such as isolated rural areas. To overcome this, researchers are investigating lightweight AI models and hardware acceleration strategies [13].

14.4.3 Ethical Implications of AI-driven Automation

AI decision-making processes can potentially introduce bias, leading to unintended consequences, especially in critical fields like healthcare and finance. To ensure fairness and accountability in AI-powered IoT solutions, it is necessary to establish regulatory frameworks [14].

14.5 Future Directions and Innovations

14.5.1 Self-healing IoT Networks Powered by AI

In the coming years, IoT networks will incorporate AI to enable self-healing capabilities. These systems will automatically identify and fix network issues without human assistance, thereby improving overall system durability [15].

14.5.2 IoT Security Enhanced by AI and Quantum Computing

The integration of quantum computing is anticipated to transform IoT security. By facilitating the development of highly secure encryption algorithms, it will strengthen AI-driven IoT networks against cybersecurity threats [16].

14.5.3 Swarm Intelligence in IoT Communication Through AI

Research is underway to apply AI-powered swarm intelligence, inspired by natural swarm behaviors, to IoT device coordination. This approach shows promise for applications such as networks of drones and systems for monitoring the environment [17].

14.5.4 Autonomous IoT Management via Zero-touch AI

The implementation of zero-touch AI for IoT management will enable networks to operate autonomously with minimal human oversight. This advancement is expected to enhance both scalability and operational efficiency [18].

14.6 Conclusion

The integration of AI and ML into IoT communication is revolutionizing sustainability efforts by creating smart, adaptable, and highly efficient systems across various sectors. This fusion allows devices to analyze data instantaneously, optimize energy consumption, and enhance decision-making processes without human input. Cutting-edge technologies such as Edge AI, FL, and AI-powered 6G networks are transforming intelligent connectivity by decreasing response times, bolstering security measures, and facilitating seamless automation. These advancements are particularly vital in areas like smart urban planning, industrial automation, precision farming, and telemedicine, where optimized IoT communication can result in substantial resource conservation and environmental advantages.

Nevertheless, several obstacles must be overcome to fully harness the potential of AI-enhanced IoT communication in achieving SDGs. A primary concern is the security and privacy of the vast amounts of sensitive data generated by AI-powered IoT systems, necessitating robust encryption and decentralized authentication protocols. Additionally, infrastructure limitations, particularly in developing regions, hinder the widespread implementation of AI-driven IoT solutions due to insufficient computational capabilities and network coverage. Ethical issues related to AI model bias, transparency in decision-making processes, and regulatory compliance further complicate the deployment of AI-powered IoT communication systems.

Upcoming technological breakthroughs will be instrumental in addressing these challenges and further improving sustainability and efficiency. Self-repairing networks will enable IoT systems to independently identify and resolve communication issues, thereby reducing maintenance expenses and enhancing system resilience. Quantum-secured IoT networks will employ advanced cryptographic methods to provide unparalleled security, ensuring data integrity and trust in critical applications. Moreover, zero-touch AI will facilitate fully autonomous IoT management, minimizing human intervention and maximizing operational efficiency. As AI and IoT technologies continue to progress, these innovations will contribute to a more sustainable, secure, and interconnected world, aligning with global sustainability objectives and paving the way for an intelligent future.

References

1 Zhang, Y., Liu, H., Wang, K., et al. (2023). AI-powered IoT for sustainable smart cities. *IEEE Internet of Things Journal* 10 (5): 4212–4225.

2 Sharma, R., Yadav, A., Kumar, V., et al. (2022). Machine learning in IoT-based smart agriculture: a review. *Computers and Electronics in Agriculture* 199: 107069.

3 Gupta, V., Mehta, P., Ranjan, R., et al. (2021). Predictive maintenance in IIoT using AI/ML. *IEEE Transactions on Industrial Informatics* 17 (2): 1234–1245.

4 Kim, J., Lee, S., Park, H., et al. (2023). Edge AI for real-time IoT analytics: challenges and opportunities. *Future Generation Computer Systems* 143: 324–336.

5 Li, X., Zhang, M., Xu, Y., et al. (2022). Federated learning for secure IoT applications. *ACM Transactions on Internet Technology* 22 (1): 1–18.

6 Nguyen, T., Pham, Q., Le, T., et al. (2023). AI-driven communication protocols for 6G IoT networks. *IEEE Network* 37 (1): 67–76.

7 Ahmed, A., Khan, R., Rehman, A., et al. (2024). 6G networks: AI and IoT convergence. *IEEE Wireless Communications* 31 (3): 44–52.

8 Kaur, M., Sharma, N., Thakur, S., et al. (2022). Digital twin for IoT-enabled smart manufacturing. *Journal of Manufacturing Systems* 62: 123–138.

9 Patel, S., Joshi, R., Mahajan, R., et al. (2023). Blockchain and AI for IoT security: a survey. *Computer Networks* 223: 109470.

10 Ribeiro, M.T., Singh, S., Gadepally, V., et al. (2022). Explainable AI for IoT decision support systems. *ACM Computing Surveys* 54 (4): 1–35.

11 Singh, P., Verma, R., Yadav, S., et al. (2021). AI-driven energy optimization in smart grids. *IEEE Transactions on Smart Grid*, 12 (6): 5432–5445.

12 Jisasoftech. (2025). Protecting AI-Enabled IoT Ecosystems from Cyber Threats [Internet]. Jisasoftech.com; 2025 [cited 2025 Aug 6]. Available from: https://www.jisasoftech.com/protecting-ai-enabled-iot-ecosystems-from-cyber-threats/

13 HeyCoach. (2025). AI for IoT Security: The Future of Cyber Defense [Internet]. HeyCoach.in [cited 2025 Aug 6]. Available from: https://heycoach.in/blog/ai-for-iot-security/

14 BeyondKey. (2024). AI Security: Risks, Privacy, Future of Cybersecurity & Safety [Internet]. BeyondKey.com [cited 2025 Aug 6]. Available from: https://www.beyondkey.com/blog/ai-security-risk-and-safety/

15 Palo Alto Networks. (2024). Made for Each Other: AI and IoT [Internet]. PaloAltoNetworks.com [cited 2025 Aug 6]. Available from: https://www.paloaltonetworks.com/blog/security-operations/made-for-each-other-ai-and-iot/

16 Forescout. (2024). The Future of Encryption in a Quantum Cryptography World [Internet]. Forescout.com [cited 2025 Aug 6]. Available from: https://www.forescout.com/blog/the-future-of-encryption-in-a-quantum-cryptography-world/

17 Cisco Blogs. (2024). A Generational Industrial Experience Is in Store at Cisco Live San Diego [Internet]. Blogs.Cisco.com [cited 2025 Aug 6]. Available from: https://blogs.cisco.com/industrial-iot/a-generational-industrial-experience-is-in-store-at-cisco-live-san-diego

18 Splunk. (2024). What Is Zero Touch in IT? [Internet]. Splunk.com [cited 2025 Aug 6]. Available from: https://www.splunk.com/en_us/blog/learn/zero-touch.html

Index

AI and ML Techniques in IoT-based Communication: A Path to Sustainable Development Goals, First Edition. Edited by Sumita Mishra, Nishu Gupta and Polat Goktas.

m

n

o